CONTROL AND DYNAMIC SYSTEMS

Advances in Theory and Applications

Volume 43

CONTRIBUTORS TO THIS VOLUME

MO-SHING CHEN
TSAI-HSIANG CHEN
HSIAO-DONG CHIANG
A. A. FOUAD
JESUS A. JATIVA
WEI-JEN LEE
WERNER LEONHARD
YOUNG-HWAN MOON
M. A. PAI
ARUN G. PHADKE
P. W. SAUER
RAYMOND R. SHOULTS
JAMES S. THORP
V. VITTAL

CONTROL AND DYNAMIC SYSTEMS

ADVANCES IN THEORY AND APPLICATIONS

Edited by
C. T. LEONDES

Department of Electrical Engineering
University of Washington
Seattle, Washington
and
School of Engineering and Applied Science
University of California, Los Angeles
Los Angeles, California

VOLUME 43: ANALYSIS AND CONTROL SYSTEM
TECHNIQUES FOR ELECTRIC
POWER SYSTEMS
Part 3 of 4

ACADEMIC PRESS, INC.
Harcourt Brace Jovanovich, Publishers
San Diego   New York   Boston
London   Sydney   Tokyo   Toronto

Academic Press Rapid Manuscript Reproduction

Academic Press, Inc.
San Diego, California 92101

United Kingdom Edition published by
ACADEMIC PRESS LIMITED
24-28 Oval Road, London NW1 7DX

Library of Congress Catalog Card Number: 64-8027

ISBN   0-12-012743-1   (alk. paper)

PRINTED IN THE UNITED STATES OF AMERICA
91 92 93 94   9 8 7 6 5 4 3 2 1

CONTENTS

CONTRIBUTORS

Numbers in parentheses indicate the pages on which the authors' contributions begin.

Mo-Shing Chen (61, 185), *The Energy Systems Research Center, University of Texas, Arlington, Texas 76019*

Tsai-Hsiang Chen (61), *The Energy Systems Research Center, University of Texas, Arlington, Texas 76019*

Hsiao-Dong Chiang (275), *School of Electrical Engineering, Cornell University, Ithaca, New York 14853*

A. A. Fouad (115), *College of Engineering, Department of Electrical Engineering and Computer Engineering, Iowa State University, Ames, Iowa 50011*

Jesus A. Jativa (377), *Energy Systems Research Center, University of Texas, Arlington, Texas 76019*

Wei-Jen Lee (185), *The Energy Systems Research Center, University of Texas, Arlington, Texas 76019*

Werner Leonhard (407), *Technical University Braunschweig, 3300 Braunschweig, Germany*

Young-Hwan Moon (185), *The Energy Systems Research Center, University of Texas, Arlington, Texas 76019*

M. A. Pai (1), *Department of Electrical and Computer Engineering, University of Illinois, Urbana, Illinois 61801*

Arun G. Phadke (335), *Department of Electrical Engineering, Virginia Polytechnic Institute and State University, Blacksburg, Virginia 24061*

P. W. Sauer (1), *Department of Electrical and Computer Engineering, University of Illinois, Urbana, Illinois 61801*

Raymond R. Shoults (377), *Energy Systems Research Center, University of Texas, Arlington, Texas 76019*

James S. Thorp (335), *School of Electrical Engineering, Cornell University, Ithaca, New York 14853*

V. Vittal (115), *College of Engineering, Department of Electrical Engineering and Computer Engineering, Iowa State University, Ames, Iowa 50011*

PREFACE

Research and development in electric power systems analysis and control techniques has been an area of significant activity for decades. However, because of increasingly powerful advances in techniques and technology, the activity in electric power systems analysis and control techniques has increased significantly over the past decade and continues to do so at an expanding rate because of the great economic significance of this field. Major centers of research and development in electrical power systems continue to grow and expand because of the great complexity, challenges, and significance of this field. These centers have become focal points for the brilliant research efforts of many academicians and industrial professionals and for the exchange of ideas between these individuals. As a result, this is a particularly appropriate time to treat advances in the many issues and modern techniques involved in electric power systems in this international series. Thus, this is the third volume of a four volume sequence in this series devoted to the significant theme of "Analysis and Control System Techniques for Electric Power Systems." The broad topics involved include transmission line and transformer modeling. Since the issues in these two fields are rather well in hand, although advances continue to be made, this four volume sequence will focus on advances in areas including power flow analysis, economic operation of power systems, generator modeling, power system stability, voltage and power control techniques, and system protection, among others.

The first contribution in this volume is "Modeling and Simulation of Multimachine Power System Dynamics," by P.W. Sauer and M.A. Pai. Modeling and simulation of multimachine power system dynamics has had a long history beginning with network analyzers and analog computers. Many of the models in use today were developed for these earlier computational techniques. The numerical methods for digital simulation have evolved separately within a basic mathematical framework. This contribution introduces fundamental component modeling and presents it in a conceptually coherent framework. As it turns out, in most cases the final models conform to what the power industry has been using for some time. Reduced-order models are presented from an integral manifold approach where the simplest approximation usually coincides with traditional practice. The computation of initial conditions for the set of differential-algebraic equations is presented in a step-by-step manner for typical simulations. The actual numerical solution methods are discussed in terms of standard integration methods and equivalent industry practice. The formulation is done in such a way that the interface between standard load-flow analysis and dynamic simulation is clear.

The next contribution is "Computer Simulation Techniques in Electric Distribution Systems," by Mo-Shing Chen and Tsai-Hsiang Chen. It has become increasingly clear that utilities and consumers need demand-side management. The least-cost planning as well as demand-side strategies are being widely adopted and mandated. The demand-side management, least-cost planning, network transformer placement study, and many other distribution analyses need rigorous operational-type analysis rather than planning-oriented analysis. The difference between these two types of analyses should be properly emphasized, otherwise the misuse of the planning-type method to analyze the operational behavior of the system will distort the explanation of the calculated results and lead to incorrect conclusions. The concept and methodologies for detailed electric distribution system simulation are discussed here. The features of general distribution systems are also addressed.

The next contribution is "Power System Transient Stability Assessment Using the Transient Energy Function Method," by A.A. Fouad and V. Vittal. Power system transient stability deals with the power system's response to disturbances. The period of interest is the transient period before the new steady-state conditions are reached. If the power system is subjected to a large disturbance and it is able to reach an acceptable steady-state condition, it is (transiently) stable. A large disturbance is a sudden change in a power system parameter or operating condition such that linearization of the system equations, for the purpose of analysis, cannot be justified. Transient stability studies are made by electric utilities in North America (and in other parts of the world) to ensure the reliability of the bulk power supply. The conventional, and still the standard, method to determine the transient stability of a power system is to solve for the system variables and parameters for a given sequence of events. Power system transient stability assessment using energy functions has been around for many years. In the late 1970s interest in the energy functions was revived by the development of a new energy function. The transient energy function described in this contribution is based on that function.

The next contribution is "Dynamic Stability Analysis and Control of Power Systems," by Young-Hwan Moon, Wei-Jen Lee, and Mo-Shing Chen. In order to maintain integrity, power systems should be designed to withstand the specified contingencies with finite probability of occurring. The stability study of electric power systems can be separated into two distinct problems, the transient (short-term) problem and the dynamic (long-term) problem. In the transient stability analysis, the performance of the power system when subjected to sudden severe impacts is studied. The concern is whether the system is able to maintain synchronism during and following these disturbances. The period of interest is relatively short (a few seconds), with the first swing being of primary importance. The stability depends strongly upon the magnitude and location of the disturbance and to a lesser extent upon the initial state or operating condition of the system. In contrast to transient stability, dynamic stability tends to be a property of the state of the system. Dynamic stability indicates the ability of all machines in a system to adjust to small load changes or impacts. The concern is whether the system has growing oscillation phenomena and their damping. As modern power systems become equipped with more fast control systems and become more complicated, dynamic stability becomes more important. The stability strongly depends on the initial state or operating condition of the system. Without question, both transient stability and dynamic stability must be secured for successful planning and operation of the system. This contribution is an in-depth presentation of issues and techniques in this area of great significance.

The next contribution, "Analytical Results on Direct Methods for Power System Transient Stability Analysis," by Hsiao-Dong Chiang, reviews recent advances in the development of analytical results for direct methods in power system transient stability analysis and provides important insight into the underlying concepts and properties of the various direct methods. Aside from a coherent view of the field, new material is presented. The exposition emphasizes fundamentals of direct methods rather than the heuristics which most direct methods are based on. In order to demonstrate one of the advantages of employing the analytical approach instead of the heuristic approach to developing direct methods for transient stability analysis, these fundamentals are then applied to a recently developed direct method, called the boundary of stability region controlling unstable equilibrium point method (BCU method) for direct analysis of power system transient stability. The BCU method has been compared favorably with other methods on large-scale power systems, according to a recent EPRI report. One implication from the development of the BCU method is that analytical results can sometimes lead to the development of reliable, yet fast, solution algorithms for solving transient stability problems in electric power systems.

A power system is one of the largest dynamic systems in existence. Generators of electric power are interconnected with loads by means of an extensive transmission and distribution network. The size of the power network is usually on the scale of continents. The secure and economic operation of the electric power network is a complex problem, involving a distributed, hierarchical, and multicentered structure. Very sophisticated monitoring, protection, and control systems have been devised to achieve a system which provides electric power of high quality and reliability. Recent developments in the field of computer-based substation functions and satellite based dissemination of high-accuracy time reference signals have provided power system engineers with a new tool for achieving even better overall system performance more economically. The next contribution, "Improved Control and Protection of Power Systems through Synchronized Phasor Measurements," by Arun G. Phadke and James S. Thorpe, describes the techniques and uses of synchronizing phasor measurements for improved monitoring, protection, and control of an electric system power network. Recent field trials for synchronized phasor measurements, based on new technology, and directions for future significant research are presented in this contribution.

The next contribution is "Real Time Power System Control: Issues Related to Variable Nonlinear Tie-line Frequency Bias for Load Frequency Control," by Raymond R. Shoults and Jesus A. Jativa. The interconnection of power systems has allowed various economic and technical advantages that were otherwise unavailable. Incremental increases in system reliability; ability to sell, buy, or exchange energy; feasibility of installing larger power plants; incremental increases in system stability; sharing of spinning reserve capacities; and taking advantage of load diversity for economy of operation are the main achievements. However, the interconnection carries with it some difficulties and obligations such as incremental increases in operational complexity, reduced ability to control steady-state power flows, propagation of the effect of faults through the entire system, propagation of steady-state oscillations, responsibility of matching generation to load within each control area, and shared responsibility to maintain frequency and time error within certain established limits. The area secondary control, referred to as automatic generation control (AGC), is responsible for regulating the system frequency within acceptable error bounds, maintaining correct interchange schedules, and distributing generation within each area according to minimum operating cost criteria. Even though tie-line frequency-bias control has been in practice for many years, it still represents the state of the art in AGC. It is based on an area control error (ACE) defined as the generation change required to restore frequency and net interchange to desired values. The derivation of ACE is based upon the assumption of steady-state conditions. A nonzero ACE represents the load-generation-net interchange unbalance within an area. It has been observed that frequency bias should follow the variable and nonlinear frequency response of a given control area. This contribution is an in-depth presentation of the role of the frequency bias parameter in the ACE calculation. The area control principle of an interconnected power system requires only the measurements of area frequency and area net interchange to calculate the ACE. This principle leads to a straightforward and efficient method of decentralized control. The input to ACE, which is assumed steady-state calculation, represents a continua of dynamic behavior. During dynamic conditions, ACE comprises the key input to load-frequency control (LFC). The final controller design for LFC may be carried out by either on-line tuning or some sort of control theory approach.

The control necessary to maintain stability in electric power systems can be achieved through either field voltage control or mechanical power control. The next contribution, "High Dynamic Performance Microcomputer Control of Electrical

Drives," by Werner Leonhard, is an in-depth presentation of the modern technology for achieving effective electric drive control necessary to achieve mechanical power control through servo control techniques.

This volume is a particularly appropriate one as the third of a companion set of four volumes on analysis and control techniques in electric power systems. The authors are all to be commended for their superb contributions, which will provide a significant reference source for workers on the international scene for years to come.

Modeling and Simulation of Multimachine Power System Dynamics[†]

P. W. Sauer and M. A. Pai

University of Illinois at Urbana-Champaign
Department of Electrical and Computer Engineering
Urbana, IL 61801

1 Introduction

Modeling and simulation of multimachine power system dynamics has had a long history beginning with network analyzers and analog computers. Many of the models in use today were developed for these earlier computational techniques. The numerical methods for digital simulation have evolved separately within a basic mathematical framework.

In this chapter we introduce fundamental component modeling and present it in a conceptually coherent framework. As it turns out, in most cases the final models conform to what the power industry has been using for some time. Reduced-order models are presented from an integral manifold approach where the simplest approximation usually coincides with traditional practice. The computation of initial conditions for the set of differential-algebraic equations is presented in a step-by-step manner for typical simulations.

The actual numerical solution methods are discussed both in terms of standard integration methods and equivalent industry practice. The formulation is done in such a way that the interface between standard load-flow analysis and dynamic simulation is clear.

[†]Major portions of the material contained in this chapter have been excerpted from the book entitled *Power System Dynamics and Stability* by P. W. Sauer and M. A. Pai, to be published by Prentice-Hall in 1992.

CONTROL AND DYNAMIC SYSTEMS, VOL. 43

2   Synchronous Machine Modeling

There is probably more literature on synchronous machines than on any other device in electrical engineering. Unfortunately, it is this vast amount of material which often makes the subject complex and confusing. In addition, the largest portion of work on reduced-order modeling is based primarily on physical intuition, practical experience and years of experimentation. The evolution of dynamic analysis has caused some problem in notation as it relates to common symbols which eventually require data from manufacturers. This chapter uses the conventions and notations of [1], which essentially follows that of many publications on synchronous machines [2]-[11]. The original Park's transformation is used together with the "X_{ad}" per-unit system [12, 13].

2.1   Three Damper Winding Model

This subsection presents the basic dynamic equations for a balanced symmetrical three-phase, synchronous machine with a field winding and three damper windings on the rotor. The simplified schematic of Figure 1 shows the coil orientation, assumed polarities, and rotor position reference. The stator windings have axes 120 electrical degrees apart and are assumed to have an equivalent sinusoidal distribution [1]. While a 2-pole machine is shown, all equations are written for a P-pole machine with $\omega = \frac{P}{2}\omega_{shaft}$ expressed in electrical radians per second. The circles with dots and x's indicate the windings. Current flow is assumed to be in the "x" and out of the "dot". The voltage polarity of the coils is assumed to be plus to minus from the "x" to the "dot".

This notation uses "motor" current notation for all the windings at this time. The transformed stator currents will be changed to "generator" current notation at the time when per-unit scaling is introduced. The fundamental Kirchhoff, Faraday and Newton's laws give:

$$v_a = i_a r_s + \frac{d\lambda_a}{dt} \tag{1}$$

$$v_b = i_b r_s + \frac{d\lambda_b}{dt} \tag{2}$$

$$v_c = i_c r_s + \frac{d\lambda_c}{dt} \tag{3}$$

$$v_{fd} = i_{fd} r_{fd} + \frac{d\lambda_{fd}}{dt} \tag{4}$$

$$v_{1d} = i_{1d} r_{1d} + \frac{d\lambda_{1d}}{dt} \tag{5}$$

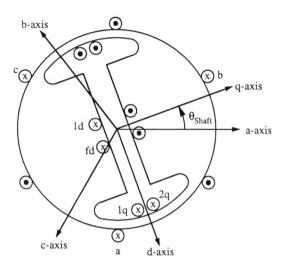

Figure 1: Synchronous machine schematic

$$v_{1q} = i_{1q}r_{1q} + \frac{d\lambda_{1q}}{dt} \tag{6}$$

$$v_{2q} = i_{2q}r_{2q} + \frac{d\lambda_{2q}}{dt} \tag{7}$$

$$\frac{d\theta_{shaft}}{dt} = \frac{2}{P}\omega \tag{8}$$

$$J\frac{2}{P}\frac{d\omega}{dt} = T_m - T_{elec} - T_{fw} \tag{9}$$

where λ is the flux linkage, r is the winding resistance, J is the inertia constant, P is the number of magnetic poles per phase, T_m is the mechanical torque to the shaft, $-T_{elec}$ is the torque of electrical origin, and $-T_{fw}$ is a friction windage torque. A major modeling challenge is to obtain the relationships between flux linkage and current. These relationships will be presented in a later subsection.

2.2 Transformations and Scaling

The first major step in synchronous machine modeling is to transform the stator variables into a reference frame fixed in the machine rotor. The general form of the transformation which accomplishes this is [1],

$$v_{dqo} \stackrel{\Delta}{=} T_{dqo}v_{abc}, \quad i_{dqo} \stackrel{\Delta}{=} T_{dqo}i_{abc}, \quad \lambda_{dqo} \stackrel{\Delta}{=} T_{dqo}\lambda_{abc} \tag{10}$$

where,

$$v_{abc} \stackrel{\Delta}{=} [v_a v_b v_c]^t, \quad i_{abc} \stackrel{\Delta}{=} [i_a i_b i_c]^t, \quad \lambda_{abc} \stackrel{\Delta}{=} [\lambda_a \lambda_b \lambda_c]^t \tag{11}$$

$$v_{dqo} \stackrel{\Delta}{=} [v_d v_q v_o]^t, \quad i_{dqo} \stackrel{\Delta}{=} [i_d i_q i_o]^t, \quad \lambda_{dqo} \stackrel{\Delta}{=} [\lambda_d \lambda_q \lambda_o]^t \tag{12}$$

and,

$$T_{dqo} \stackrel{\Delta}{=} \frac{2}{3} \begin{bmatrix} \sin\frac{P}{2}\theta_{shaft} & \sin(\frac{P}{2}\theta_{shaft} - \frac{2\pi}{3}) & \sin(\frac{P}{2}\theta_{shaft} + \frac{2\pi}{3}) \\ \cos\frac{P}{2}\theta_{shaft} & \cos(\frac{P}{2}\theta_{shaft} - \frac{2\pi}{3}) & \cos(\frac{P}{2}\theta_{shaft} + \frac{2\pi}{3}) \\ \frac{1}{2} & \frac{1}{2} & \frac{1}{2} \end{bmatrix} \tag{13}$$

with the inverse,

$$T_{dqo}^{-1} = \begin{bmatrix} \sin\frac{P}{2}\theta_{shaft} & \cos\frac{P}{2}\theta_{shaft} & 1 \\ \sin(\frac{P}{2}\theta_{shaft} - \frac{2\pi}{3}) & \cos(\frac{P}{2}\theta_{shaft} - \frac{2\pi}{3}) & 1 \\ \sin(\frac{P}{2}\theta_{shaft} + \frac{2\pi}{3}) & \cos(\frac{P}{2}\theta_{shaft} + \frac{2\pi}{3}) & 1 \end{bmatrix} \tag{14}$$

The second major step is to scale the synchronous machine equations using the traditional concept of per unit [12, 13]. The following transformed, scaled model uses new variables and parameters for all per-unit quantities. Transformed stator current direction in per unit uses generator notation. The following quantities are not scaled: ω, ω_s, δ, t.

$$\frac{1}{\omega_s}\frac{d\psi_d}{dt} = R_s I_d + \frac{\omega}{\omega_s}\psi_q + V_d \tag{15}$$

$$\frac{1}{\omega_s}\frac{d\psi_q}{dt} = R_s I_q - \frac{\omega}{\omega_s}\psi_d + V_q \tag{16}$$

$$\frac{1}{\omega_s}\frac{d\psi_o}{dt} = R_s I_o + V_o \tag{17}$$

$$\frac{1}{\omega_s}\frac{d\psi_{fd}}{dt} = -R_{fd}I_{fd} + V_{fd} \tag{18}$$

$$\frac{1}{\omega_s}\frac{d\psi_{1d}}{dt} = -R_{1d}I_{1d} + V_{1d} \tag{19}$$

$$\frac{1}{\omega_s}\frac{d\psi_{1q}}{dt} = -R_{1q}I_{1q} + V_{1q} \tag{20}$$

$$\frac{1}{\omega_s}\frac{d\psi_{2q}}{dt} = -R_{2q}I_{2q} + V_{2q} \tag{21}$$

$$\frac{d\delta}{dt} = \omega - \omega_s \tag{22}$$

$$\frac{2H}{\omega_s}\frac{d\omega}{dt} = T_M - T_{ELEC} - T_{FW} \tag{23}$$

To derive an expression for T_{ELEC}, it is necessary to look at the overall energy or power balance for the machine. This is an electromechanical system which can be divided into an electrical system, a mechanical system, and a coupling field [14]. In such a system, resistance causes real power losses in the electrical system, friction causes heat losses in the mechanical system, and hysteresis causes losses in the coupling field. Energy is stored in inductances in the electrical system, the rotating mass of the mechanical system, and the magnetic field which couples the two. Any energy which is not lost or stored must be transferred. In this chapter we assume two things about this energy balance. First, all energy stored in the electrical system inside the machine terminals is included in the energy stored in the coupling field. Secondly, the coupling field is lossless. The first assumption is arbitrary, and the second assumption neglects phenomena such as hysteresis (but not saturation). With these assumptions, it has been shown that [14]:

$$T_{ELEC} = \psi_d I_q - \psi_q I_d \tag{24}$$

To complete the dynamic model in the transformed variables, it is desirable to define an angle which is constant for rated shaft speed. We define this angle as follows,

$$\delta \triangleq \frac{P}{2}\theta_{shaft} - \omega_s t \tag{25}$$

where ω_s is a constant, normally called rated synchronous speed in electrical radians per second, giving

$$\frac{d\delta}{dt} = \omega - \omega_s \tag{26}$$

It is important to pause at this point to consider the transformation of variables. Consider a balanced set of scaled sinusoidal voltages of the form,

$$V_a = \sqrt{2}V_s cos(\omega_s t + \theta_{vs}) \tag{27}$$

$$V_b = \sqrt{2}V_s cos\left(\omega_s t + \theta_{vs} - \frac{2\pi}{3}\right) \qquad (28)$$

$$V_c = \sqrt{2}V_s cos\left(\omega_s t + \theta_{vs} + \frac{2\pi}{3}\right) \qquad (29)$$

Using the transformation (13) and the definition of δ,

$$V_d = V_s sin(\delta - \theta_{vs}) \qquad (30)$$

$$V_q = V_s cos(\delta - \theta_{vs}) \qquad (31)$$

These two real equations can be written as one complex equation,

$$(V_d + jV_q)e^{j(\delta-\pi/2)} = V_s e^{j\theta_{vs}} \qquad (32)$$

This is recognized as the per-unit RMS phasor of (27).

It is also important at this point to note that the model of (15)-(23) was derived using essentially four general assumptions. These assumptions are summarized below:

(a) The stator has three coils in a balanced, symmetrical configuration centered 120 electrical degrees apart.

(b) The rotor has four coils in a balanced symmetrical configuration located in pairs 90 electrical degrees apart.

(c) The relationships between the flux linkages and currents must reflect a conservative coupling field.

(d) The relationships between the flux linkages and currents must be independent of θ_{shaft} when expressed in the dqo coordinate system.

The following subsection gives a general flux linkage/current relationship which completes the dynamic model.

2.3   General Magnetic Circuit Model

In this section we propose a fairly generalized treatment of nonlinearities in the magnetic circuit. The generalization is motivated by the multitude of various representations of saturation which have appeared in the literature. Virtually all methods proposed to date involve the addition of one or more nonlinear terms to a linear model. The following treatment returns to the original "abc" variables so that any assumptions or added terms can be traced through the transformation and scaling processes. The flux linkage/current relationships must satisfy assumptions c and d at the end of the last subsection if the results here are to be valid for the general model of (15)-(23). Toward this end, we propose a flux linkage/current relationship of the following form:

$$\lambda_{abc} = L_{ss}(\theta_{shaft})i_{abc} + L_{sr}(\theta_{shaft})i_{rotor}$$

$$-S_{abc}(i_{abc}, \lambda_{rotor}, \theta_{shaft}) \qquad (33)$$

$$\lambda_{rotor} = L_{sr}^t(\theta_{shaft})i_{abc} + L_{rr}(\theta_{shaft})i_{rotor}$$

$$-S_{rotor}(i_{abc}, \lambda_{rotor}, \theta_{shaft}) \qquad (34)$$

where

$$i_{rotor} \triangleq [i_{fd}\ i_{1d}\ i_{1q}\ i_{2q}]^t \ , \ \lambda_{rotor} \triangleq [\lambda_{fd}\ \lambda_{1d}\ \lambda_{1q}\ \lambda_{2q}]^t \qquad (35)$$

with L_{ss}, L_{rr} symmetric. The structure and specific entries in each inductance matrix are given in [1]. The saturation functions S_{abc} and S_{rotor} must satisfy assumptions c and d at the end of the last subsection. With these two assumptions, S_{abc} and S_{rotor} must be such that when (33) and (34) are per unitized and transformed using (10), the following nonlinear flux linkage/current relationship is obtained,

$$\psi_d = X_d(-I_d) + X_{md}I_{fd} + X_{md}I_{1d} - S_d^{(1)}(Y_1) \qquad (36)$$

$$\psi_{fd} = X_{md}(-I_d) + X_{fd}I_{fd} + c_dX_{md}I_{1d} - S_{fd}^{(1)}(Y_1) \qquad (37)$$

$$\psi_{1d} = X_{md}(-I_d) + c_dX_{md}I_{fd} + X_{1d}I_{1d} - S_{1d}^{(1)}(Y_1) \qquad (38)$$

and

$$\psi_q = X_q(-I_q) + X_{mq}I_{1q} + X_{mq}I_{2q} - S_q^{(1)}(Y_1) \qquad (39)$$

$$\psi_{1q} = X_{mq}(-I_q) + X_{1q}I_{1q} + c_qX_{mq}I_{2q} - S_{1q}^{(1)}(Y_1) \qquad (40)$$

$$\psi_{2q} = X_{mq}(-I_q) + c_qX_{mq}I_{1q} + X_{2q}I_{2q} - S_{2q}^{(1)}(Y_1) \qquad (41)$$

and

$$\psi_o = X_{\ell s}(-I_o) - S_o^{(1)}(Y_1) \qquad (42)$$

where,

$$Y_1 \triangleq [I_d \ \psi_{fd} \ \psi_{1d} \ I_q \ \psi_{1q} \ \psi_{2q} \ I_o]^t \tag{43}$$

This system includes the possibility of coupling between all of the "d", "q", and "o" subsystems.

While several examples [15] have shown that the terms c_d and c_q are important in some simulations, it is customary to make the following simplification,

$$c_d \approx 1, \quad c_q \approx 1 \tag{44}$$

This assumption makes the respective off-diagonal entries of the decoupled inductance matrices in (36)-(41) equal. An alternative way to obtain the same structure without the above simplification would require a different choice of scaling and different definitions of leakage reactances [3]. It is customary to introduce three new variables as,

$$E'_q \triangleq \frac{X_{md}}{X_{fd}} \psi_{fd} \tag{45}$$

$$E_{fd} \triangleq \frac{X_{md}}{R_{fd}} V_{fd} \tag{46}$$

$$E'_d \triangleq -\frac{X_{mq}}{X_{1q}} \psi_{1q} \tag{47}$$

Using new variables E'_d and E'_q and rearranging (36)-(43) so that rotor currents can be eliminated gives (with $c_d = c_q = 1$)

$$\psi_d = -X''_d I_d + \frac{(X''_d - X_{\ell s})}{(X'_d - X_{\ell s})} E'_q + \frac{(X'_d - X''_d)}{(X'_d - X_{\ell s})} \psi_{1d} - S_d^{(2)}(Y_2) \tag{48}$$

$$I_{fd} = \frac{1}{X_{md}} [E'_q + (X_d - X'_d)(I_d - I_{1d}) + S_{fd}^{(2)}(Y_2)] \tag{49}$$

$$I_{1d} = \frac{X'_d - X''_d}{(X'_d - X_{\ell s})^2} [\psi_{1d} + (X'_d - X_{\ell s})I_d - E'_q + S_{1d}^{(2)}(Y_2)] \tag{50}$$

and

$$\psi_q = -X''_q I_q - \frac{(X''_q - X_{\ell s})}{(X'_q - X_{\ell s})} E'_d + \frac{(X'_q - X''_q)}{(X'_q - X_{\ell s})} \psi_{2q} - S_q^{(2)}(Y_2) \tag{51}$$

$$I_{1q} = \frac{1}{X_{mq}} [-E'_d + (X_q - X'_q)(I_q - I_{2q}) + S_{1q}^{(2)}(Y_2)] \tag{52}$$

$$I_{2q} = \frac{X_q' - X_q''}{(X_q' - X_{\ell s})^2}[\psi_{2q} + (X_q' - X_{\ell s})I_q + E_d' + S_{2q}^{(2)}(Y_2)] \tag{53}$$

and

$$\psi_o = -X_{\ell s}I_o - S_o^{(2)}(Y_2) \tag{54}$$

where,

$$Y_2 \overset{\Delta}{=} [I_d \ \ E_q' \ \ \psi_{1d} \ \ I_q \ \ E_d' \ \ \psi_{2q} \ \ I_o]^t \tag{55}$$

and $X_{\ell s}$, X_d', X_d'', X_q', X_q'' are defined in terms of X_d, X_{md}, X_{fd}, X_{1d}, X_q, X_{mq}, X_{1q}, X_{2q} [1].

Elimination of rotor currents from (15)-(23) gives the final dynamic model with general saturation functions,

$$\frac{1}{\omega_s}\frac{d\psi_d}{dt} = R_sI_d + \frac{\omega}{\omega_s}\psi_q + V_d \tag{56}$$

$$\frac{1}{\omega_s}\frac{d\psi_q}{dt} = R_sI_q - \frac{\omega}{\omega_s}\psi_d + V_q \tag{57}$$

$$\frac{1}{\omega_s}\frac{d\psi_o}{dt} = R_sI_o + V_o \tag{58}$$

$$T_{do}'\frac{dE_q'}{dt} = -E_q' - (X_d - X_d')[I_d - \frac{X_d' - X_d''}{(X_d' - X_{\ell s})^2}(\psi_{1d} + (X_d'$$
$$-X_{\ell s})I_d - E_q' + S_{1d}^{(2)}(Y_2))] - S_{fd}^{(2)}(Y_2) + E_{fd} \tag{59}$$

$$T_{do}''\frac{d\psi_{1d}}{dt} = -\psi_{1d} + E_q' - (X_d' - X_{\ell s})I_d - S_{1d}^{(2)}(Y_2) \tag{60}$$

$$T_{qo}'\frac{dE_d'}{dt} = -E_d' + (X_q - X_q')[I_q - \frac{X_q' - X_q''}{(X_q' - X_{\ell s})^2}(\psi_{2q} + (X_q'$$
$$-X_{\ell s})I_q + E_d' + S_{2q}^{(2)}(Y_2))] + S_{1q}^{(2)}(Y_2) \tag{61}$$

$$T_{qo}''\frac{d\psi_{2q}}{dt} = -\psi_{2q} - E_d' - (X_q' - X_{\ell s})I_q - S_{2q}^{(2)}(Y_2) \tag{62}$$

$$\frac{d\delta}{dt} = \omega - \omega_s \tag{63}$$

$$\frac{2H}{\omega_s}\frac{d\omega}{dt} = T_M - (\psi_dI_q - \psi_qI_d) - T_{FW} \tag{64}$$

with the three algebraic equations,

$$\psi_d = -X_d''I_d + \frac{(X_d'' - X_{ls})}{(X_d' - X_{ls})}E_q' + \frac{(X_d' - X_d'')}{(X_d' - X_{ls})}\psi_{1d}$$

$$-S_d^{(2)}(Y_2) \tag{65}$$

$$\psi_q = -X_q''I_q - \frac{(X_q'' - X_{ls})}{(X_q' - X_{ls})}E_d' + \frac{(X_q' - X_q'')}{(X_q' - X_{ls})}\psi_{2q}$$

$$-S_q^{(2)}(Y_2) \tag{66}$$

$$\psi_o = -X_{ls}I_o - S_o^{(2)}(Y_2) \tag{67}$$

$$Y_2 = [I_d \; E_q' \; \psi_{1d} \; I_q \; E_d' \; \psi_{2q} \; I_o]^t \tag{68}$$

The time constants T_{do}', T_{do}'', T_{qo}', T_{qo}'' are defined in terms of the X and R parameters [1].

One purpose for beginning this subsection by returning to the "abc" variables was to trace the nonlinearities through the transformation and scaling process. This ensures that the resulting model with nonlinearities is in some sense consistent. This was partly motivated by the proliferation of different methods to account for saturation in the literature. For example, the literature talks about "X_{md} saturating," or X_{md} being a function of the dynamic states. This could imply that many constants would change when saturation is considered. With the presentation given above, it is clear that all constants can be left unchanged while the nonlinearities are included in a set of functions to be specified based on some design calculation or test procedure.

The saturation functions cannot be arbitrarily specified. They must be such that the assumptions of a conservative coupling field are satisfied [14]. A set of saturation functions that do satisfy the assumption of a conservative coupling field is,

$$S_d^{(2)} = S_{1d}^{(2)} = S_q^{(2)} = S_{1q}^{(2)} = S_{2q}^{(2)} = S_o^{(2)} = 0 \tag{69}$$

$$S_{fd}^{(2)} = S_G(E_q') \tag{70}$$

where S_G is found from the open-circuit test,

$$X_{md}I_{fd} = V_{toc} + S_G(V_{toc}) \tag{71}$$

A typical analytical approximation of S_G is,

$$S_G(V) = A(e^{BV} - 1) \qquad (72)$$

where A and B are chosen to match the open-circuit magnetization curve at two points. This together with $S_G(0) = 0$ gives a three point fit.

3   Synchronous Machine Control Models

The primary objective of an electrical power system is to maintain balanced sinusoidal voltages with virtually constant magnitude and frequency. In the synchronous machine models of the last section, the terminal constraints (relationships between V_d, I_d, V_q, I_q, V_o, I_o) were not specified. These will be discussed in the next section. In addition, the two quantities E_{fd} and T_M were left as inputs to be specified. Specifying E_{fd} and T_M to be constants in the model means that the machine does not have voltage or speed (and hence frequency) control. If a synchronous machine is to be useful for a wide range of operating conditions, it should be capable of participating in the attempt to maintain constant voltage and frequency. This means that E_{fd} and T_M should be systematically adjusted to accommodate any change in terminal constraints. The physical device which provides the value of E_{fd} is called the exciter. The physical device which provides the value of T_M is called the prime mover. This section is devoted to basic mathematical models of these components and their associated control systems.

3.1   Exciter Model

One primary reason for using three-phase generators is the constant electrical torque developed in steady state by the interaction of the magnetic fields produced by the armature ac currents with the field dc currents. Furthermore, for balanced three-phase machines, a dc current can be produced in the field winding by a dc voltage source. In steady state, adjustment of the field voltage changes the field current and therefore the terminal voltage. Perhaps the simplest scheme for voltage control would be a battery with a rheostat adjusted voltage divider connected to the field winding through slip rings. Manual adjustment of the rheostat could be used to continuously react to changes in operating conditions to maintain a voltage magnitude at some point. Since large amounts of power are normally required for the field excitation, the control device is usually not a battery, and is referred to as the main exciter. This main exciter may be either a dc generator driven off the main shaft (with brushes and slip rings), an inverted ac generator driven off the main shaft (brushless with rotating diodes), or a static device such as an ac to dc converter fed from the synchronous machine terminals or auxilliary power (with slip rings). The main exciter may have a pilot exciter which provides the means for changing the output of the main exciter. In any case, E_{fd} is normally not manipulated directly, but is changed through the actuation of the exciter. A basic model for an exciter is [16, 17]

$$T_E \frac{dE_{fd}}{dt} = -(K_E + S_E(E_{fd})) E_{fd} + V_R \qquad (73)$$

where K_E is typically 1.0 for a separately excited dc generator and a small negative number for a self-excited dc generator. The saturation function S_E is similar to that for synchronous machines. The input V_R is normally the scaled output of the amplifier (or pilot exciter) which is applied to the field of the main exciter.

3.2   Voltage Regulator Model

The exciter provides the mechanism for controlling the synchronous machine terminal voltage magnitude. In order to automatically control terminal voltage, a transducer signal must be compared to a reference voltage and amplified to produce the exciter input V_R. The amplifier can be a pilot exciter or a solid state amplifier. In either case, the amplifier is often modeled as,

$$T_A \frac{dV_R}{dt} = -V_R + K_A V_{in} \qquad (74)$$

$$V_R^{min} \le V_R \le V_R^{max} \qquad (75)$$

where V_{in} is the amplifier input, T_A is the amplifier time constant, and K_A is the amplifier gain. The V_R limit can be multivalued to allow a higher limit during transients. The steady-state limit would be lower to reflect thermal constraints on the exciter and synchronous machine field winding. Recall that V_R is the scaled input to the main exciter. We have assumed that the amplifier data has been scaled according to our given per-unit system.

If the voltage V_{in} is simply the error voltage produced by the difference between a reference voltage and conditioned potential transformer connected to the synchronous machine terminals, the closed-loop control system can exhibit instabilities. This can be seen by noting that the self excited dc exciter can have a negative K_E such that its open loop eigenvalue is positive for small saturation S_E. Even without this potential instability, there is always a need to shape the regulator response to achieve desirable dynamic performance. In many standard excitation systems, this is accomplished through a stabilizing transformer whose input is connected to the output of the exciter and whose output voltage is subtracted from the amplifier input. For V_F the scaled output of the stabilizing transformer, the dynamic model is,

$$T_F \frac{dV_F}{dt} = -V_F + K_F \left(-\frac{K_E + S_E(E_{fd})}{T_E} E_{fd} + \frac{V_R}{T_E} \right) \qquad (76)$$

Another form of this model is often used by defining,

$$R_f \triangleq \frac{K_F}{T_F} E_{fd} - V_F \qquad (77)$$

With R_f (called rate feedback) as the dynamic state,

$$T_F \frac{dR_f}{dt} = -R_f + \frac{K_F}{T_F} E_{fd} \qquad (78)$$

This form will be used throughout the remainder of the chapter.

In keeping with the philosophy of this chapter as one of fundamental dynamic modeling, we conclude this section with a summary of a fundamental model of an excitation system:

$$T_E \frac{dE_{fd}}{dt} = -(K_E + S_E(E_{fd}))E_{fd} + V_R \qquad (79)$$

$$T_F \frac{dR_f}{dt} = -R_f + \frac{K_F}{T_F} E_{fd} \qquad (80)$$

$$T_A \frac{dV_R}{dt} = -V_R + K_A R_f - \frac{K_A K_F}{T_F} E_{fd}$$

$$+ K_A(V_{ref} - V_t) \qquad (81)$$

$$V_R^{min} \leq V_R \leq V_R^{max} \qquad (82)$$

The above excitation system model is often referred to as an IEEE type I model [16].

3.3   Turbine Model

The frequency of the ac voltage at the terminals of a synchronous machine is determined by its shaft speed and the number of magnetic poles of the machine. The steady-state speed of a synchronous machine is determined by the speed of the prime mover which drives its shaft. Typical prime movers are diesel engines, gasoline engines, steam turbines, hydro turbines (water wheels) and gas turbines. The prime mover output affects the input T_M in the model of section 2. This subsection presents basic models for steam turbines.

Steam plants consist of a fuel supply to a steam boiler which supplies a steam chest. The steam chest contains pressurized steam which enters a high-pressure (HP) turbine through a steam valve. The power into the high-pressure turbine is proportional to the valve opening. A nonreheat system would then terminate in the condensor and cooling systems, with the HP turbine shaft connected to the synchronous machine. It is common to include additional stages such as the intermediate-(IP) and low-(LP) pressure turbines. The steam is reheated upon leaving the high-pressure turbine and either reheated or simply crossed over between the IP and LP turbines. The dynamics which are normally represented are the steam chest delay, the reheat delay and/or the crossover piping delay. In a tandem connection, all stages are on the same shaft. In a cross-compound system the different stages may be connected on different shafts. These two shafts then supply the torque for two generators. In this analysis, we model the steam chest dynamics, the single reheat dynamics, and the mass dynamics for a two-stage (HP and LP) turbine tandem mounted. In this model we are interested in the effect of steam valve position (power P_{SV}) on synchronous machine torque T_M. The incremental steam chest dynamic model is a simple linear single time constant with unity gain, written in scaled variables as,

$$T_{CH}\frac{d\Delta P_{CH}}{dt} = -\Delta P_{CH} + \Delta P_{SV} \qquad (83)$$

where ΔP_{CH} is the change in output power of the steam chest. This output is either converted into torque on the high-pressure turbine or passed on to the reheat cycle. Let the fraction which is converted into torque be

$$\Delta T_{HP} = K_{HP}\Delta P_{CH} \qquad (84)$$

and the fraction passed on to the reheater be $(1-K_{HP})\Delta P_{CH}$. The dynamics of the HP turbine mass in incremental scaled variables are,

$$\frac{d\Delta\delta_{HP}}{dt} = \Delta\omega_{HP} \qquad (85)$$

$$\frac{2H_{HP}}{\omega_s}\frac{d\Delta\omega_{HP}}{dt} = \Delta T_{HP} - \Delta T_{HL} \qquad (86)$$

where ΔT_{HL} is the incremental change in torque transmitted through the shaft to the low-pressure turbine. This is modeled as a stiff spring,

$$\Delta T_{HL} = -K_{HL}(\Delta\delta_{LP} - \Delta\delta_{HP}) \qquad (87)$$

The reheat process has a time delay which can be modeled similarly as,

$$T_{RH}\frac{d\Delta P_{RH}}{dt} = -\Delta P_{RH} + (1 - K_{HP})\Delta P_{CH} \qquad (88)$$

where ΔP_{RH} is the change in output power of the reheater. Assuming this output is totally converted into torque on the low-pressure (LP) turbine,

$$\Delta T_{LP} = \Delta P_{RH} \qquad (89)$$

the dynamics of the LP turbine mass in incremental scaled variables are,

$$\frac{d\Delta\delta_{LP}}{dt} = \Delta\omega_{LP} \qquad (90)$$

$$\frac{2H_{LP}}{\omega_s}\frac{d\Delta\omega_{LP}}{dt} = \Delta T_{HL} + \Delta T_{LP} - \Delta T_M \qquad (91)$$

where the torque to the connection of the LP turbine to the synchronous machine is assumed to be transmitted through a stiff spring as,

$$\Delta T_M = -K_{LM}(\Delta\delta - \Delta\delta_{LP}) \qquad (92)$$

For operation near an equilibrium point (denoted by superscript o) with

$$P_{CH}^o = P_{SV}^o, \quad T_{HP}^o = T_{HL}^o = K_{HP}P_{CH}^o = -K_{HL}(\delta_{LP}^o - \delta_{HP}^o),$$

$$T_{LP}^o = P_{RH}^o = (1 - K_{HP})P_{CH}^o, \quad \omega_{LP}^o = \omega_s, \omega_{HP}^o = \omega_s,$$

$$T_M^o = P_{CH}^o = -K_{LM}(\delta^o - \delta_{LP}^o) \qquad (93)$$

the actual variables are,

$$P_{CH} = P_{CH}^o + \Delta P_{CH} \qquad (94)$$

$$P_{SV} = P_{SV}^o + \Delta P_{SV} \qquad (95)$$

$$\delta_{HP} = \delta_{HP}^o + \Delta\delta_{HP} \qquad (96)$$

$$\omega_{HP} = \omega_{HP}^o + \Delta\omega_{HP} \qquad (97)$$

$$P_{RH} = P_{RH}^o + \Delta P_{RH} \qquad (98)$$

$$\delta_{LP} = \delta_{LP}^o + \Delta\delta_{LP} \qquad (99)$$

$$\omega_{LP} = \omega_{LP}^o + \Delta\omega_{LP} \qquad (100)$$

$$\delta = \delta^o + \Delta\delta \qquad (101)$$

$$T_M = T_M^o + \Delta T_M \qquad (102)$$

The steam turbine model is written as,

$$T_{CH}\frac{dP_{CH}}{dt} = -P_{CH} + P_{SV} \tag{103}$$

$$\frac{d\delta_{HP}}{dt} = \omega_{HP} - \omega_s \tag{104}$$

$$\frac{2H_{HP}}{\omega_s}\frac{d\omega_{HP}}{dt} = K_{HP}P_{CH} + K_{HL}(\delta_{LP} - \delta_{HP}) \tag{105}$$

$$T_{RH}\frac{dP_{RH}}{dt} = -P_{RH} + (1 - K_{HP})P_{CH} \tag{106}$$

$$\frac{d\delta_{LP}}{dt} = \omega_{LP} - \omega_s \tag{107}$$

$$\frac{2H_{LP}}{\omega_s}\frac{d\omega_{LP}}{dt} = -K_{HL}(\delta_{LP} - \delta_{HP}) + P_{RH} + K_{LM}(\delta - \delta_{LP}) \tag{108}$$

and T_M in the synchronous machine speed equation must be replaced with

$$T_M = -K_{LM}(\delta - \delta_{LP}) \tag{109}$$

The steam valve position P_{SV} will become a dynamic state when the steam governor equations are added.

For rigid shaft couplings,

$$K_{HP}P_{CH} = T_M - P_{RH} \tag{110}$$

and the two turbine masses are added into the synchronous machine inertia to give the following steam turbine model with T_M as a dynamic state,

$$T_{RH}\frac{dT_M}{dt} = -T_M + (1 - \frac{K_{HP}T_{RH}}{T_{CH}})P_{CH}$$
$$+\frac{K_{HP}T_{RH}}{T_{CH}}P_{SV} \tag{111}$$

$$T_{CH}\frac{dP_{CH}}{dt} = -P_{CH} + P_{SV} \tag{112}$$

and T_M remains as a state in the synchronous machine model. It is possible to add more reheat stages and additional detail. It is also possible to further simplify.

For a nonreheat system, simply set $T_{RH} = 0$ in (111)-(112) and the following model is obtained,

$$T_{CH}\frac{dT_M}{dt} = -T_M + P_{SV} \tag{113}$$

where again T_M remains a state in the synchronous machine model and P_{SV} will become a state when the governor is added. The above non-reheat model is often referred to as a type-A steam turbine model [18, 19].

3.4 Speed Governor Model

The prime mover provides the mechanism for controlling the synchronous machine speed and hence terminal voltage frequency. In order to automatically control speed (and therefore frequency), a device must sense either speed or frequency in such a way that comparison with a desired value can be used to create an error signal to take corrective action. A basic model for a single time constant governor reflecting valve position dynamics with speed regulation R (typically .05) is [19],

$$T_{SV}\frac{dP_{SV}}{dt} = -P_{SV} + P_C - \frac{1}{R}\frac{\omega}{\omega_s} \tag{114}$$

$$0 \le P_{SV} \le P_{SV}^{max} \tag{115}$$

In addition to the limit on the valve position P_{SV}, it may also be important to constrain the derivative of P_{SV} as rate limits. The quantity P_C is a control input which can be either a constant, or the output of an automatic generation control (AGC) scheme. In order to provide zero steady-state error in speed (and therefore frequency), an integral control is needed. In multimachine power systems this load frequency control (LFC) is used together with economic dispatch to maintain frequency at minimum cost on an area and systemwide basis. In this case, P_C would be the output of a load reference motor which is driven by an AGC signal based on a unit control error.

4 Integral Manifolds for Model Reduction

This section presents a mathematical technique to reduce the order of dynamic models. The concept of integral manifolds is used later to systematically develop lower-order models. The term manifold in this section refers to a functional relationship between variables. For example, a manifold for z as a function of x is simply another term for the expression,

$$z = h(x) \tag{116}$$

When x is a scalar, the manifold is a line when plotted in the z, x space. When x is two dimensional, the manifold is a surface.

To define an integral manifold and the concept of reduced-order modeling, we need to introduce a multidimensional dynamic model of the form,

$$\frac{dx}{dt} = f(x, z) \qquad\qquad x(o) = x^o \tag{117}$$

$$\frac{dz}{dt} = g(x, z) \qquad\qquad z(o) = z^o \tag{118}$$

An integral manifold for z as a function of x is a manifold

$$z = h(x) \tag{119}$$

which satisfies the differential equation for z. Thus $h(x)$ is an integral manifold of (117)-(118) if it satifies,

$$\frac{\partial h}{\partial x} f(x, h) = g(x, h) \tag{120}$$

If the initial conditions of x and z lie on the manifold ($z^o = h(x^o)$), then the integral manifold is an exact solution of the differential equation (118), and the following reduced-order model is exact

$$\frac{dx}{dt} = f(x, h(x)) \qquad\qquad x(o) = x^o \tag{121}$$

4.1   Integral Manifolds for Linear Systems

The concept of integral manifolds is illustrated in this subsection through a series of examples. Consider the oversimplified system

$$\frac{dx}{dt} = -x + z \qquad\qquad x(o) = x^o \tag{122}$$

$$\frac{dz}{dt} = -10z + 10 \qquad\qquad z(o) = z^o \tag{123}$$

This system clearly has eigenvalues -1 and -10. Suppose we are interested only in the mode associated with the -1 eigenvalue. We make the observation that if z is one at any time, then z remains equal to one for all time. This means that $z = 1$ is an exact integral manifold of this dynamic system. If this integral manifold is substituted into the x differential equation, the reduced-order model (valid only for $z^o = 1$) is,

$$\frac{dx}{dt} = -x + 1 \qquad\qquad x(o) = x^o \qquad (124)$$

which clearly exhibits the exact eigenvalue of interest. If z^o is equal to 1.0 then (124) gives the exact response of x for any x^o. If z^o is not equal to 1.0 then (124) will not give the exact response of x. For this case we define the off-manifold variable η as,

$$\eta \overset{\Delta}{=} z - 1 \qquad (125)$$

which has the dynamics

$$\frac{d\eta}{dt} = -10\eta \qquad\qquad \eta(o) = z^o - 1 \qquad (126)$$

Note that $\eta = 0$ is an exact integral manifold because if $\eta = 0$ at any time, $\eta = 0$ for all time. The exact solution when $\eta(0) \neq 0$ is,

$$\eta(t) = (z^o - 1)e^{-10t} \qquad (127)$$

The integral manifold plus off-manifold solution for z as a function of x and t is,

$$z = 1 + (z^o - 1)e^{-10t} \qquad (128)$$

This gives the exact reduced-order model (for any z^o),

$$\frac{dx}{dt} = -x + 1 + (z^o - 1)e^{-10t} \qquad x(o) = x^o \qquad (129)$$

While this may have been obvious from the beginning, the steps which led to this result are important for cases where it is not obvious.

We now extend this concept to the more general case with coupling in both equations as,

$$\frac{dx}{dt} = -x + z \qquad (130)$$

$$\frac{dz}{dt} = -10x - 10z \qquad (131)$$

The eigenvalues of this system are -2.3 and -8.7. If we are only interested in capturing the mode with eigenvalue -2.3, we propose that this mode is associated with the state x and seek to eliminate z from (130)-(131) through an integral manifold of the general linear form

$$z = hx + c \tag{132}$$

Substitution into (131) (and using (130)) gives,

$$h(-x + hx + c) = -10x - 10hx - 10c \tag{133}$$

For arbitrary x, this has the solutions,

$$c = 0 \qquad\qquad h = -1.3 \tag{134}$$

$$c = 0 \qquad\qquad h = -7.7 \tag{135}$$

Thus there are two integral manifolds for z

$$IM\#1: \quad z = -1.3x \tag{136}$$

$$IM\#2: \quad z = -7.7x \tag{137}$$

Substitution of IM#1 into (130) gives the slow part of x,

$$\frac{dx}{dt} = -2.3x \tag{138}$$

and substitution of IM#2 into (130) gives the fast part of x,

$$\frac{dx}{dt} = -8.7x \tag{139}$$

Several important points are illustrated here. First, integral manifolds can be used to decompose systems. Second, integral manifolds are not unique. Third, the choice of the integral manifold determines the phenomena retained in the reduced-order model. To understand more about this technique, consider the same system with the introduction of the small parameter ϵ ($1/10$ in the above example), and the inclusion of initial conditions,

$$\frac{dx}{dt} = -x + z \qquad\qquad x(o) = x^o \tag{140}$$

$$\epsilon \frac{dz}{dt} = -x - z \qquad\qquad z(o) = z^o \tag{141}$$

Again we seek a linear manifold,

$$z = h(\epsilon)x + c(\epsilon) \tag{142}$$

where we presume that h and c would normally depend on the small parameter ϵ. Substitution of (142) into (140)-(141) gives,

$$\epsilon h(\epsilon)(-x + h(\epsilon)x + c(\epsilon)) = -x - h(\epsilon)x - c(\epsilon) \tag{143}$$

Again for arbitrary x, the solutions are,

$$c(\epsilon) = 0, \quad h(\epsilon) = -\frac{1-\epsilon}{2\epsilon} + \frac{1}{2\epsilon}\sqrt{(1-\epsilon)^2 - 4\epsilon}, \quad h(o) = -1 \tag{144}$$

$$c(\epsilon) = 0, \quad h(\epsilon) = -\frac{1-\epsilon}{2\epsilon} - \frac{1}{2\epsilon}\sqrt{(1-\epsilon)^2 - 4\epsilon}, \quad h(o) = -\infty \tag{145}$$

For positive $\epsilon < 1$, these solutions exist only for,

$$0 \leq \epsilon \leq 0.17157 \tag{146}$$

This makes sense since the original system has complex eigenvalues for ϵ less than one but greater that 0.17157. It would not be possible to capture a complex mode from a single real state equation.

This simple example illustrates that integral manifolds may not exist, and may not be unique. If we are interested in the "slow mode" which we propose is associated with the variable x, we need a systematic way to compute the correct integral manifold. If z is infinitely fast ($\epsilon = 0$) the integral manifold of interest is $z = -x$ ($h = -1$). When ϵ is near zero, we propose that the integral manifold of interest should be near -1. To systematically compute h for ϵ near zero, we expand $h(\epsilon)$ in a power series in ϵ as,

$$h(\epsilon) = h_o + \epsilon h_1 + \epsilon^2 h_2 + \dots \tag{147}$$

and return to the example by substituting into (141) and (142) (using $c(\epsilon) = 0$)

$$\epsilon(h_o + \epsilon h_1 + \epsilon^2 h_2 + \dots)(-x + (h_o + \epsilon h_1 + \epsilon^2 h_2 + \dots)x) =$$
$$-x - (h_o + \epsilon h_1 + \epsilon^2 h_2 + \dots)x \tag{148}$$

For arbitrary x, we solve for $h_o, h_1, h_2, \dots$ by equating coefficients of powers of ϵ:

$$\epsilon^o \quad : \quad 0 = -1 - h_o \quad or \quad h_o = -1 \tag{149}$$

$$\epsilon^1 \quad : \quad h_o(-1 + h_o) = -h_1 \quad or \quad h_1 = h_o - h_o^2 = -2 \tag{150}$$

Stopping with these terms,

$$h(\epsilon) \approx -1 - 2\epsilon \tag{151}$$

This approximates the exact integral manifold of (140)-(141) as

$$z \approx -(1 + 2\epsilon)x \qquad (152)$$

Using this in (140) gives the approximate reduced-order model (valid when the initial conditions satisfy $z^o = h(\epsilon)x^o$),

$$\frac{dx}{dt} \approx -(2 + 2\epsilon)x \qquad x(o) = x^o \qquad (153)$$

This model could be improved to any degree of accuracy by including additional terms of $h(\epsilon)$. Since these terms were computed from a power series near the integral manifold of interest (slow manifold), the correct mode has been captured. While computation of the integral manifold is the primary task in model reduction, it may not make sense to find a good approximation and then ignore the possibility that z does not start on the manifold $(z^o \neq h(\epsilon)x^o)$.

4.2   Integral Manifolds for Nonlinear Systems

While there are many reduction techniques which can be applied to linear systems, the primary advantage of the integral manifold approach is its straightforward extension to nonlinear systems. We begin by considering the general form,

$$\frac{dx}{dt} = f(x, z) \qquad x(o) = x^o \qquad (154)$$

$$\frac{dz}{dt} = g(x, z) \qquad z(o) = z^o \qquad (155)$$

In order to analyze the dynamics of x it is necessary to also compute the dynamics of z. A reduced-order model involving only x requires the elimination of z from (154). An integral manifold for z as a function of x has the form

$$z = h(x) \qquad (156)$$

and must satisfy (155)

$$\frac{\partial h}{\partial x} f(x, h) = g(x, h) \qquad (157)$$

While in general it is very difficult to find such an integral manifold, there are several very important cases where h can either be found exactly or approximated to any degree of accuracy. We begin with a generic example which closely resembles the single synchronous machine connected to an infinite bus with stator transients,

$$\frac{dx_1}{dt} = x_2 - 1 \tag{158}$$

$$\frac{dx_2}{dt} = f(x_1, x_2, z_1, z_2) \tag{159}$$

$$\frac{dz_1}{dt} = -\sigma z_1 + x_2 z_2 + V \sin x_1 \tag{160}$$

$$\frac{dz_2}{dt} = -\sigma z_2 - x_2 z_1 + V \cos x_1 \tag{161}$$

We suppose that only x_1 and x_2 are of interest and look for an integral manifold of the form,

$$z_1 = h_1(x_1, x_2) \tag{162}$$

$$z_2 = h_2(x_1, x_2) \tag{163}$$

This two-dimensional integral manifold must satisfy,

$$\frac{\partial h_1}{\partial x_1}(x_2 - 1) + \frac{\partial h_1}{\partial x_2} f(x_1, x_2, h_1, h_2) =$$
$$-\sigma h_1 + x_2 h_2 + V \sin x_1 \tag{164}$$

$$\frac{\partial h_2}{\partial x_1}(x_2 - 1) + \frac{\partial h_2}{\partial x_2} f(x_1, x_2, h_1, h_2) =$$
$$-\sigma h_2 - x_2 h_1 + V \cos x_1 \tag{165}$$

These partial differential equations can be solved by first assuming that h_1 and h_2 are independent of x_2 and then equating coefficients of x_2 to give,

$$\frac{\partial h_1}{\partial x_1} = h_2 \quad , \quad \frac{\partial h_1}{\partial x_1} = \sigma h_1 - V \sin x_1 \tag{166}$$

$$\frac{\partial h_2}{\partial x_1} = -h_1 \quad , \quad \frac{\partial h_2}{\partial x_1} = \sigma h_2 - V \cos x_1 \tag{167}$$

Eliminating the partials gives the solution,

$$h_1 = V \cos \alpha \cos(\alpha - x_1) \tag{168}$$

$$h_2 = V \cos \alpha \sin(\alpha - x_1) \tag{169}$$

where $tan \: \alpha = \sigma$. Thus if the initial conditions on z_1, z_2 and x_1 satisfy (162)-(163), then substitution of (162)-(163) into (159) gives an exact reduced-order model. Such exact integral manifolds are rare in dynamic systems. The synchronous machine with zero stator resistance connected to an infinite bus stands out as a unique device with this property [22].

A very broad class of systems where integral manifolds can often be found or approximated to any degree of accuracy is the class of two-time-scale systems of the form,

$$\frac{dx}{dt} = f(x,z) \quad x(o) = x^o \tag{170}$$

$$\epsilon \frac{dz}{dt} = g(x,z) \quad z(o) = z^o \tag{171}$$

These systems are called two-time-scale because when ϵ is small, the z variables are predominantly fast and the x variables are predominantly slow. This is clear because the derivative of z with respect to time is proportional to $1/\epsilon$ which is large for small ϵ. As in the linear case of the last subsection, we propose an integral manifold for z as a function of x and ϵ,

$$z = h(x, \epsilon) \tag{172}$$

We assume that ϵ is sufficiently small so that the manifold can be expressed as a power series in ϵ as,

$$h(x, \epsilon) = h_o(x) + \epsilon h_1(x) + \epsilon^2 h_2(x) + ... \tag{173}$$

Substitution into (172) and then (171) gives,

$$\epsilon \left(\frac{\partial h_o}{\partial x} + \epsilon \frac{\partial h_1}{\partial x} + ... \right) f(x,h) = g(x,h) \tag{174}$$

Expanding f and g about $\epsilon = 0$,

$$f(x,h) = f(x,h_o) + \epsilon \frac{\partial f}{\partial z} |_{z=h_o} h_1 + ... \tag{175}$$

$$g(x,h) = g(x,h_o) + \epsilon \frac{\partial g}{\partial z} |_{z=h_o} h_1 + ... \tag{176}$$

The partial differential equation to be solved is,

$$\epsilon \left(\frac{\partial H_o}{\partial x} + \epsilon \frac{\partial h_1}{\partial x} + ... \right) \: (f(x,h_o) + \epsilon \frac{\partial f}{\partial z} |_{z=h_o} h_1 + ...) =$$

$$g(x,h_o) + \epsilon \frac{\partial g}{\partial z} |_{z=h_o} h_1 + ... \tag{177}$$

Equating coefficients of powers of ϵ produces a set of algebraic equations to be solved for h_o, h_1, h_2...,

$$\epsilon^o : \qquad 0 = g(x, h_o) \tag{178}$$

$$\epsilon^1 : \qquad \frac{\partial h_o}{\partial x} f(x, h_o) = \frac{\partial g}{\partial z} |_{z=z_o} h_1 \tag{179}$$

$$etc.$$

Clearly the most important equation is (178) which requires the solution of the nonlinear equation for h_o. Once this is found, the solution for h_1 simply requires nonsingular $\partial g / \partial z$. Normally if (178) can be solved, the nonsingularity of $\partial g / \partial z$ follows.

As in the linear case, the use of an integral manifold in a reduced-order model can give exact results only if it is found exactly and if the initial conditions start on it. If the initial conditions do not start on the manifold (do not satisfy (172)), an error will be introduced. To eliminte this error, it is necessary to compute the off-manifold dynamics. This is done by introducing the off-manifold variable,

$$\eta \overset{\Delta}{=} z - h(x, \epsilon) \tag{180}$$

with the following dynamics,

$$\epsilon \frac{d\eta}{dt} = g(x, \eta + h) - \epsilon \frac{\partial h}{\partial x} f(x, \eta + h) \tag{181}$$

and the initial condition,

$$\eta(o) = z^o - h(x^o, \epsilon) \tag{182}$$

These off-manifold dynamics are normally difficult to compute because they require x. As a first approximation, Equation (181) could be solved using x as a constant equal to its initial condition. This is a reasonably good approximation because the off-manifold dynamics should decay rapidly (if they are stable) before x changes significantly.

It can be shown that if z is stable, the use of h_o as an approximation for h and neglecting off-manifold dynamics only introduces "order ϵ" error into the slow variable x response [23, 24]. If further accuracy is desired and h is approximated by $h_o + \epsilon h_1$, there will still be "order ϵ" error if the off-manifold dynamics are neglected. In order to reduce the error to "order ϵ^2", it is necessary to include h_1 and approximate η to order ϵ^2. This can be done by approximating the off-manifold dynamics as,

$$\epsilon \frac{d\eta}{dt} \approx g(x^o, \eta + h_o + \epsilon h_1) - \epsilon \frac{\partial h_o}{\partial x} f(x^o, \eta + h_o) \tag{183}$$

with

$$\eta(o) = z^o - h_o(x^o) - \epsilon h_1(x^o) \tag{184}$$

and h_o, h_1, $\partial h_o / \partial x$ evaluated at $x = x^o$.

In many cases, the first approximation of integral manifolds corresponds to well known "quasi-steady-state" models. The advantage of an integral manifold approach lies in the ability to generate improved reduced-order models without resorting to increased model order. In addition, it provides a sound theoretical basis for reduced-order modeling and can in some cases give criteria for determining when reduced-order models are valid [23, 24].

5   Multimachine Dynamic Models

The models presented in the earlier sections did not specify the generator terminal constraints. In this section we interconnect the synchronous machine models through models of transformers, transmission lines and loads. This interconnection is greatly simplified from a theoretical point of view if all elements have pure resistive/inductive models. In this case it is easy to formulate the dynamic state equations of the network and load transients in terms of loop flux linkages [20]. For notation, we adopt the following "number" symbols,

   m = number of synchronous machines
          (numbered i=1,...,m)

   n = total number of system three-phase buses
          (excluding the reference or ground bus)

   b = total number of machines plus transformers
          plus lines plus loads (total three-phase branches)

It is convenient to transform all synchronous machine stator and network variables into the synchronously-rotating reference frame. Variables in the synchronously-rotating reference frame are denoted with D,Q,0 subscripts. For synchronous machine i, the relationship between the two reference frames is,

$$(V_{Di} + jV_{Qi}) = (V_{di} + jV_{qi})e^{j(\delta_i - \pi/2)} \quad i = 1, ..., m \tag{185}$$

$$V_{0i} = V_{oi} \quad\quad\quad\quad i = 1, ..., m \tag{186}$$

and similarily for current and flux linkage. For all transformers, transmission lines and loads modeled as balanced three-phase R-L elements, the differential equations for the synchronous machine stators, transformers, transmission lines and R-L loads in terms of branch variables are,

$$\epsilon \frac{d\psi_{Di}}{dt} = R_i I_{Di} + \psi_{Qi} + V_{Di} \quad i = 1, ..., b \tag{187}$$

$$\epsilon \frac{d\psi_{Qi}}{dt} = R_i I_{Qi} - \psi_{Di} + V_{Qi} \quad i = 1, ..., b \tag{188}$$

$$\epsilon \frac{d\psi_{Oi}}{dt} = R_i I_{Oi} + V_{Oi} \quad i = 1, ..., b \tag{189}$$

where $\epsilon = 1/\omega_s$, $R_i = R_{si}$ i=1,...,m and the branch-flux linkages are related to the branch currents by algebraic equations. These differential equations are not normally independent. To properly present the model, we need to add the topology connection that converts the b branch equations into independent loops. When this is done, the independent differential equations in vector form are,

$$\epsilon \frac{d\psi_{Dloop}}{dt} = R_{Dloop} I_{Dloop} + \psi_{Qloop} \tag{190}$$

$$\epsilon \frac{d\psi_{Qloop}}{dt} = R_{Qloop} I_{Qloop} - \psi_{Dloop} \tag{191}$$

$$\epsilon \frac{d\psi_{Oloop}}{dt} = R_{Oloop} I_{Oloop} \tag{192}$$

where the loop-flux-linkage vectors are related to the loop-current vectors by algebraic equations.

5.1 Elimination of Stator/Network/Load Transients

While it is possible to perform simulations with full detail, virtually all system studies are done with the so-called stator/network transients neglected. This is done because the above dynamics are very fast compared to the remaining dynamics. This is due to the smallness of ϵ compared to other time constants in the model. A formal reduction of this model has been given in [20]-[22]. Additional background is given in [25] and [26]. The reduction uses the basic concepts of integral manifolds as introduced in the last section. We summarize this result here and begin by noting that if ϵ is sufficiently small (we assume $\epsilon = 1/\omega_s$ is sufficiently small), there exists an integral manifold for ψ_{Dloop}, ψ_{Qloop}, ψ_{Oloop} in terms of the remaining slower dynamic states. A first approximation of the integral manifold is obtained by setting ϵ equal to zero. In terms of the branch variables this gives,

$$0 = R_i I_{Di} + \psi_{Qi} + V_{Di} \quad i = 1, ..., b \qquad (193)$$

$$0 = R_i I_{Qi} - \psi_{Di} + V_{Qi} \quad i = 1, ..., b \qquad (194)$$

$$0 = R_i I_{Oi} + V_{Oi} \quad i = 1, ..., b \qquad (195)$$

The algebraic equations which relate the branch-flux linkages to branch currents (with $S_{di}^{(2)} = S_{qi}^{(2)} = S_{oi}^{(2)} = 0$) are,

$$\psi_{di} = -X_{di}'' I_{di} + \frac{(X_{di}'' - X_{\ell si})}{(X_{di}' - X_{\ell si})} E_{qi}' + \frac{(X_{di}' - X_{di}'')}{(X_{di}' - X_{\ell si})} \psi_{1di} \qquad (196)$$

$$\psi_{qi} = -X_{qi}'' I_{qi} - \frac{(X_{qi}'' - X_{\ell si})}{(X_{qi}' - X_{\ell si})} E_{di}' + \frac{(X_{qi}' - X_{qi}'')}{(X_{qi}' - X_{\ell si})} \psi_{2qi}$$

$$\psi_{oi} = -X_{\ell si} I_{oi} \qquad (197)$$

all for i=1,...,m where

$$(V_{Di} + jV_{Qi}) = (V_{di} + jV_{qi})e^{j(\delta_i - \frac{\pi}{2})} \quad i = 1, ..., m \qquad (198)$$

$$(I_{Di} + jI_{Qi}) = (I_{di} + jI_{qi})e^{j(\delta_i - \frac{\pi}{2})} \quad i = 1, ..., m \qquad (199)$$

$$(\psi_{Di} + j\psi_{Qi}) = (\psi_{di} + j\psi_{qi})e^{j(\delta_i - \frac{\pi}{2})} \quad i = 1, ..., m \qquad (200)$$

and,

$$\psi_{Di} = -X_{epi} I_{Di} \quad i = m + 1, ..., b \qquad (201)$$

$$\psi_{Qi} = -X_{epi} I_{Qi} \quad i = m + 1, ..., b \qquad (202)$$

$$\psi_{Oi} = -X_{eoi} I_{Oi} \quad i = m + 1, ..., b \qquad (203)$$

with X_{epi} and X_{eoi} being the positive- and zero- sequence reactances respectively. For three-wire or balanced-four-wire systems, all of the "0" variables are zero. The D and Q equations can be combined to form a single complex equation as,

$$0 = R_i(I_{Di} + jI_{Qi}) - j(\psi_{Di} + j\psi_{Qi}) + (V_{Di} + jV_{Qi})$$

$$i = 1, ..., b \qquad (204)$$

and

$$\psi_{Di} + j\psi_{Qi} = -X''_{di}(I_{di} + jI_{qi})e^{j(\delta_i - \pi/2)} - j\overline{E}''_i \quad i = 1, ..., m \quad (205)$$

$$\psi_{Di} + j\psi_{Qi} = -X_{epi}(I_{Di} + jI_{Qi}) \quad i = m+1, ..., b \quad (206)$$

where,

$$\overline{E}''_i = \left[(X''_{qi} - X''_{di})I_{qi} + \frac{(X''_{qi} - X_{\ell si})}{(X'_{qi} - X_{\ell si})} E'_{di} - \frac{(X'_{qi} - X''_{qi})}{(X'_{qi} - X_{\ell si})} \psi_{2qi} \right.$$

$$\left. + j\frac{(X''_{di} - X_{\ell si})}{(X'_{di} - X_{\ell si})} E'_{qi} + j\frac{(X'_{di} - X''_{di})}{(X'_{di} - X_{\ell si})} \psi_{1di} \right] e^{j(\delta_i - \pi/2)}$$

$$i = 1, ..., m \quad (207)$$

The algebraic equations for the m synchronous machines have the circuit representation of Figure 2. The algebraic equations for the $b - m$ branches have the circuit representation of Figure 3.

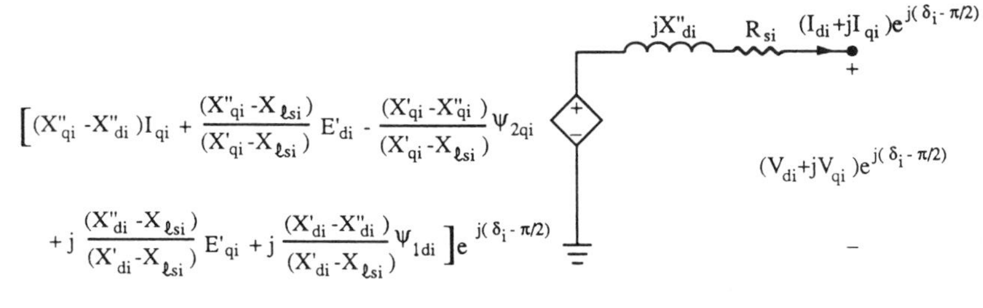

Figure 2: Multimachine subtransient dynamic circuit (i=1,...,m)

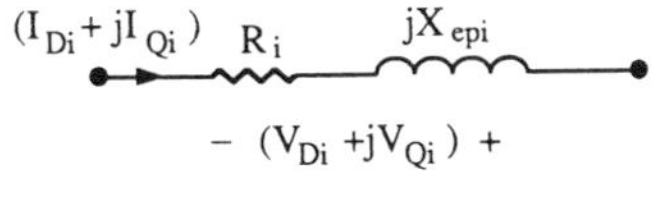

Figure 3: Network plus R-L load dynamic circuit (i=m+1,...,b)

The interconnection of each individual machine is done by simply interconnecting the b branches into the multimachine plus network dynamic circuit.

We emphasize that while these circuits look like steady-state phasor circuits, they are simply circuit representations of the algebraic equations obtained by the elimination of the stator/network transients using an approximation of the integral manifold for the fast dynamics.

Before presenting the multimachine dynamic model without stator/network transients, we now make a very big assumption. We assume that an integral manifold also exists for all dynamic states associated with network elements and loads that are not simple R-L elements. We assume that the integral manifold for the network states can be approximated by the same representation as above (allowing X_{ep} to be negative). We also assume that the integral manifold for the load electrical dynamic states can be approximated by sets of two real algebraic equations which can be written as sets of complex equations of the following general form

$$(V_{Di} + jV_{Qi})(I_{LDi} - jI_{LQi}) = P_{Li}(V_i) + jQ_{Li}(V_i) \quad i = 1, ..., n \quad (208)$$

where,

$$V_i = \sqrt{V_{Di}^2 + V_{Qi}^2} \quad i = 1, ..., n \tag{209}$$

and $P_{Li}(V_i)$, $Q_{Li}(V_i)$ are specified functions of V_i, that would normally be negative for passive loads.

With this generalized model for the network and loads, we write the algebraic equations for the interconnection of m machine circuits with all transformers, lines and loads using the standard network bus admittance matrix. In order to convert from branch subscript notation to bus subscript notation, we assume that loads are present at all n buses and connected to the reference bus. These n branch voltages are then the n bus voltages denoted as,

$$V_i e^{j\theta_i} = (V_{Di} + jV_{Qi}) \quad i = 1, ..., n \tag{210}$$

with this notation, the network interconnection of machines is done by,

$$(I_{di} + jI_{qi})e^{j(\delta_i - \pi/2)} + (I_{LDi} + jI_{LQi}) = \sum_{k=1}^{n} Y_{ik} e^{j\alpha_{ik}} V_k e^{j\theta_k}$$

$$i = 1, ..., m \tag{211}$$

$$(I_{LDi} + jI_{LQi}) = \sum_{k=1}^{n} Y_{ik} e^{j\alpha_{ik}} V_k e^{j\theta_k}$$

$$i = m + 1, ..., n \tag{212}$$

where $Y_{ik}e^{j\alpha_{ik}}$ is the ik^{th} entry of the network bus admittance matrix. It has the same formulation as the admittance matrix used for standard load-flow analysis.

Choosing a single time constant, rigid shaft turbine model and neglecting saturation ($S_{fd}^{(2)} = S_{1d}^{(2)} = S_{1q}^{(2)} = S_{2q}^{(2)} = 0$) the multimachine dynamic model without stator/network transients is,

$$T'_{doi}\frac{dE'_{qi}}{dt} = -E'_{qi} - (X_{di} - X''_{di})[I_{di} - \frac{(X'_{di} - X''_{di})}{(X'_{di} - X_{\ell si})^2}$$

$$(\psi_{1di} + (X'_{di} - X_{\ell si})I_{di} - E'_{qi})] + E_{fdi}$$

$$i = 1, ..., m \qquad (213)$$

$$T''_{doi}\frac{d\psi_{1di}}{dt} = -\psi_{1di} + E'_{qi} - (X'_{di} - X_{\ell si})I_{di} \quad i = 1, ..., m \qquad (214)$$

$$T'_{qoi}\frac{dE'_{di}}{dt} = -E'_{di} + (X_{qi} - X'_{qi})\left[I_{qi} - \frac{(X'_{qi} - X''_{qi})}{(X'_{qi} - X_{\ell si})^2}\right.$$

$$\left.(\psi_{2qi} + (X'_{qi} - X_{\ell si})I_{qi} + E'_{di})\right] \quad i = 1, ..., m \qquad (215)$$

$$T''_{qoi}\frac{d\psi_{2qi}}{dt} = -\psi_{2qi} - E'_{di} - (X'_{qi} - X_{\ell si})I_{qi} \quad i = 1, ..., m \qquad (216)$$

$$\frac{d\delta_i}{dt} = \omega_i - \omega_s \quad i = 1, ..., m \qquad (217)$$

$$\frac{2H_i}{\omega_s}\frac{d\omega_i}{dt} = T_{Mi} - \frac{(X''_{di} - X_{\ell si})}{(X'_{di} - X_{\ell si})}E'_{qi}I_{qi} - \frac{(X'_{di} - X''_{di})}{(X'_{di} - X_{\ell si})}\psi_{1di}I_{qi}$$

$$-\frac{(X''_{qi} - X_{\ell si})}{(X'_{qi} - X_{\ell si})}E'_{di}I_{di} + \frac{(X'_{qi} - X''_{qi})}{(X'_{qi} - X_{\ell si})}\psi_{2qi}I_{di}$$

$$-(X''_{qi} - X''_{di})I_{di}I_{qi} - T_{FWi} \quad i = 1, ..., m \qquad (218)$$

$$T_{Ei}\frac{dE_{fdi}}{dt} = -(K_{Ei} + S_{Ei}(E_{fdi}))E_{fdi} + V_{Ri} \quad i = 1, ..., m \qquad (219)$$

$$T_{Fi}\frac{dR_{fi}}{dt} = -R_{fi} + \frac{K_{Fi}}{T_{Fi}}E_{fdi} \quad i = 1, ..., m \qquad (220)$$

$$T_{Ai}\frac{dV_{Ri}}{dt} = -V_{Ri} + K_{Ai}R_{fi} - \frac{K_{Ai}K_{Fi}}{T_{Fi}}E_{fdi}$$

$$+K_{Ai}(V_{refi} - V_i) \quad i = 1, ..., m \qquad (221)$$

$$T_{CHi}\frac{dT_{Mi}}{dt} = -T_{Mi} + P_{SVi} \quad i = 1, ..., m \qquad (222)$$

$$T_{SVi}\frac{dP_{SVi}}{dt} = -P_{SVi} + P_{Ci} - \frac{1}{R_i}\frac{\omega_i}{\omega_s} \quad i = 1, ..., m \tag{223}$$

with the limit constraints,

$$V_{Ri}^{min} \le V_{Ri} \le V_{Ri}^{max} \quad i = 1, ..., m \tag{224}$$

$$0 \le P_{SVi} \le P_{SVi}^{max} \quad i = 1, ..., m \tag{225}$$

and the algebraic constraints,

$$0 = V_i e^{j\theta_i} + (R_{si} + jX_{di}'')(I_{di} + jI_{qi})e^{j(\delta_i - \frac{\pi}{2})}$$
$$- \left[(X_{qi}'' - X_{di}'')I_{qi} + \frac{(X_{qi}'' - X_{\ell si})}{(X_{qi}' - X_{\ell si})}E_{di}' \right.$$
$$- \frac{(X_{qi}' - X_{qi}'')}{(X_{qi}' - X_{\ell si})}\psi_{2qi} + j\frac{(X_{di}'' - X_{\ell si})}{(X_{di}' - X_{\ell si})}E_{qi}'$$
$$\left. +j\frac{(X_{di}' - X_{di}'')}{(X_{di}' - X_{\ell si})}\psi_{1di} \right] e^{j(\delta_i - \frac{\pi}{2})}$$
$$i = 1, ..., m \tag{226}$$

$$V_i e^{j\theta_i}(I_{di} - jI_{qi})e^{-j(\delta_i - \frac{\pi}{2})} + P_{Li}(V_i) + jQ_{Li}(V_i) =$$
$$\sum_{k=1}^{n} V_i V_k Y_{ik} e^{j(\theta_i - \theta_k - \alpha_{ik})} \quad i = 1, ..., m \tag{227}$$

$$P_{Li}(V_i) + jQ_{Li}(V_i) = \sum_{k=1}^{n} V_i V_k Y_{ik} e^{j(\theta_i - \theta_k - \alpha_{ik})}$$
$$i = m + 1, ..., n \tag{228}$$

The voltage regulator input voltage is V_i which is automatically defined through the network algebraic constraints. Also, for given functions $P_{Li}(V_i)$ and $Q_{Li}(V_i)$, the $n + m$ complex algebraic equations must be solved for V_i, θ_i (i=1,...,n), I_{di}, I_{qi} (i=1,...,m) in terms of the states δ_i, E_{di}', ψ_{2qi}, E_{qi}', ψ_{1di} (i=1,...,m). The currents can clearly be explicitly eliminated by solving either (226) or (227) and substituting into the differential equations and remaining algebraic equations. This would leave n nonlinear complex algebraic equations to be solved for the n complex voltages $V_i e^{j\theta_i}$.

The Special Case of "Impedance Loads"

In some cases, the above dynamic model can be put in explicit closed form without algebraic equations. We make the special assumptions about the load representations,

$$P_{Li}(V_i) = k_{P2i}V_i^2 \quad i = 1, ..., n \tag{229}$$

$$Q_{Li}(V_i) = k_{Q2i}V_i^2 \quad i = 1, ..., n \tag{230}$$

In this case, the loads and $R_{si} + jX_{di}''$ can be added into the bus admittance matrix diagonal entries to obtain the following algebraic equations

$$
(I_{di} + jI_{qi})e^{j(\delta_i - \frac{\pi}{2})} = \frac{1}{(R_{si} + jX_{di}'')} \left[(X_{qi}'' - X_{di}'')I_{qi} + \frac{(X_{qi}'' - X_{\ell si})}{(X_{qi}' - X_{\ell si})} E_{di}' \right.
$$
$$
- \frac{(X_{qi}' - X_{qi}'')}{(X_{qi}' - X_{\ell si})}\psi_{2qi} + j\frac{(X_{di}'' - X_{\ell si})}{(X_{di}' - X_{\ell si})} E_{qi}'
$$
$$
\left. +j\frac{(X_{di}' - X_{di}'')}{(X_{di}' - X_{\ell si})}\psi_{1di} \right] e^{j(\delta_i - \frac{\pi}{2})} - \left(\frac{1}{R_{si} + jX_{di}''} \right) V_i e^{j\theta_i}
$$
$$
i = 1, ..., m \tag{231}
$$

$$
0 = - \left(\frac{1}{R_{si} + jX_{di}''} \right) \left[(X_{qi}'' - X_{di}'')I_{qi} + \frac{(X_{qi}'' - X_{\ell si})}{(X_{qi}' - X_{\ell si})} E_{di}' \right.
$$
$$
- \frac{(X_{qi}' - X_{qi}'')}{(X_{qi}' - X_{\ell si})}\psi_{2qi} + j\frac{(X_{di}'' - X_{\ell si})}{(X_{di}' - X_{\ell si})} E_{qi}'
$$
$$
\left. +j\frac{(X_{di}' - X_{di}'')}{(X_{di}' - X_{\ell si})}\psi_{1di} \right] e^{j(\delta_i - \frac{\pi}{2})}
$$
$$
+ \sum_{k=1}^{n} (G_{ik}'' + jB_{ik}'')V_k e^{j\theta_k} \quad i = 1, ..., m \tag{232}
$$

$$
0 = \sum_{k=1}^{n} (G_{ik}'' + jB_{ik}'')V_k e^{j\theta_k} \quad i = m+1, ..., n \tag{233}
$$

where $G_{ik}'' + jB_{ik}''$ is the ik^{th} entry of the bus admittance matrix including all "constant-impedance" loads and $1/(R_{si} + jX_{di}'')$ on the ith diagonal.

Clearly all of the $V_k e^{j\theta_k}$ (k=1,...,n) can be eliminated by solving (232) and (233) and substituting into (231) to obtain m complex equations which are linear in I_{di}, I_{qi} (i=1,...,m). These can in principle be solved through the inverse of a matrix which would be a function of δ_i, E'_{di}, ψ_{2qi}, E'_{qi}, ψ_{1di} (i=1,...,m). This inverse can be avoided if the following simplification is made,

$$X''_{qi} = X''_{di} \quad i = 1, ..., m \tag{234}$$

With this assumption, (232) and (233) can be solved for $V_k e^{j\theta_k}$ and substituted into (231) to obtain,

$$
\begin{aligned}
I_{di} + jI_{qi} = \sum_{k=1}^{m} (G''_{red} + jB''_{red}) & \left[\frac{(X''_{qk} - X_{\ell sk})}{(X'_{qk} - X_{\ell sk})} E'_{dk} \right. \\
& \quad - \frac{(X'_{qk} - X''_{qk})}{(X'_{qk} - X_{\ell sk})} \psi_{2qk} + j \frac{(X''_{dk} - X_{\ell sk})}{(X'_{dk} - X_{\ell sk})} E'_{qk} \\
& \quad \left. + j \frac{(X'_{dk} - X''_{dk})}{(X'_{dk} - X_{\ell sk})} \psi_{1dk} \right] e^{j(\delta_k - \delta_i)} \quad i = 1, ..., m \tag{235}
\end{aligned}
$$

where $G''_{red} + jB''_{red}$ is the ik^{th} entry of an $m \times m$ admittance matrix (often called the matrix reduced to "internal nodes"). This can easily be solved for I_{di}, I_{qi} (i=1,...,m) and substituted into the differential equations (213)-(223). Substitution of I_{di} I_{qi} into (226) also gives V_i as a function of the states (needed for (221)) so the resulting dynamic model is in explicit form without algebraic equations.

5.2   Multimachine Two-Axis Model

The reduced-order model of the last subsection still contains the damper winding dynamics of ψ_{1di} and ψ_{2qi}. If T''_{doi} and T''_{qoi} are sufficiently small, there is an integral manifold for these dynamic states. A first approximation of the fast damper winding integral manifold is found by setting T''_{doi} and T''_{qoi} equal to zero in (214) and (216) to obtain

$$0 = -\psi_{1di} + E'_{qi} - (X'_{di} - X_{\ell si})I_{di} \quad i = 1, ..., m \tag{236}$$

$$0 = -\psi_{2qi} - E'_{di} - (X'_{qi} - X_{\ell si})I_{qi} \quad i = 1, ..., m \tag{237}$$

This first approximation of the damper winding integral manifolds is equivalent to a machine without the $1d$ and $2q$ damper windings. An improved approximation of these damper winding integral manifolds would lead to explicit damping torques of the form [21],

$$T_{Damp\,i} = D_i(\omega_i - \omega_s) \tag{238}$$

It is customary to use such an approximation when damper windings are eliminated. When (236) and (237) are used to eliminate ψ_{1di} and ψ_{2qi}, the synchronous machine dynamic circuit is changed from Figure 2 to Figure 4, giving the following multimachine two-axis model (with a damping torque term),

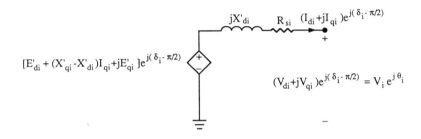

Figure 4: Synchronous machine two-axis model dynamic circuit (i=1,...,m)

$$T'_{doi}\frac{dE'_{qi}}{dt} = -E'_{qi} - (X_{di} - X'_{di})I_{di} + E_{fdi} \quad i = 1, ..., m \qquad (239)$$

$$T'_{qoi}\frac{dE'_{di}}{dt} = -E'_{di} + (X_{qi} - X'_{qi})I_{qi} \quad i = 1, ..., m \qquad (240)$$

$$\frac{d\delta_i}{dt} = \omega_i - \omega_s \quad i = 1, ..., m \qquad (241)$$

$$\frac{2H_i}{\omega_s}\frac{d\omega_i}{dt} = T_{Mi} - E'_{di}I_{di} - E'_{qi}I_{qi} - (X'_{qi} - X'_{di})I_{di}I_{qi}$$

$$-T_{FWi} - D_i(\omega_i - \omega_s) \quad i = 1, ..., m \qquad (242)$$

$$T_{Ei}\frac{dE_{fdi}}{dt} = -(K_{Ei} + S_{Ei}(E_{fdi}))E_{fdi} + V_{Ri} \quad i = 1, ..., m \qquad (243)$$

$$T_{Fi}\frac{dR_{fi}}{dt} = -R_{fi} + \frac{K_{Fi}}{T_{Fi}}E_{fdi} \quad i = 1, ..., m \qquad (244)$$

$$T_{Ai}\frac{dV_{Ri}}{dt} = -V_{Ri} + K_{Ai}R_{fi} - \frac{K_{Ai}K_{Fi}}{T_{Fi}}E_{fdi}$$

$$+K_{Ai}(V_{refi} - V_i) \quad i = 1, ..., m \qquad (245)$$

$$T_{CHi}\frac{dT_{Mi}}{dt} = -T_{Mi} + P_{SVi} \quad i = 1, ..., m \tag{246}$$

$$T_{SVi}\frac{dP_{SVi}}{dt} = -P_{SVi} + P_{Ci} - \frac{1}{R_i}\frac{\omega_i}{\omega_s} \quad i = 1, ..., m \tag{247}$$

with the limit constraints,

$$V_{Ri}^{min} \le V_{Ri} \le V_{Ri}^{max} \quad i = 1, ..., m \tag{248}$$

$$0 \le P_{svi} \le P_{svi}^{max} \quad i = 1, ..., m \tag{249}$$

and the algebraic constraints,

$$0 = V_i e^{j\theta_i} + (R_{si} + jX'_{di})(I_{di} + jI_{qi})e^{j(\delta_i - \frac{\pi}{2})}$$
$$- [E'_{di} + (X'_{qi} - X'_{di})I_{qi} + jE'_{qi}]e^{j(\delta_i - \frac{\pi}{2})}$$
$$i = 1, ..., m \tag{250}$$

$$V_i e^{j\theta_i}(I_{di} - jI_{qi})e^{-j(\delta_i - \frac{\pi}{2})} + P_{Li}(V_i) + jQ_{Li}(V_i)$$
$$= \sum_{k=1}^{n} V_i V_k Y_{ik} e^{j(\theta_i - \theta_k - \alpha_{ik})} \quad i = 1, ..., m \tag{251}$$

$$P_{Li}(V_i) + jQ_{Li}(V_i) = \sum_{k=1}^{n} V_i V_k Y_{ik} e^{j(\theta_i - \theta_k - \alpha_{ik})}$$
$$i = m + 1, ..., n \tag{252}$$

As before, for given functions $P_{Li}(V_i)$ and $Q_{Li}(V_i)$, the $n + m$ complex algebraic equations must be solved for V_i, θ_i (i=1,...,m), I_{di}, I_{qi} (i=1,...,m) in terms of the states δ_i, E'_{di}, E'_{qi} (i=1,...,m). The currents can clearly be explicitly eliminated by solving either (250) or (251) and substituting into the differential equations and remaining algebraic equations. This would leave only n complex algebraic equations to be solved for the n complex voltages $V_i e^{j\theta_i}$.

The Special Case of "Impedance Loads"

In some cases, this two-axis dynamic model can be put in explicit closed form without algebraic equations. We make the special assumptions about the load representations,

$$P_{Li}(V_i) = k_{P2i}V_i^2 \quad i = 1, ..., n \tag{253}$$

$$Q_{Li}(V_i) = k_{Q2i}V_i^2 \quad i = 1, ..., n \tag{254}$$

In this case, the loads and $R_{si} + jX'_{di}$ can be added into the bus admittance matrix diagonal entries to obtain the following simplified algebraic equations for the two-axis model with "constant-impedance" loads,

$$(I_{di} + jI_{qi})e^{j(\delta_i - \frac{\pi}{2})} = \left(\frac{1}{R_{si} + jX'_{di}}\right)[E'_{di} + (X'_{qi} - X'_{di})I_{qi}$$

$$+ jE'_{qi}]e^{j(\delta_i - \frac{\pi}{2})} - \left(\frac{1}{R_{si} + jX'_{di}}\right)V_i e^{j\theta_i}$$

$$i = 1, ..., m \qquad (255)$$

$$0 = -\left(\frac{1}{R_{si} + jX'_{di}}\right)[E'_{di} + (X'_{qi} - X'_{di})I_{qi}$$

$$+ jE'_{qi}]e^{j(\delta_i - \frac{\pi}{2})} + \sum_{k=1}^{n}(G'_{ik} + jB'_{ik})V_k e^{j\theta_k}$$

$$i = 1, ..., m \qquad (256)$$

$$0 = \sum_{k=1}^{n}(G'_{ik} + jB'_{ik})V_k e^{j\theta_k}$$

$$i = m + 1, ..., n \qquad (257)$$

where $G'_{ik} + jB'_{ik}$ is the ik^{th} entry of the admittance matrix including all "constant-impedance" loads and $1/(R_{si} + jX'_{di})$ on the ith diagonal.

Clearly all of the $V_k e^{j\theta_k}$ (k=1,...,n) can be eliminated by solving (256) and (257) and substituting into (255) to obtain m complex equations which are linear in I_{di}, I_{qi} (i=1,...,m). These can in principle be solved through the inverse of a matrix which would be a function of the states δ_i, E'_{qi}, E'_{di} (i=1,...,m). This inverse can be avoided if the following simplification is made,

$$X'_{qi} = X'_{di} \qquad i = 1, ..., m \qquad (258)$$

With this assumption, (256) and (257) can be solved for $V_k e^{j\theta_k}$ and substituted into (255) to obtain,

$$I_{di} + jI_{qi} = \sum_{k=1}^{m}\left(G'_{red} + jB'_{red}\right)[E'_{dk} + jE'_{qk}]e^{j(\delta_k - \delta_i)}$$

$$i = 1, ..., m \qquad (259)$$

where $G'_{red_{ik}} + jB'_{red_{ik}}$ is the ik^{th} entry of an $m \times m$ admittance matrix (often called the matrix reduced to "internal nodes"). This can easily be solved for I_{di}, I_{qi} (i=1,...,m) and substituted into the differential equations (239)-(247). Substitution of I_{di}, I_{qi} into (250) also gives V_i as a function of the states (needed for (245)) so the resulting dynamic model is in explicit form without algebraic equations.

5.3   Multimachine Flux-Decay Model

The reduced-order model of the last subsection still contains the damper winding dynamics of E'_{di}. If T'_{qoi} for all i=1,...,m are sufficiently small, there is an integral manifold for these dynamic states. A first approximation of the remaining fast damper winding integral manifold is found by setting T'_{qoi} equal to zero in (240) to obtain

$$0 = -E'_{di} + (X_{qi} - X'_{qi})I_{qi} \quad i = 1, ..., m \tag{260}$$

When equation (260) is used to eliminate E'_{di}, the synchronous machine dynamic circuit is changed from Figure 4 to Figure 5, giving the following multimachine one-axis or flux-decay model (with speed damping term),

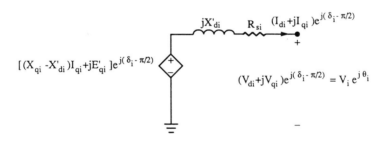

Figure 5: Synchronous machine flux-decay model dynamic circuit (i=1,...,m)

$$T'_{doi}\frac{dE'_{qi}}{dt} = -E'_{qi} - (X_{di} - X'_{di})I_{di} + E_{fdi} \quad i = 1, ..., m \tag{261}$$

$$\frac{d\delta_i}{dt} = \omega_i - \omega_s \quad i = 1, ..., m \tag{262}$$

$$\frac{2H_i}{\omega_s}\frac{d\omega_i}{dt} = T_{Mi} - E'_{qi}I_{qi} - (X_{qi} - X'_{di})I_{di}I_{qi} - T_{FWi}$$

$$- D_i(\omega_i - \omega_s) \quad i = 1, ..., m \tag{263}$$

$$T_{Ei}\frac{dE_{fdi}}{dt} = -(K_{Ei} + S_{Ei}(E_{fdi}))E_{fdi} + V_{Ri} \quad i = 1, ..., m \tag{264}$$

$$T_{Fi}\frac{dR_{fi}}{dt} = -R_{fi} + \frac{K_{Fi}}{T_{Fi}}E_{fdi} \quad i = 1, ..., m \tag{265}$$

$$T_{Ai}\frac{dV_{Ri}}{dt} = -V_{Ri} + K_{Ai}R_{fi} - \frac{K_{Ai}K_{fi}}{T_{Fi}}E_{fdi}$$

$$K_{Ai}(V_{refi} - V_i) \quad i = 1, ..., m \tag{266}$$

$$T_{CHi}\frac{dT_{Mi}}{dt} = -T_{Mi} + P_{SVi} \quad i = 1, ..., m \tag{267}$$

$$T_{SVi}\frac{dP_{SVi}}{dt} = -P_{SVi} + P_{Ci} - \frac{1}{R_i}\frac{\omega_i}{\omega_s} \quad i = 1, ..., m \tag{268}$$

with the limit constraints,

$$V_{Ri}^{min} \le V_{Ri} \le V_{Ri}^{max} \quad i = 1, ..., m \tag{269}$$

$$0 \le P_{svi} \le P_{svi}^{max} \quad i = 1, ..., m \tag{270}$$

and the algebraic constraints,

$$0 = V_i e^{j\theta_i} + (R_{si} + jX'_{di})(I_{di} + jI_{qi})e^{j(\delta_i - \frac{\pi}{2})}$$

$$-[(X_{qi} - X'_{di})I_{qi} + jE'_{qi}]e^{j(\delta_i - \frac{\pi}{2})}$$

$$i = 1, ..., m \tag{271}$$

$$V_i e^{j\theta_i} (I_{di} - jI_{qi})e^{-j(\delta_i - \frac{\pi}{2})} + P_{Li}(V_i) + jQ_{Li}(V_i)$$

$$= \sum_{k=1}^{n} V_i V_k Y_{ik} e^{j(\theta_i - \theta_k - \alpha_{ik})} \quad i = 1, ..., m \tag{272}$$

$$P_{Li}(V_i) + jQ_{Li}(V_i) = \sum_{k=1}^{n} V_i V_k Y_{ik} e^{j(\theta_i - \theta_k - \alpha_{ik})}$$

$$i = m + 1, ..., n \tag{273}$$

As before, for given functions $P_{Li}(V_i)$ and $Q_{Li}(V_i)$, the $n + m$ complex algebraic equations must be solved for V_i, θ_i (i=1,...,n), I_{di}, I_{qi} (i=1,...,m) in terms of the states δ_i, E'_{qi} (i=1,...,m). The currents can clearly be explicitly eliminated by solving either (271) or (272) and substituting into the differential equations and remaining algebraic equations. This would leave only n complex algebraic equations to be solved for the n complex voltages $V_i e^{j\theta_i}$.

The Special Case of "Impedance Loads"

In some cases, this flux-decay dynamic model can be put in explicit closed form without algebraic equations. We make the special assumptions about the load representations,

$$P_{Li}(V_i) = k_{P2i}V_i^2 \quad i = 1,...,n \tag{274}$$

$$Q_{Li}(V_i) = k_{Q2i}V_i^2 \quad i = 1,...,n \tag{275}$$

In this case, the loads and $R_{si} + jX'_{di}$ can be added into the bus admittance matrix diagonal entries to obtain the following simplified algebraic equations for the flux-decay model with "constant-impedance" loads,

$$(I_{di} + jI_{qi})e^{j(\delta_i - \frac{\pi}{2})} = \frac{1}{(R_{si} + jX'_{di})}[(X_{qi} - X'_{di})I_{qi} + jE'_{qi}]e^{j(\delta_i - \frac{\pi}{2})}$$

$$-\frac{1}{(R_{si} + jX'_{di})}V_i e^{j\theta_i} \quad i = 1,...,m \tag{276}$$

$$0 = -\frac{1}{(R_{si} + jX'_{di})}[(X_{qi} - X'_{di})I_{qi} + jE'_{qi}]e^{j(\delta_i - \frac{\pi}{2})}$$

$$+ \sum_{k=1}^{n}(G'_{ik} + jB'_{ik})V_k e^{j\theta_k} \quad i = 1,...,m \tag{277}$$

$$0 = \sum_{k=1}^{n}(G'_{ik} + jB'_{ik})V_k e^{j\theta_k} \quad i = m+1,...,n \tag{278}$$

where $G'_{ik} + jB'_{ik}$ is the ik^{th} entry of the admittance matrix including all "constant-impedance" loads and $1/(R_{si} + jX'_{di})$ on the ith diagonal.

Clearly all of the $V_k e^{j\theta_k}$ (k=1,...,n) can be eliminated by solving (277) and (278) and substituting into (276) to obtain m complex equations which are linear in I_{di}, I_{qi} (i=1,...,m). These can in principle be solved through the inverse of a matrix which would be a function of the states δ_i, E'_{qi} (i=1,...,m). This inverse can be avoided if the following simplification (usually not considered valid) is made,

$$X_{qi} = X'_{di} \quad i = 1,...,m \tag{279}$$

With this simplification, (277) and (278) can be solved for $V_k e^{j\theta_k}$ and substituted into (276) to obtain,

$$I_{di} + jI_{qi} = \sum_{k=1}^{m} (G'_{red_{ik}} + jB'_{red_{ik}})E'_{qk}e^{j(\delta_k - \delta_i)} \quad i = 1, ..., m \qquad (280)$$

where $G'_{red_{ik}} + jB'_{red_{ik}}$ is an $m \times m$ admittance matrix (often called the matrix reduced to "internal nodes"). This can easily be solved for I_{di}, I_{qi} (i=1,...,m) and substituted into the differential equations (261)-(268). Substitution of I_{di}, I_{qi} into (271) also gives V_i as a function of the states (needed for (266)) so the resulting dynamic model is in explicit form without algebraic equations. We emphasize that the simplification of (279) is usually not considered valid for most machines.

5.4   Multimachine Classical Model

The derivation of the classical model requires assumptions which cannot be rigorously supported. Returning to the multimachine two-axis model, rather than assuming $T'_{qoi} = 0$ (i=1,..,m), we assume that an integral manifold exists for $E'_{di}, E'_{qi}, E_{fdi}, R_{fi}, V_{Ri}$ (i=1,...,m) which as a first approximation gives each E'_{qi} equal to a constant and each $(E'_{di} + (X'_{qi} - X'_{di})I_{qi})$ equal to a constant. For this constant based on initial values $E'^o_{di}, I^o_{qi}, E'^o_{qi}$, we define the constant voltage

$$E'^o_i \triangleq \sqrt{(E'^o_{di} + (X'_{qi} - X'_{di})I^o_{qi})^2 + (E'^o_{qi})^2} \qquad (281)$$

and the constant angle,

$$\delta'^o_i \triangleq tan^{-1}\left(\frac{E'^o_{qi}}{E'^o_{di} + (X'_{qi} - X'_{di})I^o_{qi}}\right) - \frac{\pi}{2} \qquad (282)$$

The classical model dynamic circuit is then shown in Figure 6. The classical model is usually used with the assumption of constant shaft torque, so we assume,

$$T_{CHi} = \infty \qquad (283)$$

The classical model (with speed damping) is then a 2m order system (obtained from (239)-(252))

$$\frac{d\delta_i}{dt} = \omega_i - \omega_s \quad i = 1, ..., m \qquad (284)$$

$$\frac{2H_i}{\omega_s}\frac{d\omega_i}{dt} = T^o_{Mi} - Real[(E'^o_i e^{j(\delta_i + \delta'^o_i)})(I_{di} - jI_{qi})e^{-j(\delta_i - \frac{\pi}{2})}]$$

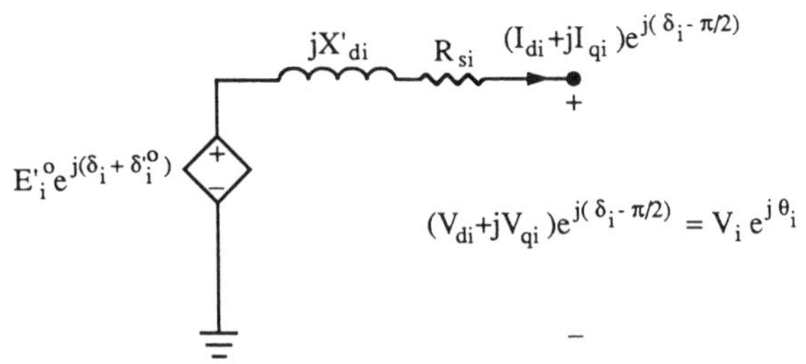

Figure 6: Synchronous machine classical model dynamic circuit (i=1,...,m)

$$-T_{FWi} - D_i(\omega_i - \omega_s) \quad i = 1, ..., m \tag{285}$$

and the algebraic constraints,

$$0 = V_i e^{j\theta_i} + (R_{si} + jX'_{di})(I_{di} + jI_{qi})e^{j(\delta_i - \frac{\pi}{2})}$$

$$-E_i^{\prime o}e^{j(\delta_i + \delta_i^{\prime o})} \quad i = 1, ..., m \tag{286}$$

$$V_i e^{j\theta_i} \ (I_{di} - jI_{qi})e^{-j(\delta_i - \frac{\pi}{2})} + P_{Li}(V_i) + jQ_{Li}(V_i) =$$

$$\sum_{k=1}^{n} V_i V_k Y_{ik} e^{j(\theta_i - \theta_k - \alpha_{ik})} \quad i = 1, ..., m \tag{287}$$

$$P_{Li}(V_i) + jQ_{Li}(V_i) = \sum_{k=1}^{n} V_i V_k Y_{ik} e^{j(\theta_i - \theta_k - \alpha_{ik})}$$

$$i = m + 1, ..., n \tag{288}$$

As before, for given functions $P_{Li}(V_i)$ and $Q_{Li}(V_i)$, the $n + m$ complex algebraic equations must be solved for V_i, θ_i (i=1,...,n), I_{di}, I_{qi}, (i=1,...,m) in terms of the states δ_i. The currents can easily be explicitly eliminated by solving either (286) or (287) and substituting into the differential equations and remaining algebraic equations. This would leave only n complex equations to be solved for the n complex voltages $V_i e^{j\theta_i}$.

The Special Case of "Impedance Loads"

In some cases, this classical model can be put in explicit closed form without algebraic equations. We make the special assumptions about the load representations,

$$P_{Li}(V_i) = k_{P2i}V_i^2 \quad i = 1, ..., n \tag{289}$$

$$Q_{Li}(V_i) = k_{Q2i}V_i^2 \quad i = 1, ..., n \tag{290}$$

In this case, the loads and $R_{si} + jX'_{di}$ can be added into the bus admittance matrix diagonal entries to obtain the following simplified algebraic equations for the classical model with "constant-impedance" loads,

$$(I_{di} + jI_{qi})e^{j(\delta_i - \frac{\pi}{2})} = \frac{1}{(R_{si} + jX'_{di})}E_i'^o e^{j(\delta_i + \delta_i'^o)}$$

$$- \frac{1}{(R_{si} + jX'_{di})}V_i e^{j\theta_i} \quad i = 1, ..., m \tag{291}$$

$$0 = -\frac{1}{(R_{si} + jX'_{di})}E_i'^o e^{j(\delta_i + \delta_i'^o)}$$

$$+ \sum_{k=1}^{n}(G'_{ik} + jB'_{ik})V_k e^{j\theta_k} \quad i = 1, ..., m \tag{292}$$

$$0 = \sum_{k=1}^{n}(G'_{ik} + jB'_{ik})V_k e^{j\theta_k}$$

$$i = m+1, ..., n \tag{293}$$

where $G'_{ik} + jB'_{ik}$ is the ik^{th} entry of the admittance matrix including all "constant-impedance" loads <u>and</u> $1/(R_{si} + jX'_{di})$ on the ith diagonal.

Clearly all of the $V_k e^{j\theta_k}$ (k=1,...,n) can be eliminated by solving (292) and (293) and substituting into (291) to obtain m complex equations of the form,

$$(I_{di} + jI_{qi})e^{j(\delta_i - \frac{\pi}{2})} = \sum_{k=1}^{m}(G'_{red_{ik}} + jB'_{red_{ik}})E_k'^o e^{j(\delta_k + \delta_k'^o)} \tag{294}$$

where $G'_{red_{ik}} + jB'_{red_{ik}}$ is the ik^{th} entry of an $m \times m$ admittance matrix (often called the matrix reduced to "internal nodes"). Defining,

$$\delta_{classical} \triangleq \delta_i + \delta_i'^o \quad i = 1, ..., m \tag{295}$$

the multimachine classical model with constant-impedance loads is found by substituting (294) into (285),

$$\frac{d\delta_{classical}}{dt} = \omega_i - \omega_s \quad i = 1, ..., m \tag{296}$$

$$\frac{2H_i}{\omega_s}\frac{d\omega_i}{dt} = T_{Mi} - \sum_{k=1}^{m} E_i^{\prime o} E_k^{\prime o} G_{red_{ik}}^{\prime} cos(\delta_{classical_i} - \delta_{classical_k})$$

$$- \sum_{k=1}^{m} E_i^{\prime o} E_k^{\prime o} B_{red_{ik}}^{\prime} sin(\delta_{classical_i} - \delta_{classical_k})$$

$$-T_{FWi} - D_i(\omega_i - \omega_s) \quad i = 1, ..., m \tag{297}$$

The classical model can also be obtained formally from the two-axis model by setting $X_{qi}' = X_{di}'$, $T_{CHi} = T_{qoi}' = T_{doi}' = \infty$ (i=1,...,m), or from the flux-decay model by setting $X_{qi} = X_{di}'$, $T_{CHi} = T_{doi}' = \infty$ (i=1,...,m). In this latter case, $\delta_i^{\prime o}$ is equal to zero so that $\delta_{classical}$ is equal to δ_i and $E_i^{\prime o}$ is equal to $E_{qi}^{\prime o}$ (i=1,...,m).

5.5 Alternative Formulations

Examination of the four models in Eq. (213)-(228), (239)-(252), (261)-(273), (284)-(288) shows the differential plus algebraic equations to be of the form,

$$\frac{dx_i}{dt} = f_i(x_i, I_{di}, I_{qi}, V_i) + B_i u_i \quad i = 1, ..., m \tag{298}$$

$$I_{di} = h_{di}(x_i, V_i, \theta_i) \quad i = 1, ..., m \tag{299}$$

$$I_{qi} = h_{qi}(x_i, V_i, \theta_i) \quad i = 1, ..., m \tag{300}$$

$$(I_{di} + jI_{qi})e^{j(\delta_i - \frac{\pi}{2})} + \frac{P_{Li}(V_i) - jQ_{Li}(V_i)}{\overline{V}_i^*} = \sum_{k=1}^{n} \overline{Y}_{ik}\overline{V}_k$$
$$i = 1, ..., m \tag{301}$$

$$\frac{P_{Li}(V_i) - jQ_{Li}(V_i)}{\overline{V}_i^*} = \sum_{k=1}^{n} \overline{Y}_{ik}\overline{V}_k$$
$$i = m+1, ..., n \tag{302}$$

where $*$ denotes complex conjugation. This formulation solves the synchronous machine circuit equations for I_{di}, I_{qi} in terms of dynamic states and the machine terminal-voltage magnitude and angle. The network equations are written as current equations. For example, in the two-axis model,

$$x_i = [E'_{qi} E'_{di} \delta_i \omega_i E_{fdi} R_{fi} V_{Ri} T_{Mi} P_{SVi}]^t \quad i = 1, ..., m \qquad (303)$$

$$u_i = [V_{refi} P_{Ci}]^t \quad i = 1, ..., m \qquad (304)$$

and,

$$\begin{bmatrix} I_{di} \\ I_{qi} \end{bmatrix} = \begin{bmatrix} R_{si} & -X'_{qi} \\ X'_{di} & R_{si} \end{bmatrix}^{-1} \begin{bmatrix} E'_{di} - V_i \sin(\delta_i - \theta_i) \\ E'_{qi} - V_i \cos(\delta_i - \theta_i) \end{bmatrix}$$
$$i = 1, ..., m \qquad (305)$$

The model (298)-(302) can clearly be put in the form,

$$\frac{dx_i}{dt} = F_i(x_i, V_i, \theta_i) + B_i u_i \quad i = 1, ..., m \qquad (306)$$

$$0 = G(x, V, \theta) \qquad (307)$$

where G is a vector of 2n real equations to be solved for the 2n real variables V and θ in terms of x. This can also be written in terms of V_{Di} and V_{Qi}, where

$$\overline{V}_i = V_i e^{j\theta_i} = (V_{di} + jV_{qi}) e^{j(\delta_i - \frac{\pi}{2})} = (V_{Di} + jV_{Qi}) \quad i = 1, ..., m \qquad (308)$$

Another formulation introduces the variable,

$$P_{Ei} \triangleq E'_{di} I_{di} + E'_{qi} I_{qi} + (X'_{qi} - X'_{di}) I_{di} I_{qi} \quad i = 1, ..., m \qquad (309)$$

and the following structure,

$$\frac{dx_i}{dt} = A_i(x_i) x_i + A'_i w_i + B_i u_i \quad i = 1, ..., m \qquad (310)$$

$$w_i = h_i(x_i, V_{Di}, V_{Qi}) \quad i = 1, .., m \qquad (311)$$

$$\overline{I}(x, w, \overline{V}) = \overline{YV} \qquad (312)$$

where

$$x \triangleq [x_1, ..., x_m]^t \qquad (313)$$

$$w \triangleq [w_1, ..., w_m]^t \qquad (314)$$

$$w_i \triangleq [I_{di} I_{qi} P_{Ei} V_i]^t \quad i = 1, ..., m \qquad (315)$$

The only state dependence in the matrix A_i is due to the exciter saturation. With their respective simplifying assumptions ($X''_{di} = X''_{qi}$, etc.) and
constant-impedance-load models, each model can be put in the form,

$$\frac{dx_i}{dt} = f'_i(x) + B_i u_i \quad i = 1, ..., m \tag{316}$$

6   Initial Condition Computations

While we are free to arbitrarily specify the initial conditions of all dynamic
states and control inputs, most dynamic studies begin from steady-state
operation. In power systems, the most widely used calculation for steady-
state operation is the standard load flow. We begin this section with a
definition and review of this calculation.

6.1   Standard Load Flow

Using the previously defined numbering notation, we define standard load
flow as the following algorithm. Given the bus admittance matrix,

a) Specify bus-voltage magnitudes numbered 1 to m ($V_1, ..., V_m$).

b) Specify bus-voltage angle number 1 (θ_1).

c) Specify net injected real power P_i at buses numbered 2 to m.

d) Specify load powers P_{Li} and Q_{Li} at all buses numbered 1 to n.

e) Solve the following "standard load-flow" equations for $V_i(i = m+1, ..., n)$,
$\theta_i(i = 2, ..., n)$:

$$0 = -P_i + \sum_{k=1}^{n} V_i V_k Y_{ik} \cos(\theta_i - \theta_k - \alpha_{ik}) \quad \begin{matrix} \text{i=2,...,m} \\ \text{(PV buses)} \end{matrix} \tag{317}$$

$$0 = -P_{Li} + \sum_{k=1}^{n} V_i V_k Y_{ik} \cos(\theta_i - \theta_k - \alpha_{ik}) \quad \begin{matrix} \text{i=m+1,...,n} \\ \text{(PQ buses)} \end{matrix} \tag{318}$$

$$0 = -Q_{Li} + \sum_{k=1}^{n} V_i V_k Y_{ik} \sin(\theta_i - \theta_k - \alpha_{ik}) \quad \begin{matrix} \text{i=m+1,...,n} \\ \text{(PQ buses)} \end{matrix} \tag{319}$$

where $P_i(i = 2, ..., m)$, $V_i(i = 1, ..., m)$, $P_{Li}(i = m + 1, ..., n)$, $Q_{Li}(i = m + 1, ..., n)$, and θ_1 are specified numbers. The standard load-flow Jacobian matrix is the linearization of (317)-(319) with respect to $\theta_2, ..., \theta_n$,
$V_{m+1}, ..., V_n$ [27]. After this solution, compute

$$P_1 + jQ_1 = \sum_{k=1}^{n} V_1 V_k Y_{1k} e^{j(\theta_1 - \theta_k - \alpha_{1k})} \tag{320}$$

$$Q_i = \sum_{k=1}^{n} V_i V_k Y_{ik} \sin(\theta_i - \theta_k - \alpha_{ik}) \quad i = 2, ..., m \tag{321}$$

This standard load flow has many variations including the addition of other devices such as tap-changing-under-load (TCUL) transformers, switching var sources and high-voltage-dc (HVDC) converters. It can also include inequality constraints on quantities such as Q_i, and be revised to distribute the slack power between all generators.

We would like to make one important point about load flow. Load flow is normally used to evaluate operation at a specific load level (specified by a given set of powers). For a specified load and generation schedule, the solution is independent of the actual load model. That is, it is certainly possible to evaluate the voltage at a constant-impedance load for a specific case where that impedance load consumes a specific amount of power. Thus the use of "constant power" in load-flow analysis does not require or even imply that the load is truly a constant-power device. It merely gives the voltage at the buses when the loads (any type) consume a specific amount of power. The load characteristic is important when the analyst wants to study the system in response to a change such as contingency analysis or dynamic analysis. For these purposes, standard load flow usually provides the "initial conditions."

6.2   Initial Dynamic States and Control Inputs

For the formulation of Eq. (298)-(302) we need to systematically compute the initial values of $x_1, ..., x_m$ and $u_1, ..., u_m$ from a standard load-flow solution. For illustration purposes, we choose the two-axis model of (239)-(252). The first step in computing initial conditions is normally the calculation of generator currents from

$$I_{Gi} e^{j\gamma_i} = ((P_i - P_{Li}(V_i)) - j(Q_i - Q_{Li}(V_i))/(V_i e^{-j\theta_i})$$

$$i = 1, ..., m \tag{322}$$

where all V_i, θ_i are known from the load-flow solution. This current is in the network reference frame and is equal to $(I_{di} + jI_{qi})e^{j(\delta_i - \pi/2)}$.

The machine rotor angles are obtained from equation (250) and the algebraic equation obtained from (240), where we assume all time derivatives are zero in steady state,

$$E'_{di} = (X_{qi} - X'_{qi})I_{qi} \quad i = 1, ..., m \tag{323}$$

Using this in (250)

$$V_i e^{j\theta_i} + (R_{si} + jX'_{di})(I_{di} + jI_{qi})e^{j(\delta_i - \pi/2)}$$
$$- [(X_{qi} - X'_{qi})I_{qi} + (X'_{qi} - X'_{di})I_{qi} + jE'_{qi}]e^{j(\delta_i - \pi/2)} = 0$$
$$i = 1, ..., m \tag{324}$$

or,

$$V_i e^{j\theta_i} + (R_{si} + jX_{qi})(I_{di} + jI_{qi})e^{j(\delta_i - \pi/2)}$$
$$- j[(X_{qi} - X'_{di})I_{di} + E'_{qi}]e^{j(\delta_i - \pi/2)} = 0 \quad i = 1, ..., m \tag{325}$$

Now replacing $(I_{di} + jI_{qi})e^{j(\delta_i - \pi/2)} = I_{Gi}e^{j\gamma_i}$ which is already calculated in (322) we get

$$V_i e^{j\theta_i} + (R_{si} + jX_{qi})I_{Gi}e^{j\gamma_i}$$
$$= [(X_{qi} - X'_{di})I_{di} + E'_{qi}]e^{j\delta_i} \quad i = 1, ..., m \tag{326}$$

The right-hand side of (326) is a voltage behind the impedance $(R_{si} + jX_{qi})$ and has an angle δ_i. Therefore,

$$\delta_i = \text{ angle of } [V_i e^{j\theta_i} + (R_{si} + jX_{qi})I_{Gi}e^{j\gamma_i}] \tag{327}$$

With these quantities, the remaining dynamic and algebraic states can be found by,

$$I_{di} + jI_{qi} = I_{Gi}e^{j(\gamma_i - \delta_i + \pi/2)} \quad i = 1, ..., m \tag{328}$$

$$V_{di} + jV_{qi} = V_i e^{j(\theta_i - \delta_i + \pi/2)} \quad i = 1, ..., m \tag{329}$$

followed by

$$E'_{qi} = |V_i e^{j\theta_i} + (R_{si} + jX_{qi})I_{Gi}e^{j\gamma_i}| - (X_{qi} - X'_{di})I_{di}$$

$$i = 1, ..., m \qquad (330)$$

where $|\cdot|$ denotes complex absolute value (or magnitude), and

$$E'_{di} = (X_{qi} - X'_{qi})I_{qi} \quad i = 1, ..., m \qquad (331)$$

From (239) by setting the derivative of E'_{qi} to zero,

$$E_{fdi} = E'_{qi} + (X_{di} - X'_{di})I_{di} \qquad (332)$$

With this field voltage, R_{fi}, V_{Ri} and V_{refi} can be found from (243)-(245) by setting the derivatives to zero,

$$R_{fi} = \frac{K_{Fi}}{T_{Fi}}E_{fdi} \qquad i = 1, ..., m \qquad (333)$$

$$V_{Ri} = (K_{Ei} + S_{Ei}(E_{fdi}))E_{fdi} \qquad i = 1, ..., m \qquad (334)$$

$$V_{refi} = V_i + (V_{Ri}/K_{Ai}) \qquad i = 1, ..., m \qquad (335)$$

The mechanical states and P_{Ci} are found from (241), (242) and (246) and (247) as,

$$\omega_i = \omega_s \qquad i = 1, ..., m \qquad (336)$$

$$T_{Mi} = E'_{di}I_{di} + E'_{qi}I_{qi} + (X'_{qi} - X'_{di})I_{di}I_{qi} + T_{FW} \qquad i = 1, ..., m \quad (337)$$

$$P_{SVi} = T_{Mi} \qquad i = 1, ..., m \qquad (338)$$

$$P_{Ci} = P_{SVi} + (1/R_i) \qquad i = 1, ..., m \qquad (339)$$

This completes the computation of all dynamic-state initial conditions and fixed inputs. When synchronous machine saturation is included, this process may be iterative, with zero saturation as the initial guess.

For a given disturbance, the inputs remain fixed throughout the simulation. If the disturbance occurs in the algebraic equations, the algebraic states must change instantaneously to satisfy the initial condition of the dynamic states and the new algebraic equations. Thus it may be necessary to resolve the algebraic equations with the dynamic states specified at their initial conditions to determine the new initial values of the algebraic states.

From the above description it is clear that once a standard load-flow solution is found, the remaining dynamic states and inputs can be found in a straightforward way. The machine relative angles δ_i can always be found provided,

$$V_i e^{j\theta_i} + (R_{si} + jX_{qi})I_{Gi}e^{j\gamma_i} \neq 0 \qquad i = 1, ..., m \qquad (340)$$

If control limits are enforced, a solution satisfying these limits may not exist. In this case, the state which is limited would need to be fixed at its limiting value and a corresponding new steady-state solution would have to be found. This would require a new load flow specifying either different values of generator voltages, different generator real powers, or possibly specifying generator reactive power injections, thus allowing generator voltage to be a part of the load-flow solution. In fact, the use of reactive power limits in load flow can usually be traced back to an attempt to consider excitation system limits or generator capability limits.

6.3   Angle Reference

In any rotational system, the reference for angles is usually arbitrary. Examination clearly shows that the order of this dynamic system can be reduced from 9m to 9m-1 by introducing the new relative angles (arbitrarily selecting δ_1 as reference)

$$\delta_i' = \delta_i - \delta_1 \qquad i = 1, ..., m \qquad (341)$$

$$\theta_i' = \theta_i - \delta_1 \qquad i = 1, ..., n \qquad (342)$$

The full system remains exactly the same as before with each δ_i replaced by δ_i', each θ_i replaced by θ_i' and ω_s replaced by ω_1 in each $d\delta_i'/dt$ equation. During a transient, the angle δ_1 still changes from its initial condition as ω_1 changes, so that each original δ_i and θ_i can be easily recovered if needed. The angle δ_1' remains at zero for all time. Thus for dynamic simulation, the differential equation for δ_1 is normally replaced by the algebraic equation which simply states $\delta_1' = 0$. Notice that θ_1 is normally arbitrarily selected as zero for the load-flow analysis. This means that the initial value of δ_1 is normally not zero. During a transient, θ_1' and θ_1 change as well as all angles except δ_1'. If the inertia of machine 1 is set to infinity, ω_1 and δ_1 remain constant for all time.

7 Simulation of Power System Dynamics

Several algorithms have been proposed for use in power system dynamic analysis. There are basically two approaches used today [28].

1. The Simultaneous-Implicit (SI) method

2. The Partitioned-Explicit (PE) method

These two methods and several related issues are presented in this section.

7.1 Simultaneous-Implicit (SI) Method

Since the inputs u_i are constant in these formulations, we can suppress them in our presentation. The formulation of equations (306)-(307) can be written as

$$\frac{dx}{dt} = f(x,y) \quad , \quad x(o) = x_o \tag{343}$$

$$0 = g(x,y) \quad , \quad y(o) = y_o \tag{344}$$

In the SI method (343) and (344) are algebraized using either the implicit Euler's method or a trapezoidal integration method. These resulting algebraic equations are solved simultaneously using Newton's method at each time step.

Choosing the trapezoidal rule to find x and y at time $t_k + \Delta t$ from known values at t_k, we introduce the vector notation

$$x_k \stackrel{\Delta}{=} x(t_k) \tag{345}$$

$$y_k \stackrel{\Delta}{=} y(t_k) \tag{346}$$

$$x_{k+1} \stackrel{\Delta}{=} x(t_k + \Delta t) \tag{347}$$

$$y_{k+1} \stackrel{\Delta}{=} y(t_k + \Delta t) \tag{348}$$

and solve the following nonlinear equations,

$$x_{k+1} = x_k + \frac{\Delta t}{2}[f(x_{k+1}, y_{k+1}) + f(x_k, y_k)] \tag{349}$$

$$0 = g(x_{k+1}, y_{k+1}) \tag{350}$$

Rearranging the equations

$$F_1(x_{k+1}, y_{k+1}, x_k, y_k) \triangleq [x_{k+1} - \frac{\Delta t}{2} f(x_{k+1}, y_{k+1})] - [x_k + \frac{\Delta t}{2} f(x_k, y_k)] = 0$$

$$(351)$$

$$F_2(x_{k+1}, y_{k+1}) \triangleq g(x_{k+1}, y_{k+1}) = 0 \qquad (352)$$

At each time step (351) and (352) are solved by Newton's method for Δx_{k+1}, Δy_{k+1}, where at the pth iteration,

$$\begin{bmatrix} J_1 & J_2 \\ J_3 & J_4 \end{bmatrix}^{(p)} \begin{bmatrix} \Delta x_{k+1} \\ \Delta y_{k+1} \end{bmatrix}^{(p)} = - \begin{bmatrix} F_1 \\ F_2 \end{bmatrix}^{(p)} \qquad (353)$$

$$x_{k+1}^{(p+1)} = x_{k+1}^{(p)} + \Delta x_{k+1}^{(p)} \qquad (354)$$

$$y_{k+1}^{(p+1)} = y_{k+1}^{(p)} + \Delta y_{k+1}^{(p)} \qquad (355)$$

The structure of the Jacobian is as follows

$$J_1 = I - \frac{\Delta t}{2} \frac{\partial f}{\partial x_{k+1}} \qquad (356)$$

$$J_2 = -\frac{\Delta t}{2} \frac{\partial f}{\partial y_{k+1}} \qquad (357)$$

$$J_3 = \frac{\partial g}{\partial x_{k+1}} \qquad (358)$$

$$J_4 = \frac{\partial g}{\partial y_{k+1}} \qquad (359)$$

7.2   Disturbance Simulation

The typical disturbance corresponds to a network disturbance such as a fault where the parameters in the algebraic equations change at $t = 0$. The algebraic variables change instantaneously whereas the state variables do not. Hence at $t = 0^+$, with the network disturbance, one solves the algebraic equations first as

$$g^f(x(o), y(o^+)) = 0 \qquad (360)$$

where the superscript f indicates the algebraic equations in the faulted state. With the value of $y(o^+)$ so obtained, the trapezoidal method is then applied. Note that the initial guess for the vector $\begin{bmatrix} x \\ y \end{bmatrix}$ at time t_{k+1} is the converged value at the previous time instant, i.e.,

$$\begin{bmatrix} x_{k+1} \\ y_{k+1} \end{bmatrix}^{(o)} = \begin{bmatrix} x_k \\ y_k \end{bmatrix} \tag{361}$$

To clarify the nature of $g^f(x, y)$, if there is a change in load, this is simply taken care of by changing P_{Li} and Q_{Li}. If there is a short circuit at bus i, then set $\overline{V}_i \equiv 0$ and delete the injected current equations at bus i from $g(x, y)$ to obtain $g^f(x, y)$.

7.3 Partitioned-Explicit (PE) Method

Still using the formulation of equations (343)-(344), the PE scheme solves two smaller problems but introduces an interface error. Choosing the trapezoidal rule and the notation of (345)-(348), the partitioned-explicit method solves the following nonlinear equations in sequence. First solve,

$$x_{k+1} = x_k + \frac{\Delta t}{2}[f(x_{k+1}, y_k) + f(x_k, y_k)] \tag{362}$$

for x_{k+1}, and then solve

$$0 = g(x_{k+1}, y_{k+1}) \tag{363}$$

for y_{k+1}. The first solution solves m sets of decoupled differential equations with the algebraic variables treated as constants. The second solution solves the algebraic equations with the dynamic states treated as constants.

7.4 Methodology of a Commerical Program

In this section we expand on the formulation of (310)-(315) as used in the EPRI RP1208 package known as the Extended-Transient-Midterm-Stability-Package (ETMSP) [29]. Consider the set of equations

$$\frac{dx}{dt} = A(x)x + A'w + Bu \tag{364}$$

$$w = h(x, V_D, V_Q) \tag{365}$$

and the complex set of nodal equations

$$\overline{I}(x, w, \overline{V}) = \overline{Y}\,\overline{V} \tag{366}$$

We first separate (366) into real and imaginary parts using rectangular coordinates

$$\overline{Y}_{ij} = G_{ij} + jB_{ij}, \quad \overline{V}_i = V_{Di} + jV_{Qi} \tag{367}$$

We now denote (366) as

$$I_w^e(x, w, V^e) = Y^e V^e \tag{368}$$

where the superscript e stands for "expanded" and each complex admittance element in $\overline{Y}$ has been replaced by a 2×2 block in Y^e. The components of V^e are $(V_{D1}, V_{Q1}, \cdots, V_{Dn} V_{Qn})$. We can now formally substitute w from (365) into (364) and (366) to get

$$\frac{dx}{dt} = A(x)x + A'h'(x, V^e) + Bu \tag{369}$$

$$I^e(x, V^e) = Y^e V^e \tag{370}$$

where h' is the same as h, but written here as a function of one $2n$ vector (V^e) rather than two n vectors (V_D, V_Q). Application of the trapezoidal rule to (369) and (370) will result in (for constant u)

$$x_{k+1} - x_k - \frac{\Delta t}{2}[A(x_{k+1})x_{k+1} + A'h'(x_{k+1}, V_{k+1}^e) + Bu$$
$$+ \ A(x_k)x_k + A'h'(x_k, V_k^e) + Bu] = 0 \tag{371}$$

$$Y^e V_{k+1}^e - I^e(x_{k+1}, V_{k+1}^e) = 0 \tag{372}$$

Applying Newton's method to (371), (372) gives the linear set of equations as

$$\begin{bmatrix} A_{GG1} & & 0 & A_{GV1} \\ & \ddots & & \vdots \\ 0 & & A_{GGm} & A_{GVm} \\ \hline A_{VG1} & \cdots & A_{VGm} & Y^e + Y_L^e \end{bmatrix} \begin{bmatrix} \Delta x_{k+1,1}^{(p)} \\ \vdots \\ \Delta x_{k+1,m}^{(p)} \\ \hline \Delta V_{k+1}^{e(p)} \end{bmatrix} = \begin{bmatrix} R_{G1} \\ \vdots \\ R_{Gm} \\ \hline R_V^e \end{bmatrix} \tag{373}$$

where

$$Y_L^e \triangleq - \left[\frac{\partial I^e}{\partial V^e} \right] \tag{374}$$

The right hand side of (373) is the vector of residuals of (371)-(372) at the pth iteration. The solution method to solve the linear equation (373) is as follows. We recognize that

$$\Delta x_{k+1,i}^{(p)} = A_{GGi}^{-1}(R_{Gi} - A_{GVi}\Delta V_{k+1}^{e(p)}) \quad i = 1, ..., m \tag{375}$$

$$[Y^e + Y_L^e]\Delta V_{k+1}^{e(p)} + \sum_{i=1}^{m}(A_{VGi}\Delta x_{k+1,i}^{(p)}) = [R_V^e] \tag{376}$$

or,

$$[Y^e + Y_L^e + Y_G^e]\Delta V_{k+1}^{e(p)} = [R_V^{e'}] \tag{377}$$

where,

$$R_V^{e'} = R_V^e - \sum_{i=1}^{m} A_{VGi}A_{GGi}^{-1}R_{Gi} \tag{378}$$

and

$$Y_G^e = - \sum_{i=1}^{m} A_{VGi}A_{GGi}^{-1}A_{GVi} \tag{379}$$

The Jacobian in (377) is expensive to compute at each iteration. Let $J_{k+1}^{(p)}$ be the Jacobian evaluated at $t = t_{k+1}$ and p represent the iteration at that time instant. As a result of solving (377) we get

$$\begin{bmatrix} x_{k+1}^{(p+1)} \\ V_{k+1}^{e(p+1)} \end{bmatrix} = \begin{bmatrix} x_{k+1}^{(p)} \\ V_{k+1}^{e(p)} \end{bmatrix} + \begin{bmatrix} \Delta x_{k+1}^{(p)} \\ \Delta V_{k+1}^{e(p)} \end{bmatrix}$$

The very-dishonest-Newton method [29] holds the Jacobian fixed for a period of time. This means that the initial Jacobian at $t = t_{k+1}$ is held constant for some time steps after t_{k+1}. This reduces the overall cost of computation. The choice of when to re-evaluate the Jacobian is based on experience. Of course the Jacobian must be re-evaluated at any major system change. Between time steps it is re-evaluated if the previous time step iteration was considered too slow (took three or more iterations). Within a time step, a maximum of five iterations are taken using the same Jacobian.

8   Conclusions

Modeling and simulation of multimachine power system dynamics involves essentially five stages:

(a) Development of basic component models from fundamental laws

(b) Interconnection of individual component models into a system model

(c) Model reduction

(d) Initial condition computation

(e) Solution of the final model

Stage (a) is somewhat flexible and can include a wide range of possible detail. The specific component models used will depend on the phenomena of interest. This chapter has included the fundamental models needed to study most basic phenomena. Stage (b) includes the interface between individual unit component models and the network interconnection. This chapter has presented the basic methodology for this interconnection. Stage (c) recognizes that different models may be used to study different phenomena. Integral manifold theory provides a systematic approach to the model reduction process. It was used in the chapter to provide the theoretical basis for various model simplifications. Stage (d) specifies the initial state of the system. In most cases this state is calculated from standard steady-state load flow. This chapter has shown how load flow is incorporated in dynamic analysis. Stage (e) is the most computationally expensive, and remains a fruitful area for research. The power system structure is special, with the decoupling between individual unit differential equations and the system algebraic equations.

9 Acknowledgements

The authors gratefully acknowledge Professor Petar V. Kokotovic for many valuable discussions and for the contribution of many ideas on the use of singular perturbation and integral manifolds in power system dynamic modeling. Portions of this work were supported by the National Science Foundation Grant #NSF ECS 87-19055, and the Grainger Foundation Endowment to the University of Illinois.

References

[1] P. C. Krause, *Analysis of Electric Machinery*, McGraw-Hill Book Co., New York, NY, 1986.

[2] A. E. Fitzgerald, C. Kingsley, and S. D. Umans, *Electric Machinery*, McGraw-Hill Book Co., New York, NY, 1983.

[3] E. W. Kimbark, *Power System Stability: Synchronous Machines*, Dover Publications, Inc., New York, NY, 1956.

[4] B. Adkins, *The General Theory of Electrical Machines*, Chapman and Hall Ltd., London, England, 1962.

[5] F. P. DeMello and L. N. Hannett, "Determination of Synchronous Machine Stability Study Constants," EPRI Report EL-1424, vol. 3, Electric Power Research Institute, Palo Alto, CA, June 1980.

[6] D. P. Gelopulos, "Midterm Simulation of Electric Power Systems," EPRI Report EL-596, Electric Power Research Institute, Palo Alto, CA, June 1979.

[7] D. W. Olive, "New Techniques for the Calculation of Dynamic Stability," IEEE Trans. on Power Apparatus and Systems, vol. PAS-85/No. 7, July 1966.

[8] C. Concordia, *Synchronous Machine*, John Wiley and Sons, New York, NY, 1951.

[9] P. L. Dandeno, P. Kundur and R. P. Schulz, "Recent Trends and Progress in Synchronous Machine Modeling in the Electric Utility Industry," IEEE Proceedings, vol. 62, no. 7, July 1974, pp. 941-950.

[10] IEEE, "Symposium on Adequacy and Philosophy of Modeling: Dynamic System Performance," 75CH0970-4-PWR, Tutorial at IEEE/PES 1975 Winter Meeting, New York, 1975.

[11] IEEE, "Symposium on Synchronous Machine Modeling for Power System Studies," 83TH0101-6-PWR, Tutorial at IEEE/PES 1983 Winter Meeting, New York, 1983.

[12] A. W. Rankin, "Per-unit Impedance of Synchronous Machines," AIEE Trans., vol. 64, August 1945.

[13] M. R. Harris, P. J. Lawrenson and J. M. Stephenson, *Per-unit Systems with Special Reference to Electrical Machines*, Cambridge at the University Press, Cambridge, England, 1970.

[14] P. W. Sauer, "Constraints on Saturation Modeling in Machines," Submitted for publication.

[15] I. M. Canay, "Causes of Discrepancies on Calculation of Rotor Quantities and Exact Equivalent Diagrams of the Synchronous Machine," IEEE Trans. on Power Apparatus and Systems, vol. PAS-88, no. 7, July 1969.

[16] IEEE Committee report, "Computer Representation of Excitation Systems," IEEE Trans. on Power Apparatus and Systems, vol. PAS-87, no. 6, June 1968, pp. 1460-1464.

[17] IEEE Committee report, "Excitation System Models for Power System Stability Studies," IEEE Trans. on Power Apparatus and Systems, vol. PAS-100, no. 2, Feb. 1981, pp. 494-509.

[18] C. C. Young, "Equipment and System Modeling for Large Scale Stability Studies," IEEE Trans. on Power Apparatus and Systems, vol. PAS-91, no. 1, Jan/Feb 1972, pp. 99-109.

[19] IEEE Committee report, "Dynamic Models for Steam and Hydro Turbines in Power System Studies," IEEE Trans. on Power Apparatus and Systems, vol. PAS-92, no. 6, Nov/Dec 1973, pp. 1904-1915.

[20] P. W. Sauer, D. J. LaGesse, S. Ahmed-Zaid and M. A. Pai, "Reduced Order Modeling of Interconnected Multimachine Power Systems Using Time-Scale Decomposition," IEEE Trans. on Power Systems, vol. PWRS-2, no. 2, May 1987, pp. 310-320.

[21] P. W. Sauer, S. Ahmed-Zaid and P. V. Kokotovic, "An Integral Manifold Approach To Reduced Order Dynamic Modeling of Synchronous Machines," IEEE Trans. on Power Systems, vol. 3, no. 1, Feb. 1988, pp. 17-23.

[22] P. V. Kokotovic and P. W. Sauer, "Integral Manifold as a Tool for Reduced Order Modeling of Nonlinear Systems: A Synchronous Machine Case Study," IEEE Trans. on Circuits and Systems, vol. 36, no. 3, Mar. 1989, pp. 403-410.

[23] J. H. Chow, Editor, Time-Scale Modeling of Dynamic Networks with Applications to Power Systems, vol. 46 of lecture notes in Control and Information Sciences, Springer-Verlag, New York, 1982.

[24] P. V. Kokotovic, H. K. Khalil and J. O'Reilly, *Singular Perturbation Methods in Control: Analysis and Design*, Academic Press, 1986.

[25] P. W. Sauer, "Reduced Order Dynamic Modeling of Machines and Power Systems," Proceedings of the American Power Conference, vol. 49, Chicago, IL, 1987, pp. 789-794.

[26] P. W. Sauer, S. Ahmed-Zaid and M. A. Pai, "Systematic Inclusion of Stator Transients in Reduced Order Synchronous Machine Models," IEEE Trans. on Power Apparatus and Systems, vol. PAS-103, no. 6, June 1984, pp. 1348-1354.

[27] P. W. Sauer and M. A. Pai, "Power system steady-state stability and the load-flow Jacobian," IEEE Trans. on Power Systems, vol. 5, no. 4, November 1990, pp. 1374-1383.

[28] B. Stott, "Power system dynamic response calculations," Proc. of IEEE, vol. 67, Feb. 1979, pp. 219-241.

[29] Extended Transient-Midterm Stability Package: Final Report, EPRI EL-4610, Electric Power Research Institute, Palo Alto, CA, January 1987.

COMPUTER SIMULATION TECHNIQUES
IN ELECTRIC DISTRIBUTION SYSTEMS

MO-SHING CHEN, TSAI-HSIANG CHEN

The Energy Systems Research Center
The University of Texas at Arlington
Arlington, Texas 76019

I. INTRODUCTION

It has becoming increasingly clear that utilities and consumers need demand-side management. The least-cost planning as well as demand-side strategies are being widely adopted and mandated. The demand-side management, least-cost planning, network transformer placement study, and many other distribution analyses need rigorous operational-type analysis rather than planning oriented analysis. The difference between these two types of analyses should be properly emphasized, otherwise, the misuse of the planning-type method to analyze the operational behavior of the system will distort the explanation of the calculated results and lead to incorrect conclusions. The concept and methodologies for detailed electric distribution system simulation are discussed here. The features of general distribution system are also addressed.

The distribution system is basically unbalanced. Many factors cause system imbalance such as untransposed feeders, conductor bundles, single-phase loads, unequal three-phase loads, single and double-phase "radial spurs" on primary feeders. Furthermore, even if the network is balanced, asymmetrical faults introduce imbalance. To avoid significant error arising from inherent system imbalance, rigorous distribution system analysis using detailed component models

CONTROL AND DYNAMIC SYSTEMS, VOL. 43

is required.   Based on these detailed component models and considering the
numerical stability and converge problem, execution time, data requirements, and
their impact on a extremely large-scale distribution system simulation, a suitable
solution technique has been found and is presented here.

Utilizing the phase frame representation for all network elements, a program,
entitled "Generalized Distribution Analysis System" (GDAS), with a number of
features and capabilities has been developed for large-scale distribution system
simulations and is introduced here.   These features include power flow,
contingency, system loss and short circuit analysis of numerous voltage levels
simultaneously on an individual phase basis, cogenerator and delta-wye transformer
simulation, and load modeling.   The system being analyzed can be balanced or
unbalanced and can be a radial, network, or mixed type distribution system.
Furthermore, because the individual phase representation is employed for both
system and component models, the system can comprise single, double, and three-
phase systems simultaneously.

With sufficient power and generality to handle the complexity of today's
distribution system, the distribution analysis system would provide distribution
system engineers with the capability to study and better understand actual system
conditions.   Together, the features of the system provided a framework for what
was called a generalized distribution analysis system.

The solution technique is discussed in the following section.   The sample
system results is presented in Section III.   Finally, discussion and conclusion are
drawn.

II. TECHNICAL APPROACH

In this section, basic considerations, the individual phase representation
concept, component models, solution techniques, data requirements, and the
numerical stability and convergence problem are addressed.   The methodologies
presented here and the prototype GDAS program can be used as a platform for
advanced applications.

A. BASIC CONSIDERATIONS

To perform a rigorous simulation on a large-scale distribution system combining primary and secondary networks, the following features should be considered:

- The system is basically unbalanced;
- The networks can be extremely large;
- The cogenerators (synchronous and induction) are allowed on both the primary feeders and secondary networks.

The considerations listed above are based upon the features of general electric distribution system. There are two primary considerations in the development of an effective solution technique for distribution system simulation, as shown below:

- Formulation of a mathematical description of components such as cogenerators, transformers, feeders, shunt elements, and loads;
- Selection of suitable numerical methods.

The design must also consider the inter-relationship of these two factors, and the impact of the chosen method, considering large-scale system models, convergence, execution time, and data requirements. Furthermore, the selection of suitable numerical methods should consider that the power flow and short circuit analyses can be integrated easily.

B. PHASE REPRESENTATION

The GDAS employ an individual phase, as opposed to a balanced three-phase, representation. Under the assumption that the power system is sufficiently balanced, the balanced three-phase representation adopt positive sequence network to represent a three phase system. For some transmission systems, this representation may be acceptable, but it is not suitable for a inherent unbalanced distribution system. Therefore, in the GDAS program three individual phase buses was employed to represent a three-phase bus and one individual phase bus to represent a single-phase bus. In a word, the GDAS program employed true phase coordinates for both component and system representations.

A discussion of the differences between three-phase and balanced power flow results and the effects of mutual couplings can be found in [1]. A brief summary of that discussion is given below:

- For the buses which are electrically far away from the substation, the differences between the calculated bus voltages obtained by the three-phase power flow and those by the balanced power flow are generally significant. For heavy-loaded feeders, it is expected that the result calculated by the balanced power flow approach would be unacceptable from a practical operational viewpoint.

- For an unbalanced system, the feeder losses obtained by the three-phase power flow approach method is higher than those by the balanced power flow approach method. This is consistent with the fact that system imbalance causes extra losses.

- The additional drops in bus voltages are caused by neglecting the mutual couplings in the three-phase power flow analysis. This behavior will be realized when the voltage sources at the swing bus are predominantly positive-sequence and when the feeder positive-sequence impedance is increased by the exclusion of the mutual couplings. For the purpose of accurately analyzing the operational behavior of the system, it is apparent that the consideration of the mutual couplings is necessary for the power flow procedure.

The symmetrical component approach is convenient for both power flow and short circuit analyses of a basically balanced system since no mutual coupling is assumed to exist between the sequence components. However, the basic imbalance of a distribution system reduces the benefits of the symmetrical component approach because of mutual coupling between the sequence components and detailed analysis required.

C. COMPONENT MODELS

In electric distribution analysis, individual system components are given mathematical representations that approximate their physical behavior. These mathematical representations are referred to as "models". In the GDAS program, components are modeled by their equivalent circuits in terms of inductance, capacitance, resistance, and injected current.

These mathematical component models depend upon the type of study to be performed and hence may differ for each study. For example, using GDAS in power flow and short circuit analyses, the same transformer and conductor models are used, but the cogenerator and load models are different. Furthermore, due to inherent system unbalance, distribution system analysis requires detailed models.

To develop an electric distribution analysis system, the following component models were determined to be necessary.

- Conductor - individual phase representation for both primary and secondary with capacitive line charging on primary conductor only;
- Cogenerators - three-phase representation for both synchronous and induction cogenerators;
- Transformers - principally delta-grounded wye for network transformers;
- Demands or loads - considering the voltage characteristics of the loads;
- Capacitors - principally the shunt capacitors for var compensation, power factor improvement, or voltage profile improvement.

Other components such as network protectors, fuses, automatic switches, etc., although necessary in contingency analysis, are not important in power flow and short circuit studies, and therefore, are not presented.

In the following sections, the conductor, capacitor, cogenerator, transformer, and load models are presented. For a detailed discussion of feeder models for both coupled and uncoupled feeders see references[2] and [3] and reference [4] for a more detailed discussion of the load model used.

Since most distribution system elements are three-phase, only the three-phase model is presented. Single and two-phase models can be obtained by following similar procedures. In the GDAS program, all the single, two, and three-phase models are available. The component models use the bus frame of reference and are developed for Z_{Bus}-based solution techniques, for example Optimal Ordered Bi-factorization Y_{Bus} method and direct Z_{Bus} Gauss method. These component models may require little, if any modification to apply them to other solution techniques.

1. Conductors

Conductors, both overhead and underground, are modeled on a per-phase basis. Their impedances are calculated using the methods originally developed by Carson [5] and later simplified by Clarke [6] and Lewis [7]. The presence of earth, neutral return, and conductor mutual coupling are all recognized. For a conductor, the only thing of concern is its modeling, hence, the computation steps for line impedance is, therefore, not presented.

In GDAS program, the three-phase primary feeders and secondary cables can be represented by the general π model shown in Figure 1. The general π model considers the line charging as well as the mutual coupling. The shunt capacitance part of secondary cables can be neglected because the secondary network voltage is very low.

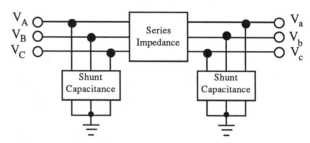

Figure 1 General Conductor Model

a. Series Impedance

Figure 2 shows the phase and neutral series impedance of the original three phase conductor. The neutral wire is assumed to have multiple grounding, and voltage drop along the neutral wire is zero. Thus, Kron's matrix reduction technique can be applied. After Kron's reduction is applied, the equivalent representation shown in Figure 3 is obtained. The effects of the neutral wire are still included in this representation. The dotted line in Figure 3 for the neutral return reflects the network reduction that has been performed. Figure 3 shows the series impedance part of a conductor, and its series admittance matrix is shown in Equation (1). By using the individual phase representation, the equivalent circuit of the series part of a conductor is shown in Figure 4.

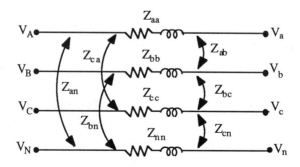

Figure 2 Original Three-Phase Conductor

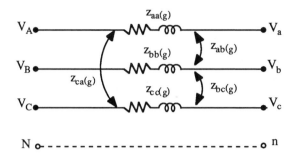

Figure 3 Series Impedance Part of a Three-phase Conductor

$$
[Y^{abc}] = \begin{bmatrix} Y_{aa(g)} & Y_{ab(g)} & Y_{ac(g)} \\ Y_{ba(g)} & Y_{bb(g)} & Y_{bc(g)} \\ Y_{ca(g)} & Y_{cb(g)} & Y_{cc(g)} \end{bmatrix} \tag{1}
$$

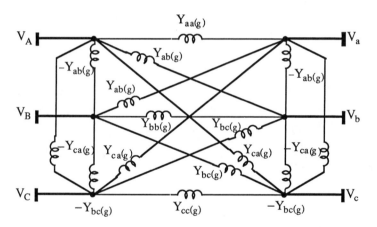

Figure 4 Equivalent Circuit of Series Admittance Part of a Three-phase Conductor

b. Shunt Capacitance

The shunt capacitance part of the original conductor is shown in Figure 5(a). Which can be represented by its equivalent injected currents shown in Figure 5(b). The injections in Figure 5(b) can be reformulated into three single-phase injection currents I_a, I_b, and I_c shown below:

$$I_a = -\frac{1}{2}(y_{ab} + y_{ac} + y_{an})V_a + \frac{y_{ab}}{2}V_b + \frac{y_{ac}}{2}V_c$$

$$I_b = -\frac{1}{2}(y_{ab} + y_{bc} + y_{bn})V_b + \frac{y_{ab}}{2}V_a + \frac{y_{bc}}{2}V_c \qquad (2)$$

$$I_c = -\frac{1}{2}(y_{ac} + y_{bc} + y_{cn})V_c + \frac{y_{ac}}{2}V_a + \frac{y_{bc}}{2}V_b$$

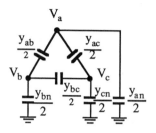

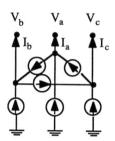

(a) Shunt Capacitances of Three-Phase Feeder

(b) Equivalent Injected Currents

Figure 5 Shunt Capacitances and Equiv. Injected Currents of 3-Phase Conductor

2. Shunt Capacitors

A three-phase shunt capacitor is schematically shown in Figure 6(a). In the GDAS program, shunt capacitors are represented by their equivalent injected currents shown in Figure 6(b). The voltage characteristics of the shunt capacitor are considered in this model.

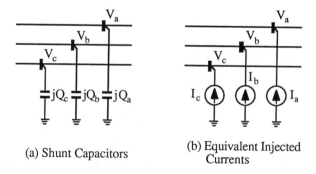

(a) Shunt Capacitors

(b) Equivalent Injected Currents

Figure 6 Three-phase Shunt Capacitor and Equivalent Injected Currents

Where

$$Q_a = Q_a^0 \left| \frac{V_a}{V_a^0} \right|^2; \quad Q_b = Q_b^0 \left| \frac{V_b}{V_b^0} \right|^2; \quad Q_c = Q_c^0 \left| \frac{V_c}{V_c^0} \right|^2 \tag{3}$$

The superscript "0" means the rating values.

If $|V| \neq 0$ then

$$I_a = \frac{-j\,Q_a}{V_a^*}; \quad I_b = \frac{-j\,Q_b}{V_b^*}; \quad I_c = \frac{-j\,Q_c}{V_c^*} \tag{4}$$

else

$$I_a = I_b = I_c = 0 \tag{5}$$

3. Cogenerators

Cogeneration is an effective means of increasing energy efficiency and reducing energy costs. The cogeneration process puts wasted heat to work. It saves energy by using the reject heat of one process as an energy input to a subsequent process, effectively using the same fuel. The favorable economics of cogeneration enable it to play a substantial role in energy development in various parts of the world[8-10].

Many utilities have seen the effects of these favorable economics and anticipate the addition of more cogenerators to their system in the furture. It is expected that the number of cogeneration facilities as well as the cogenerator capacity will increase in the future.

The impact of cogenerators upon the existing system must be studied because their contribution to the power flow and short-circuit current of the system is significant. When a request is made to generate electricity by using cogenerators in parallel with a utility system, electric distribution engineers need to study the impact that the proposed cogenerators will have upon existing facilities.

Research was conducted into the operation and characteristics of both synchronous and induction cogenerator plants available in the market. This information was used to develop a cogenerator model that was consistent with the three individual phase representation and other requirements of the GDAS, while satisfying reasonable data requirements.

This section introduces new cogenerator models for three-phase distribution power flow and short-circuit studies. The cogenerator model used in power flow studies can also be used for system loss and contingency analyses. It is significant because it represents the generator phase imbalance due to inherent distribution system imbalance. The GDAS has been developed to evaluate the performance of the distribution system with cogenerators. The derivation of these cogenerator models is presented in the following section.

a. Derivation of Cogenerator Models

A preliminary investigation of typical voltage control systems for synchronous cogenerators was done. According to the investigation results, synchronous cogenerators are not controlled to maintain constant voltage, they are controlled to maintain constant power and constant power factor. Furthermore, utilities may require that cogeneration operators provide such controls. For example, Con Edison, under some conditions, may require a power factor controller to maintain a constant power factor on the synchronous generator by controlling the voltage regulator. The power factor controller must be capable of maintaining a power factor within plus or minus one percent at any set point. As a result, the synchronous cogenerators can be represented approximately as constant complex power devices in the power flow study, i.e. cogenerators can be represented as P-Q specified devices in the power flow calculation.

As for induction cogenerators, their reactive power will vary with the terminal voltage change. Thus, the reactive power consumption of the induction cogenerators is not exactly constant. For simplification, the induction cogenerators can be treated as P-Q specified devices because the bus voltages are near 1.0 p.u. in steady state cases. As a result, both the synchronous and induction generators can be represented as P-Q specified devices, i.e. constant complex power devices.

The cogenerators are modeled as an internal voltage "E_I" behind the proper reactance; the generator subtransient reactance X_d"[11,12] as shown in the Figure 7. This is the Thevenin equivalent circuit of a cogenerator. This model is different from the traditional power flow generator bus (PV Bus) which is represented by specified injected powers at a specified voltage. Using the Figure 7 model, injected power of each phase, under an unbalanced terminal voltage condition, can be calculated in detail.

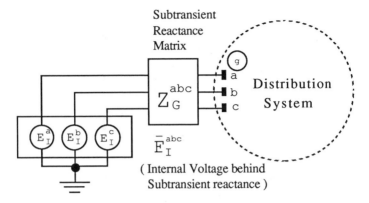

Figure 7 Thevenin Equivalent Circuit of Cogenerator

Where

$\overline{E}_I^{abc}$ are the voltages behind subtransient reactance

Z_G^{abc} is the subtransient impedance matrix

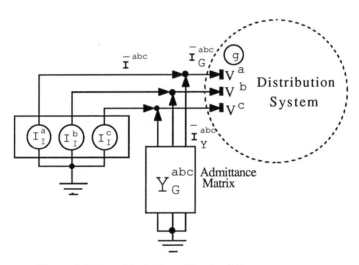

Figure 8 Norton Equivalent Circuit of Cogenerator

Based on the preliminary investigation and assumptions discussed above, one has

Total real power $= P_T = p^a + p^b + p^c = \text{Constant}$

Total reactive power $= Q_T = q^a + q^b + q^c = \text{Constant}$

The internal voltage $\overline{E}_I^{abc}$ is a balanced three-phase voltage in both magnitude and angle, assuming a balanced design of the generator windings. In the GDAS program, a Norton equivalent circuit of Figure 7 is used to represent the cogenerator model shown in Figure 8. The current vector in Figure 8 should be balanced and

$$Y_G^{abc} = \left\{ Z_G^{abc} \right\}^{-1} ; \qquad \overline{I}_I^{abc} = Y_G^{abc} \, \overline{E}_I^{abc}$$

Sequence models in symmetrical component coordinates are used to derive the phase model of cogenerator shown in Figure 8. The sequence models shown in Figure 9 are widely used in power system analysis.

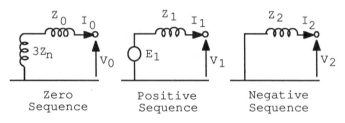

Figure 9 Sequence Models of Cogenerator

Where

Z_0 : Zero-sequence impedance

Z_n : Impedance between neutral and ground

Z_1 : Positive-sequence impedance (Subtransient reactance X_d'')

Z_2 : Negative-sequence impedance

E_1 : Generated voltage

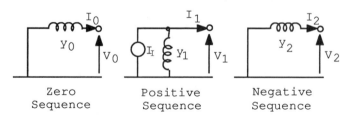

Figure 10 Norton Equivalent Circuit of Cogenerator

Where

$$y_0 = \frac{1}{z_0 + 3z_n} \qquad y_1 = \frac{1}{z_1} \qquad y_2 = \frac{1}{z_2}$$

The subtransient reactance is used not only in short circuit studies but also in power flow studies. Rotor effects should be considered even in the steady state condition of power flow analysis because of phase imbalance.

Figure 10 is the Norton equivalent circuit of Figure 9, where a current source I_I is used instead of the voltage source E_1.

The sequence models in Figure 10 are represented by

$$\begin{bmatrix} I_0 \\ I_1 \\ I_2 \end{bmatrix} = \begin{bmatrix} 0 \\ I_I \\ 0 \end{bmatrix} - \begin{bmatrix} y_0 & 0 & 0 \\ 0 & y_1 & 0 \\ 0 & 0 & y_2 \end{bmatrix} \begin{bmatrix} V_0 \\ V_1 \\ V_2 \end{bmatrix} \tag{6}$$

The phase model can be obtained by transforming Eq. (6) to phase coordinates using the transformation matrix T_s.

$$T_s \begin{bmatrix} I_0 \\ I_1 \\ I_2 \end{bmatrix} = T_s \begin{bmatrix} 0 \\ I_I \\ 0 \end{bmatrix} - T_s \begin{bmatrix} y_0 & 0 & 0 \\ 0 & y_1 & 0 \\ 0 & 0 & y_2 \end{bmatrix} T_s^{-1} T_s \begin{bmatrix} V_0 \\ V_1 \\ V_2 \end{bmatrix} \tag{7}$$

Where

$$T_s = \frac{1}{\sqrt{3}} \begin{bmatrix} 1 & 1 & 1 \\ 1 & a^2 & a \\ 1 & a & a^2 \end{bmatrix} \qquad T_s^{-1} = T_s^* = \frac{1}{\sqrt{3}} \begin{bmatrix} 1 & 1 & 1 \\ 1 & a & a^2 \\ 1 & a^2 & a \end{bmatrix} \tag{8}$$

Thus the phase model is given by Eq. (9), resulting in the physical representations shown in Figure 8.

$$\begin{bmatrix} I_a \\ I_b \\ I_c \end{bmatrix} = \frac{1}{\sqrt{3}} \begin{bmatrix} 1 \\ a^2 \\ a \end{bmatrix} I_I - Y_G^{abc} \begin{bmatrix} V_a \\ V_b \\ V_c \end{bmatrix} \tag{9}$$

Where

$$Y_G^{abc} = T_s \, Y_G^{012} \, T_s^{-1} = T_s \begin{bmatrix} y_0 & 0 & 0 \\ 0 & y_1 & 0 \\ 0 & 0 & y_2 \end{bmatrix} T_s^{-1} \tag{10}$$

The current source is balanced, thereby, by constraining the specified P_T and Q_T in power flow studies.

Multiplying Eq. (9) on both sides by V^{abc*t}

$$\begin{bmatrix} V_a^* V_b^* V_c^* \end{bmatrix} \begin{bmatrix} I_a \\ I_b \\ I_c \end{bmatrix} = \frac{1}{\sqrt{3}} \begin{bmatrix} V_a^* V_b^* V_c^* \end{bmatrix} \begin{bmatrix} 1 \\ a^2 \\ a \end{bmatrix} I_I - \begin{bmatrix} V_a^* V_b^* V_c^* \end{bmatrix} Y_G^{abc} \begin{bmatrix} V_a \\ V_b \\ V_c \end{bmatrix} \tag{11}$$

Then the constraint is represented by

$$(p_a + p_b + p_c) - j(q_a + q_b + q_c)$$
$$= \frac{1}{\sqrt{3}} I_I (V_a^* + a^2 V_b^* + a V_c^*) - \overline{V}^{abc*t} Y_G^{abc} \overline{V}^{abc} \tag{12}$$

and

$$P_T - j Q_T = (p_a + p_b + p_c) - j(q_a + q_b + q_c) \tag{13}$$

Thus, specifying total P_T and total Q_T is sufficient to guarantee a balanced current source.

The cogenerator model shown in Figure 8 can be applied to both power flow and short circuit analysis. In the power flow study, the injection current I_I will be updated using the Eq. (12) based on the update voltage at each iteration. In the short circuit study, the I_I is constant, and is calculated using the converged power flow voltages at the bus where the cogenerator is connected to the system. In power flow studies the admittance matrix Y_G^{abc}, total real power P_T and total reactive power Q_T are held constant. In short circuit studies the admittance matrix Y_G^{abc} and the internal balanced current source I_I are held constant. Although some variables are held constant during the iterative process, the voltages that are dependent on the system condition will change at each iteration. Therefore, in both power flow and short circuit studies, the injected currents will change due to the voltage change at the buses where the cogenerator is connected.

b. Cogenerator Model for Power Flow Study

Figure 8 shows the cogenerator model for power flow analysis,
Where

$$y_0 = \frac{1}{x_0} ; \qquad y_1 = \frac{1}{x_d''} ; \qquad y_2 = \frac{1}{x_2} ;$$

$$Y_G = T_s \begin{bmatrix} y_0 & 0 & 0 \\ 0 & y_1 & 0 \\ 0 & 0 & y_2 \end{bmatrix} T_s^{-1} ; \qquad T_s = \frac{1}{\sqrt{3}} \begin{bmatrix} 1 & 1 & 1 \\ 1 & a^2 & a \\ 1 & a & a^2 \end{bmatrix} ;$$

$$\overline{V}_g^{abc} = \begin{bmatrix} V_g^a & V_g^b & V_g^c \end{bmatrix}^t ; \qquad \overline{I}_g^{abc} = \begin{bmatrix} I_g^a & I_g^b & I_g^c \end{bmatrix}^t ;$$

$$I_I = \frac{(P_T - j Q_T) + \overline{V}_g^{abc*t} Y_G \overline{V}_g^{abc}}{\frac{1}{\sqrt{3}} (V_g^{a*} + a^2 V_g^{b*} + a V_g^{c*})} \tag{14}$$

$$\bar{I}_I^{abc} = \frac{1}{\sqrt{3}} \begin{bmatrix} 1 \\ a^2 \\ a \end{bmatrix} I_I \tag{15}$$

$$\bar{I}_g^{abc} = \bar{I}_I^{abc} - Y_G^{abc} \, \bar{V}_g^{abc} = \frac{1}{\sqrt{3}} \begin{bmatrix} 1 \\ a^2 \\ a \end{bmatrix} I_I - Y_G^{abc} \, \bar{V}_g^{abc} \tag{16}$$

Figure 11 shows the procedure for calculating I_a, I_b and I_c, the cogenerator injected source currents, for each iteration in the power flow study.

From the above discussion, a cogenerator is represented by three injected currents. This representation is suitable for Z_{Bus} or the factorized Y_{Bus} methods, however, for Newton-Raphson, Gauss-Seidel method or other solution methods, the representation may require modification. The basic algorithm is unchanged, however.

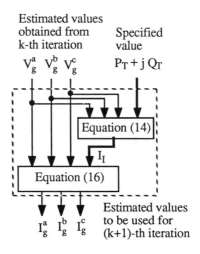

Figure 11 Procedure to Calculate I_a, I_b and I_c for Power Flow Study

c. Cogenerator Model for Short-Circuit Study

The cogenerator model for short-circuit study is the same as that of the power flow study except for the calculation of the internal current source I_I. In the short-circuit study, the initial value of the internal current source I_I, obtained from the power flow calculation, is maintained constant since the internal voltage E_I of the cogenerator is assumed constant at the instant of the fault. The constant I_I is calculated using the converged power flow voltages at the bus where the cogenerator is connected to the system. The calculation procedure of the cogenerator model for short-circuit study is shown in Figure 12.

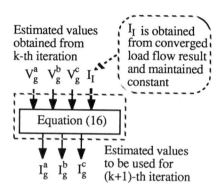

Figure 12   Procedure to Calculate I_a, I_b and I_c for Fault Study

4. Transformers

The impact of the numerous transformers in a distribution system is significant. Transformers affect system loss, zero sequence current, grounding method, and protection strategy. Although the transformer is one of the most important components of modern electric power systems, highly developed transformer models are not employed in system studies. It is the intention of this section to introduce a transformer model and its implementation method so that large-scale unbalanced distribution system problems such as power flow, short circuit, system loss, and contingency studies, can be solved.

Recognizing the fact that the system is unbalanced, the conventional transformer.models, based on a balanced three phase assumption, can no longer be considered suitable. This is done with justifiable reason. For example in the widely used, delta-grounded wye connection of distribution step-down transformers, the positive and negative sequence voltages are shifted in opposite directions, this phase shift must be included in the model to properly simulate the effects of the system imbalance.

Recent interest in unbalanced system phenomena has also produced a transformer model adaptable to the unbalanced problem which is well outlined in [13]. Further information about this model may also be found in [3,14,15]. The model developed thus far can be applied directly to distribution power flow and short-circuit analyses. However, it is still not accurate for system loss analysis because the transformer core loss contribution to total system loss is significant[16,17]. To calculate total system loss, the core loss of the transformer must be included in the model. The complete transformer model combines the unbalanced and loss models from [13] and [16] in order to integrate system loss analysis in power flow or short-circuit studies.

It is important to note that the unbalanced transformer model derived by Dillon in reference [13] cannot be applied directly to either the factorized Y_{Bus} or direct inverse Y_{Bus} method because of numerical considerations. For some connections such as grounded wye-delta, delta- grounded wye, these models make the system Y_{Bus} singular. Therefore, the application of the factorized or direct inverse Y_{Bus} methods becomes impossible. A novel implementation method is introduced to solve this problem. In this implementation method, artificial injected currents are used to make the system Y_{Bus} a nonsingular matrix.

In order to model both primary feeders and secondary systems simultaneously, on a three-phase basis, transformer models were developed by a rigorous approach. Again, engineering assumptions and trade-offs were made to arrive at a transformer model that was consistent in representation and accuracy to the remaining features of the GDAS.

A general approach is recommended whereby all transformer connections, including the three-phase common core transformers supplying the secondary network system, are represented as individual transformers connected appropriately. An overview of this approach is presented in this section, followed by a detailed description of the three-phase transformer with two most common connections, grounded wye-grounded wye and delta-grounded wye.

As used in Reference [16], a three-phase transformer is represented by two blocks shown in Figure 13. One block represents the per unit leakage admittance matrix Y_T^{abc}, and the other block models the core loss as a function of voltage on the secondary side of the transformer.

The presence of the admittance matrix block is the major distinction between the proposed model and the model used in [16,17]. In the proposed model, Dillon's model is integrated with the admittance matrix part. As a result, the copper loss, core loss, system imbalance, and phase shift characteristics are taken into account. The implementation method is introduced in the following sections.

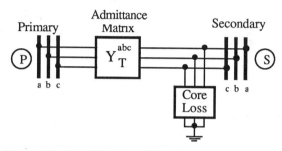

Figure 13 Overall Proposed Transformer Model

a. Core Loss

The core loss of a transformer is approximated by shunt core loss functions on each phase of the secondary terminal of the transformer. These core loss approximation functions are based on the results of EPRI load modeling research [4] which state that real and reactive power losses in the transformer core can be expressed as functions of the terminal voltage of the transformer. Transformer core loss functions represented in per unit at the system power base are[16,17]:

$$P \ (p.u.) = \frac{KVA \ Rating}{System \ Base} \ (A |V|^2 + B \ e^{C|V|^2}) \tag{17}$$

$$Q \ (p.u.) = \frac{KVA \ Rating}{System \ Base} \ (D |V|^2 + E \ e^{F|V|^2}) \tag{18}$$

Where, typically,

A=0.00267   B=0.734x10^{-9} C=13.5
D=0.00167   E=0.268x10^{-13} F=22.7
$|V|$ is the voltage magnitude in per unit.

It must be noted that the coefficients; A, B, C, D, E, and F; are machine dependent constants. For the sample system shown in Section III, core losses are represented by the functions and typical constants as indicated above. If better information becomes available, the functional representations can be easily modified.

b. Admittance Matrix

The admittance matrix part of the proposed three-phase transformer models follow the methodology derived by Dillon[13], but a novel implementation is introduced herein.

For simplification, a single three-phase transformer is approximated by three identical single-phase transformers connected appropriately. This assumption is not essential, however, it simplifies the ensuing derivation and explanation. Based upon this assumption, the characteristic submatrices used in forming the three-phase transformer admittance matrices can be developed. The matrices for the nine common connections of three-phase transformers are given in Table I.

Table I Characteristic Submatrices used in forming the
Three-phase Transformer Admittance Matrices

Transformer connection		Self Admittance		Mutual Admittance	
Bus p	Bus s	Y_p^{abc}	Y_s^{abc}	Y_{ps}^{abc}	Y_{sp}^{abc}
Wye-G	Wye-G	Y_I	Y_I	$-Y_I$	$-Y_I$
Wye-G	Wye	Y_{II}	Y_{II}	$-Y_{II}$	$-Y_{II}$
Wye-G	Delta	Y_I	Y_{II}	Y_{III}	Y_{III}^t
Wye	Wye-G	Y_{II}	Y_{II}	$-Y_{II}$	$-Y_{II}$
Wye	Wye	Y_{II}	Y_{II}	$-Y_{II}$	$-Y_{II}$
Wye	Delta	Y_{II}	Y_{II}	Y_{III}	Y_{III}^t
Delta	Wye-G	Y_{II}	Y_I	Y_{III}^t	Y_{III}
Delta	Wye	Y_{II}	Y_{II}	Y_{III}^t	Y_{III}
Delta	Delta	Y_{II}	Y_{II}	$-Y_{II}$	$-Y_{II}$

Where

$$Y_I = \begin{bmatrix} y_t & 0 & 0 \\ 0 & y_t & 0 \\ 0 & 0 & y_t \end{bmatrix} \quad Y_{II} = \frac{1}{3}\begin{bmatrix} 2y_t & -y_t & -y_t \\ -y_t & 2y_t & -y_t \\ -y_t & -y_t & 2y_t \end{bmatrix} \quad Y_{III} = \frac{1}{\sqrt{3}}\begin{bmatrix} -y_t & y_t & 0 \\ 0 & -y_t & y_t \\ y_t & 0 & -y_t \end{bmatrix} \quad (19)$$

and y_t is the leakage admittance per phase in p.u..

If the transformer has an off-nominal tap ratio $\alpha:\beta$ between the primary and secondary windings, where α and β are tappings on the primary and secondary sides, respectively, then the submatrices are modified as follows:

a) Divide the self admittance matrix of the primary by α^2

b) Divide the self admittance matrix of the secondary by β^2

c) Divide the mutual admittance matrices by $\alpha\beta$

The electrical models (equivalent circuits) and programming models (the modified equivalent circuit models for power flow and fault analysis) of the two most common connections, grounded wye-grounded wye and delta-grounded, are presented in this paper. The programming models of other connections can be derived through a similar procedure. The derivation of the characteristic submatrices of a three-phase transformer can be found in more detail in [13].

(1). Grounded Wye-Grounded Wye Transformer

A network connection diagram for the three-phase grounded wye–grounded wye transformer is shown in Figure 14.

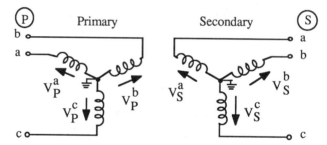

Figure 14 Network Connection Diagram for Three-phase G Wye-G Wye
Transformer

From Table I, the admittance matrix for the three-phase grounded wye-grounded wye transformer is

$$Y_T^{abc} = \begin{bmatrix} Y_I & -Y_I \\ -Y_I & Y_I \end{bmatrix} = \begin{bmatrix} y_t & 0 & 0 & -y_t & 0 & 0 \\ 0 & y_t & 0 & 0 & -y_t & 0 \\ 0 & 0 & y_t & 0 & 0 & -y_t \\ -y_t & 0 & 0 & y_t & 0 & 0 \\ 0 & -y_t & 0 & 0 & y_t & 0 \\ 0 & 0 & -y_t & 0 & 0 & y_t \end{bmatrix} \qquad (20)$$

If the tappings α and β are not equal to 1 i.e. an off-nominal tap ratio, then the admittance matrix must be modified using Eq. (21): the general form of admittance matrix for nominal and off-nominal grounded wye-grounded wye transformer. The equivalent circuit of Eq. (21) is shown in Figure 15.

$$
Y_T^{abc} = \begin{bmatrix}
\dfrac{y_t}{\alpha^2} & 0 & 0 & \dfrac{y_t}{\alpha\beta} & 0 & 0 \\[2mm]
0 & \dfrac{y_t}{\alpha^2} & 0 & 0 & \dfrac{y_t}{\alpha\beta} & 0 \\[2mm]
0 & 0 & \dfrac{y_t}{\alpha^2} & 0 & 0 & \dfrac{y_t}{\alpha\beta} \\[2mm]
\dfrac{y_t}{\alpha\beta} & 0 & 0 & \dfrac{y_t}{\beta^2} & 0 & 0 \\[2mm]
0 & \dfrac{y_t}{\alpha\beta} & 0 & 0 & \dfrac{y_t}{\beta^2} & 0 \\[2mm]
0 & 0 & \dfrac{y_t}{\alpha\beta} & 0 & 0 & \dfrac{y_t}{\beta^2}
\end{bmatrix}
\qquad (21)
$$

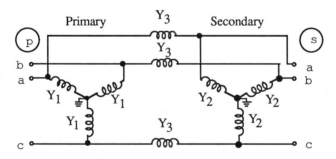

Figure 15 Equivalent Circuit for 3-phase G Wye-G Wye Transformer

Where

$$
Y_1 = \frac{y_t}{\alpha\beta}\left(\frac{\beta-\alpha}{\alpha}\right); \qquad
Y_2 = \frac{y_t}{\alpha\beta}\left(\frac{\alpha-\beta}{\beta}\right); \qquad
Y_3 = \frac{y_t}{\alpha\beta}
\qquad (22)
$$

For a nominal tap grounded wye-grounded wye transformer, both α and β are equal to unity so the shunt elements in the equivalent circuit disappear. For the off-nominal tap ratio transformer, either α or β or both α and β are not equal to unity, therefore, the shunt elements reflect the effects of the off-nominal tap ratio.

The programming model of a three-phase grounded wye-grounded wye transformer used in the GDAS program is shown in Figure 16.

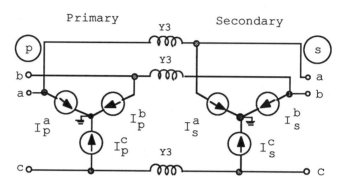

Figure 16   Programming Model for Three-phase G Wye-G Wye Transformer

Where

$$Y_3 = \frac{y_t}{\alpha\beta}$$

$$I_p^a = -\frac{y_t}{\alpha\beta}\left(\frac{\beta-\alpha}{\alpha}\right)V_p^a \qquad I_p^b = -\frac{y_t}{\alpha\beta}\left(\frac{\beta-\alpha}{\alpha}\right)V_p^b \qquad I_p^c = -\frac{y_t}{\alpha\beta}\left(\frac{\beta-\alpha}{\alpha}\right)V_p^c$$

$$I_s^a = -\frac{y_t}{\alpha\beta}\left(\frac{\alpha-\beta}{\beta}\right)V_s^a \qquad I_s^b = -\frac{y_t}{\alpha\beta}\left(\frac{\alpha-\beta}{\beta}\right)V_s^b \qquad I_s^c = -\frac{y_t}{\alpha\beta}\left(\frac{\alpha-\beta}{\beta}\right)V_s^c$$

or in vector forms

$$\overline{I}_p^{abc} = -\frac{y_t}{\alpha\beta}\left(\frac{\beta-\alpha}{\alpha}\right)\overline{V}_p^{abc} \tag{23}$$

$$\overline{I}_s^{abc} = -\frac{y_t}{\alpha\beta}\left(\frac{\alpha-\beta}{\beta}\right)\overline{V}_s^{abc} \tag{24}$$

The artificial current sources reflect the effect of the off-nominal tap ratio. If α is equal to β then

$$\overline{I}_p^{abc} = \overline{I}_s^{abc} = \overline{0}$$

and there is no injected current on either end of the transformer. Figure 16 is redrawn as Figure 17, a general programming model for three-phase transformer using in GDAS program.

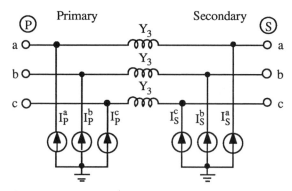

Figure 17 General Programming Model for Three-Phase Transformer

Where

$$Y_3 = \frac{y_t}{\alpha\beta} \tag{25}$$

$$\overline{I}_p^{abc} = \frac{y_t}{\alpha\beta} \left(\frac{\beta - \alpha}{\alpha} \right) \overline{V}_p^{abc} \tag{26}$$

$$\overline{I}_s^{abc} = \frac{y_t}{\alpha\beta} \left(\frac{\alpha - \beta}{\beta} \right) \overline{V}_s^{abc} \tag{27}$$

(2) Delta-Grounded Wye Transformer

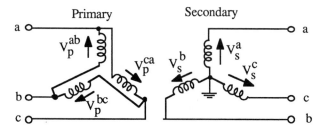

Figure 18 Network Connection Diagram for 3-phase Delta-G Wye Transformer

The network connection diagram for the three-phase delta-grounded wye transformer is shown in Figure 18.

The admittance matrix for the three-phase delta-grounded wye transformer is

$$
Y_T^{abc} = \begin{bmatrix} Y_{II} & Y_{III} \\ Y_{III} & Y_I \end{bmatrix} = \begin{bmatrix}
\frac{2}{3} y_t & -\frac{1}{3} y_t & -\frac{1}{3} y_t & \frac{-y_t}{\sqrt{3}} & \frac{y_t}{\sqrt{3}} & 0 \\
-\frac{1}{3} y_t & \frac{2}{3} y_t & -\frac{1}{3} y_t & 0 & \frac{-y_t}{\sqrt{3}} & \frac{y_t}{\sqrt{3}} \\
-\frac{1}{3} y_t & -\frac{1}{3} y_t & \frac{2}{3} y_t & \frac{y_t}{\sqrt{3}} & 0 & \frac{-y_t}{\sqrt{3}} \\
\frac{-y_t}{\sqrt{3}} & 0 & \frac{y_t}{\sqrt{3}} & y_t & 0 & 0 \\
\frac{y_t}{\sqrt{3}} & \frac{-y_t}{\sqrt{3}} & 0 & 0 & y_t & 0 \\
0 & \frac{y_t}{\sqrt{3}} & \frac{-y_t}{\sqrt{3}} & 0 & 0 & y_t
\end{bmatrix}
\tag{28}
$$

For an off-nominal delta-grounded wye transformer, the admittance matrix is modified as shown below.

$$
Y_T^{abc} = \begin{bmatrix}
\frac{2}{3}\frac{y_t}{\alpha^2} & -\frac{1}{3}\frac{y_t}{\alpha^2} & -\frac{1}{3}\frac{y_t}{\alpha^2} & \frac{-y_t}{\sqrt{3}\alpha\beta} & 0 & \frac{y_t}{\sqrt{3}\alpha\beta} \\
-\frac{1}{3}\frac{y_t}{\alpha^2} & \frac{2}{3}\frac{y_t}{\alpha^2} & -\frac{1}{3}\frac{y_t}{\alpha^2} & \frac{y_t}{\sqrt{3}\alpha\beta} & \frac{-y_t}{\sqrt{3}\alpha\beta} & 0 \\
-\frac{1}{3}\frac{y_t}{\alpha^2} & -\frac{1}{3}\frac{y_t}{\alpha^2} & \frac{2}{3}\frac{y_t}{\alpha^2} & 0 & \frac{y_t}{\sqrt{3}\alpha\beta} & \frac{-y_t}{\sqrt{3}\alpha\beta} \\
\frac{-y_t}{\sqrt{3}\alpha\beta} & \frac{y_t}{\sqrt{3}\alpha\beta} & 0 & \frac{y_t}{\beta^2} & 0 & 0 \\
0 & \frac{-y_t}{\sqrt{3}\alpha\beta} & \frac{y_t}{\sqrt{3}\alpha\beta} & 0 & \frac{y_t}{\beta^2} & 0 \\
\frac{y_t}{\sqrt{3}\alpha\beta} & 0 & \frac{-y_t}{\sqrt{3}\alpha\beta} & 0 & 0 & \frac{y_t}{\beta^2}
\end{bmatrix}
\tag{29}
$$

The equivalent circuit of Eq. (29) is shown in Figure 19.

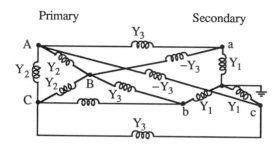

Figure 19 Equivalent Circuit for Three-phase Delta-G Wye Transformer

Where

$$Y_1 = \frac{y_t}{\beta^2} \qquad Y_2 = \frac{y_t}{3\alpha^2} \qquad Y_3 = \frac{y_t}{\sqrt{3}\alpha\beta} \qquad (30)$$

In Figure 19, the equivalent circuit of delta-grounded wye transformer, there are three elements $(Y1)$ connected to ground. These three elements must be represented by current injections because the implicit Z_{Bus} method (i.e. factorized Y_{Bus} method) is used to solve the system equations for both power flow and short circuit studies. After replacing these three admittances by current injections, the system admittance matrix (Y_{Bus}) becomes singular, if the grounded wye side is the end bus[1] or swing bus. To avoid this difficulty, three additional admittances $(-Y3)$ are replaced by current injections.

In the GDAS program, a two-step superposition method is used to implement the Z_{Bus} method. In the first step, there are no loads applied to the system and hence no current flows, therefore, the three Y2 elements in the delta side are represented by current injections. As a result, only the equivalent lines between the primary and secondary sides in the same phase are used to form the system admittance or impedance matrices $(Y_{Bus}$ or $Z_{Bus})$. All other equivalent admittances are represented by their equivalent injected currents. The programming model of a three-phase grounded wye-delta transformer used in the GDAS program is shown in Figure 20.

---

[1] The bus connected to only one bus.

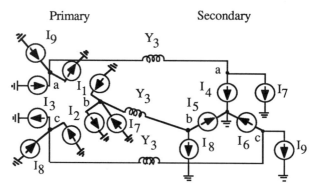

Figure 20 Programming Model for Three-phase Delta-G Wye Transformer

Where

$$Y_3 = \frac{y_t}{\sqrt{3}\alpha\beta}$$

$$I_1 = \frac{y_t}{3\alpha^2} (V_p^a - V_p^b) \qquad I_2 = \frac{y_t}{3\alpha^2} (V_p^b - V_p^c)$$

$$I_3 = \frac{y_t}{3\alpha^2} (V_p^c - V_p^a)$$

$$I_4 = \frac{y_t}{\beta^2} V_s^a \qquad\qquad I_5 = \frac{y_t}{\beta^2} V_s^b \qquad\qquad I_6 = \frac{y_t}{\beta^2} V_s^c$$

$$I_7 = - \frac{y_t}{\sqrt{3}\alpha\beta} (V_p^b - V_s^a) \qquad I_8 = - \frac{y_t}{\sqrt{3}\alpha\beta} (V_p^c - V_s^b)$$

$$I_9 = - \frac{y_t}{\sqrt{3}\alpha\beta} (V_p^a - V_s^c)$$

Another representation of programming model for three-phase grounded wye-delta transformer is shown in Figure 17.

But

$$Y_3 = \frac{y_t}{\sqrt{3}\alpha\beta}$$

$$I_p^a = \frac{y_t}{3\alpha^2} (V_p^c - V_p^a) - \frac{y_t}{3\alpha^2} (V_p^a - V_p^b) - \frac{y_t}{\sqrt{3}\alpha\beta} (V_p^a - V_s^c)$$

$$I_p^b = \frac{y_t}{3\alpha^2} (V_p^a - V_p^b) - \frac{y_t}{3\alpha^2} (V_p^b - V_p^c) - \frac{y_t}{\sqrt{3}\alpha\beta} (V_p^b - V_s^a)$$

$$I_p^c = \frac{y_t}{3\alpha^2} (V_p^b - V_p^c) - \frac{y_t}{3\alpha^2} (V_p^c - V_p^a) - \frac{y_t}{\sqrt{3}\alpha\beta} (V_p^c - V_s^b)$$

$$I_s^a = - \left\{ \frac{y_t}{\beta^2} V_s^a - \frac{y_t}{\sqrt{3}\alpha\beta} (V_p^b - V_s^a) \right\}$$

$$I_s^b = - \left\{ \frac{y_t}{\beta^2} V_s^b - \frac{y_t}{\sqrt{3}\alpha\beta} (V_p^c - V_s^b) \right\}$$

$$I_s^c = - \left\{ \frac{y_t}{\beta^2} V_s^c - \frac{y_t}{\sqrt{3}\alpha\beta} (V_p^a - V_s^c) \right\}$$

5. Demands (Loads)

There are many kinds of load components in a distribution system. Individual loads may be divided into two classes, static and dynamic. In a distribution system some of the loads consist of static or dynamic loads only, but the greater part of them consists of both static and dynamic loads, a so-called composite load. The different characteristics of loads tend to compensate one another, resulting in a composite effect.

Composite loads may be viewed as being divided into four classes: agricultural, commercial, residential, and industrial loads. Agricultural and industrial uses are generally motor loads and should be expected to have more rotating load than commercial or residential areas. On the other hand, in some countries summer heavy loads resulting from air conditioning equipment will have more dynamic load than winter heating and lighting demands. So it is not easy to determine the characteristic or the composition of a composite load.

a. Load Model for Power Flow Study

To obtain a more accurate load behavior of the whole system, the so-called "load energy models" for each individual load component have been derived. Based on these load energy models and the load window concept, energy consumption of a composite load can be obtained according to the bus voltage and ambient conditions. This allows for a more realistic modeling than with conventional power flow analysis which assumes constant power consumption at each bus.

For an unbalanced system, the positive sequence network is not adequate to estimate the power consumption of a three-phase load. Based on field test data, a detailed load energy model for different types of single-phase and three-phase loads has been determined by the Energy Systems Research Center (ESRC) at the University of Texas at Arlington. For a three-phase load, a balanced three-phase voltage source was used during the test. This implies that the unbalanced effect was not considered for the models derived. A so-called "derating factor (DF)" is introduced to solve this problem. By referring to the ANSI/NEMA standard MGI-1978, the rated load capability of polyphase induction motors is normally reduced due to voltage unbalance according to the derating factor by applying the appropriate percentage unbalance. This kind of derating factor has been considered in ESRC's load energy models because the distribution system is usually unbalanced.

The so-called load energy models for each individual load component and the load window concept are used in the GDAS program. Hence, the voltage characteristics of the load are considered in this model. The load window as illustrated in Figure 21 is a pictorial representation showing the contribution to the total load by the various components. For moderate voltage changes, the load energy models and load window concept can be used accurately, although more information about the composition of the each load bus is needed. If no information on the load composition is available, a default load window can be selected to estimate the load.

A sample of three-phase load is shown in Figure 22(a). The unbalanced load distribution is allowed. In the GDAS program, a load is actually represented by equivalent injected currents shown in Figure 22(b).

Incand. Lights	Fluor. Lights	Space Heating	Clothes Dryer	Refrig. Freezer	Elec. Range	TV	Other	Total
? %	? %	? %	? %	? %	? %	?%	? %	100%

Figure 21 A Sample of Winter Residential Load Window

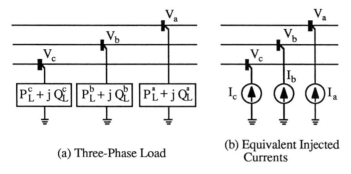

(a) Three-Phase Load

(b) Equivalent Injected Currents

Figure 22 Three-phase Loads and Equivalent Injected Currents

b. Load Model for Short Circuit Study

The so-called load energy models for each individual load component and the load window concept used in the power flow portion of the GDAS program are not suitable for short circuit simulation because of big voltage variance. The constant impedance model was used in the short circuit portion in the GDAS program. The initial power flow simulation provide pre-fault bus voltages and demand currents. Post-fault loads are calculated using this data and the post-fault voltage. The actual load impedance is not calculated explicitly, a constant impedance model is used, however. The load is represented internally as an injected current in the GDAS program.

D. SOLUTION TECHNIQUES

The analytical requirements of the GDAS, particularly the large size of the problems to be solved and the characteristics of the transformer and cogenerator models, require careful selection of a solution technique.

Independently consistent component models, when combined in a large network do not guarantee a solution. The complete network model will have different behavior when coupled with different solution techniques. Important issues in this area including numerical stability, roundoff error propagation, and convergence are discussed after the solution techniques being provided. The solution techniques for the power flow and short circuit portions of the GDAS are provided, with a discussion of the features guiding the particular choice.

1. Power Flow Solution Technique

The Z_{Bus} Gauss approach has been selected for the power flow portion of the GDAS. A brief summary of the two most common solution techniques for power flow problems, the Newton-Raphson and Z_{Bus} Gauss features is given below. A more detailed discussion can be found in [16].

The Newton-Raphson approach is known for its excellent convergence characteristics, but its major shortcoming is the requirement that the Jacobian matrix, with a rank approximately four times that of the Y_{Bus}, be recalculated for each iteration. Another drawback is that the Jacobian matrix cannot be decoupled because of the smaller X/R ratio of distribution systems. These factors, combined with the size of the proposed distribution networks to be studied makes the Newton-Raphson approach unattractive.

The Z_{Bus} Gauss methods uses the sparse bifactored Y_{Bus} matrix and equivalent current injections to solve network equations. The convergence behavior of the Z_{Bus} method is highly dependent upon the number of voltage specified buses in the system. If the only voltage specified bus in the system is the swing bus, the rate of convergence is comparable to the Newton-Raphson approach. The Z_{Bus} method meets the requirements for both rapid convergence rate and minimum memory usage. The distribution system is well suited for the Z_{Bus} method; the only voltage specified bus in the system is the substation bus and each cogenerator bus is handled as a P-Q specified bus.

a. Methodologies

The Z_{Bus} Gauss method is based upon the principle of superposition applied to the system bus voltages, the voltage of each bus is considered to arise from two different contributions, the specified source voltage and equivalent current injection. The loads, cogenerators, capacitors and reactors are modeled as current injection sources/sinks at their respective buses.

The superposition principle dictates that only one type of source will be considered at a time when calculating the bus voltages. On the one hand, when the swing bus voltage source is activated, all current injection sources are disconnected from the system. On the other hand, when all current injection sources are connected to the system, the swing bus is short-circuited to the ground. This decomposition and superposition in the analysis of a sample system is shown schematically in Figures 23,24, and 25.

The component of each bus voltage obtained by activating only the swing bus voltage source represents the no-load system voltage. This component can be determined directly as equal to the swing bus voltage for every bus in the system, however, the other component, affected by load currents and cogenerator currents, cannot be determined directly. Since load and cogenerator currents are affected by bus voltages and vice versa, these quantities must be determined in an iterative manner. The flow chart of the power flow algorithm is shown in Figure 26.

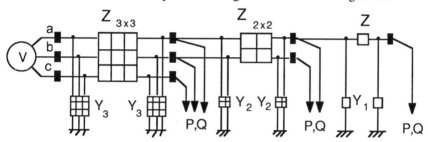

Figure 23 Distribution System Model - Full Superposition

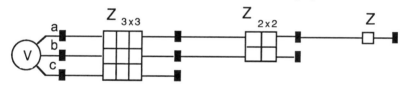

Figure 24 System Model - Decomposition, O. C. Current Sources

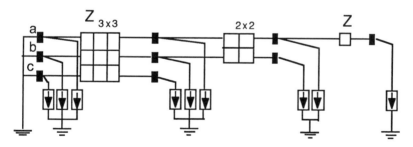

Figure 25 System Model - Decomposition, S. C. Volt. Source to Ground

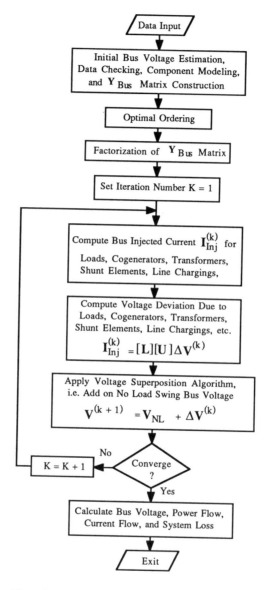

Figure 26 Flow Chart for the Factored Y_{Bus} Gauss, Power Flow Algorithm

b. Features and Capabilities

Based upon the detailed component models and the solution techniques, the power flow portion of the GDAS has the features and capabilities described below. GDAS can:

- Simultaneously analyze multivoltage primary circuits supplying a large secondary network and many spot networks in a single computer run. The electrical construction can be either radial or network with both primary and secondary at more than one voltage level;
- Model cogenerators at both primary and secondary sides. The cogenerators can be either induction or synchronous generators;
- Model three-phase transformers with various connections, for example, grounded wye-grounded wye; delta-grounded wye;
- Model composition loads. The effects of voltage change on the loads are introduced in the load models;
- Analyze system loss and detailed branch losses;
- Analyze contingency cases in order to study the effects of outage feeders on the system. The GDAS can perform these features on the combined primary-secondary models;
- Perform transformer backfeed calculations (during the contingency period).
- Analyze single, two and three-phase systems simultaneously;
- Analyze unbalanced systems;
- Analyze the effects of line chargings on a distribution system;
- Analyze the effects of shunt capacitors (or reactors) on a distribution system.

2. Short Circuit Solution Technique

The "Iterative Compensation Method", a modification of the Compensation method[18] developed for network solutions by optimally ordered triangular factorization is adopted for the short circuit analysis portion of the GDAS. Short circuit and power flow programs can use the same basic approach to component models, the system can be more tightly integrated, and the development costs of the system are decreased. In this section, the issues relating to this selection are presented.

There are three major differences between power flow and short circuit analysis:

(a) The loads are assumed to have a constant impedance characteristic immediately after a fault. The reason for this assumption is that the load models for power flow study are not accurate for a wide range of voltages. During fault the voltage at buses near the fault drop to very low values, and can also rise to abnormal high levels (for example in a single-phase-to-ground fault event).

(b) The internal voltages of cogenerator are assumed to be constant immediately after a fault.

(c) A fault current from a faulted bus to the ground is modeled as a current injection source like a load current.

The first and second differences have no baneful influence upon the application of the Z_{Bus} Gauss method to the short circuit study, but the third part causes some difficulty: the Z_{Bus} Gauss method can diverge if the initial estimated bus voltages are not accurate.

The compensation method for network solutions by optimally ordered triangular factorization was developed by Mr. William F. Tinney in 1971[18]. In this method, the compensation theorem is applied in conjunction with ordered triangular factorization of the nodal admittance matrix to simulate the effect of changes in the passive elements of the network on the solution of a problem without changing the factorization. The compensation method is suitable for GDAS short circuit analysis because sparse matrix, optimal ordering, and bi-factorization techniques are used. Furthermore, a general fault-analyzing approach program can be easily obtained.

A general fault-analyzing program using the phase co-ordinate technique and distributed source method was developed in 1979[19]. This program is capable of analyzing all kinds of series, shunt, and combining of series and shunt faults on balanced and unbalanced networks. Even the cross-country fault problem can be solved by this program. But the Y_{Bus} matrix must be modified to take into account the network changing for any fault study case. It is one of the drawbacks of this method.

The proposed method not only takes advantage of the sparsity of system equations but also benefits from the fact that no modification is needed for the factorization matrix when doing fault analyses. With this feature, the power flow and many kinds of fault studies can be performed in a single computer run, and only one factorization procedure is necessary for these analyses. It saves a lot of time by performing many jobs in a single computer run.

In additionally, the short circuit program based upon the proposed method can be easily integrated with the power flow analysis program. Hence, with the original power flow program based upon the optimally ordered triangular factorization Y_{Bus} method, only a small procedure need be added to perform the short circuit simulation. Using this method, a general approach program can also be obtained. Therefore, single-line-to-ground; double-line-to-ground; and three-phase fault can be performed by the same program. Most of the conventional short circuit programs[20,21] use the method of symmetrical components. In this way, each kind of fault is solved separately. Therefore, there are as many computer procedures needed as types of fault being analyzed. Furthermore, many problems, such as multiple unbalanced faults on unbalanced polyphase systems, cannot be solved by this method.

The GDAS short circuit program allows six individual nodes (a three-phase node equal to three individual nodes) to ground fault simultaneously. Therefore, cross-country fault, double three-phase fault, and triple double-line-to-ground fault can be analyzed by this program. Actually, any combinations of these six individual nodes are allowed. For example, a single-line-to-ground fault occurs at bus #x, on phase A; a double-line-to-ground fault at bus #y, on phase B and C; and a three-phase fault at bus #z simultaneously can be analyzed. Using the proposed method, more than six nodes to ground fault can be easily achieved. However, from a practical point of view, a six individual nodes to ground fault simultaneously is adequate for today's analysis. For this reason, this program allows only six individual nodes to ground fault simultaneously.

a. Methodologies

The methodologies for GDAS short circuit analysis are listed below:

- Transformers are represented by more detailed models considering the core and copper losses, tap-changing, phase-shifting, and the connection of the transformers;
- Distribution loads are represented by constant impedance. The impedance is obtained from the power flow results. Although this approach is also one of the approximation methods, it is much more accurate than conventional methods

that neglect the load or use the rated load to estimate the impedance to represented load;

- Primary and secondary networks are represented by the actual phase quantities. (The primary feeders are represented by series impedance and shunt capacitance, but the secondary cables are represented by series admittance only);
- Cogenerators are represented properly. The inherent imbalance characteristics of cogenerators due to the system unbalance are considered;
- "Iterative Compensation Method" used for short circuit calculation, superposition and sparsity techniques is also employed.

The techniques employed in company with the compensation theory, are:

- Using the power flow analysis program to obtain the prefault conditions;
- Modeling components in detail;
- Using the method of phase co-ordinates;
- Employing superposition theory. The voltage on each bus can be considered to have a contribution from two different types of sources. One is the voltage specified substation bus. Another is load, cogenerator, shunt capacitor or inductor, line charging and fault. All of them are modeled as current injection sources at their respective buses;
- Using the iterative method to take care of the voltage characteristics of load and other components. In this way, the fault condition can be simulated accurately.

b. Features and Capabilities

Based upon the detailed component models and the solution techniques, the short circuit portion of the GDAS has the features and capabilities described below. It can:

- Analyze multiple faults, up to six individual nodes to ground fault simultaneously. Therefore, the cross-country fault, double three-phase fault, and triple double-line-to-ground fault can be analyzed by this program;
- Simultaneously analyze multivoltage primary circuits supplying a large secondary network and many spot networks in a single computer run. The electrical construction can be either radial or network with both primary and secondary at more than one voltage level;
- Model cogenerators at both primary and secondary sides. The cogenerators can be either induction or synchronous generators;

- Model three-phase transformers with various connections, for example, grounded wye-grounded wye; delta-grounded wye;
- Calculate the transformer backfeeding current of contingency cases in order to study the effects of feeder outages upon the system. The GDAS can perform these features on the combined primary-secondary models;
- Consider the load effect. The effects of voltage change on the loads are introduced in the load models;
- Analyze single, two and three phase systems simultaneously;
- Analyze unbalanced systems;
- Consider the effects of line chargings on a distribution system during the fault;
- Consider the effects of shunt capacitors (or reactors).

A brief flow chart of the "Iterative Compensation Method" is shown in Figure 27.

E. NUMERICAL STABILITY AND CONVERGENCE

The purpose of the GDAS is to perform power flow and short circuit studies on a large distribution system combining primary and secondary networks. After converting the impedances of elements for the entire system to per unit values, the p.u. impedances are spread out over a large range. This can cause numerical instability, especially for a very large ungrounded system. Hence, a double precision representation of numbers is needed.

The power flow portion of the GDAS using the Z_{Bus} method, therefore, more loosely tied systems will converge to a solution with fewer iterations than for tightly knit systems. Generally speaking, this method has excellent convergence characteristics, and is not sensitive to initial values of the voltage profile.

The convergence characteristic described above are true for the grounded wye-grounded wye connected transformers, but for other connections, the convergence speed will be reduced considerably, especially for cases in which the system is ungrounded. For these cases, a small acceleration factor must be used to avoid divergence. For a severely unbalanced system, cogenerator models will also reduce convergence speed. The following list contains factors that affect convergence characteristics.

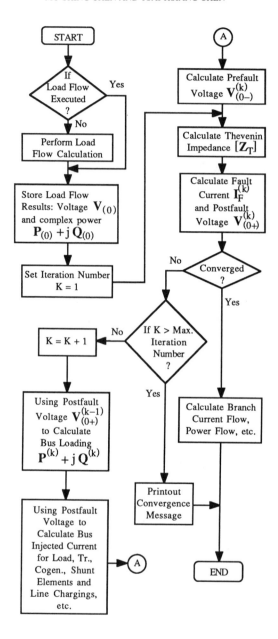

Figure 27 Flow Chart of "Iterative Compensation Method"

(a) All transformers except the grounded wye-grounded wye type connection.

(b) The degree of system unbalance.

(c) Cogenerators in unbalanced systems.

(d) Acceleration factor.

(e) System grounding.

(f) Other factors including load model, transformer tap, and shunt element. These factors, however, are not as sensitive as the first five factors.

For the short circuit analysis, the fault type also affect the converge characteristic.

F. DATA REQUIREMENTS

Considering the scope of the GDAS programs' functionality, the data requirements could become prohibitive, making the program unusable. Adopting a practical point of view, simplified data requirements were determined to be an additional necessary program constraint. In this manner, all the data needed for the GDAS to arrive the features mentioned before can be available from distribution engineers without any other effort. If more detailed information is available, the system can handle all the data, the only thing required is modification of the relevant data base files, then the system can be simulated using a more rigid approach. The simplified data requirements are listed below:

(a) Conductors

(1) Positive sequence impedance (Z_1)

(2) Negative sequence impedance (Z_2)

(3) Zero sequence impedance (Z_0)

(4) Line Charging (KVAR)

(b) Transformers

(1) Capacity (KVA)

(2) Impedance (in p.u. or %)

(3) X/R ratio

(4) Primary and secondary voltages (KV)

(5) Connection

(c) Cogenerators

(1) Subtransient reactance (X_d'')

  (2) Negative sequence reactance (X_2)

  (3) Zero sequence reactance (X_0)

  (4) Capacity (KVA)

  (5) Power Factor

  (6) Rated voltage (KV)

 (d) Loads

  (1) Demand (KVA) (and power factor)

  (2) Load composition (Load window type)

 (e) Shunt elements

  (1) Capacity (KVAR)

  (2) Rated voltage (KV)

III. SAMPLE SYSTEM RESULTS

The 12-bus distribution system shown in Figure 28 is used as a sample system. Included in the system is a 400 KVA cogenerator operating in parallel with a utility system with parameters: power factor = 0.85, X_0 = 0.086 p.u., X_d" = 0.191 p.u., $X_2 = 0.175$ p.u.. The total complex power supplied by the cogenerator to the secondary network of the system is 340 KW + j 211 KVAR.

Although the most popular connection of transformer is delta-grounded wye, many different connections will be considered for the transformers of the sample system to compare the results and corroborate the transformer models.

Many load flow and short circuit study cases are addressed here. The line and system loss problem are discussed in some of the load flow study cases for both balanced and unbalanced systems.

A. POWER FLOW STUDY RESULTS

Eight different cases described in Table II are used to demonstrate the features of the GDAS power flow program. The results from these cases corroborate the cogenerator and transformer models and illustrate the effect of the line charging of the primary feeders, the shunt capacitors, the cogenerator, the winding connection of the transformers, and the load behavior on the distribution system.

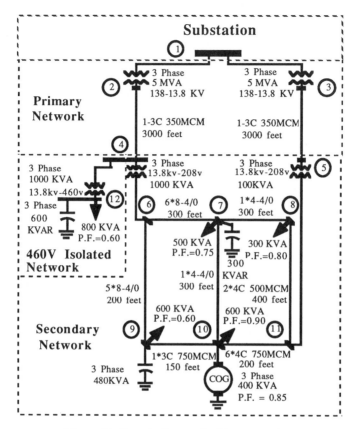

Figure 28 Sample System for Demonstration

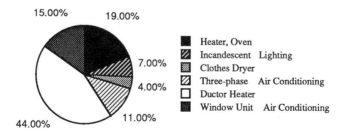

Figure 29 Pie Chart for Load Composition of Type 1

Table II Cases for Power Flow Algorithm demonstration

	Cases								
	1	2	3	4.1	4.2	5	6	7	8
A	Without	Without	Without	Without		With	With	With	With
B	1:1	1:1.025	1:1.025	1:1.025		1:1.025	1:1.025	1:1.025	1:1.025
C	GY-GY	GY-GY	GY-GY	GY-GY		GY-GY	Various	D-GY	Various
D	Without	Without	With	W	W/O	With	With	With	With
E	Heavy, Balanced	Heavy, Balanced	Heavy, Balanced	Light, Balanced		Heavy, Balanced	Heavy, Balanced	Heavy, Balanced	Heavy, UnB.
F	Type 0	Type 0	Type 0	Type 0		Type 0	Type 0	Type 1	Type 1

Note:

    A : With/Without cogenerator
    B : Off-nominal tap ratios of primary transformers
    C : Transformer connections for whole system transformers
       GY : Grounded Wye
       D : Delta
    D : With/Without shunt capacitors
    E : Loads:
       Heavy : Full load
       Light : 10% of full load
    F : Load types:
       Type 0 : Constant P, Q loads
       Type 1 : Shown in Figure 29

The voltage comparison for Cases 1, 2, 3, and 4.1 are shown in Figure 30. Case 4.2 is the same as Case 4.1 but three capacitors are removed. The voltages of Case 4.2 are more moderate than that of Case 4.1, and the power factor is improved from 0.19 leading to 0.84 lagging. The voltage comparison for Case 4.1 and Case 4.2 is shown in Figure 31. Case 4 shows that automatic switching of shunt capacitors can prevent overvoltage and a leading power factor during periods of light load.

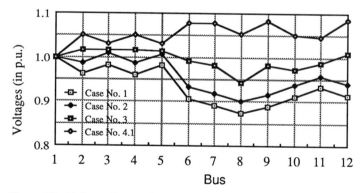

Figure 30 Voltage Comparison for Cases No. 1, 2, 3, and 4.1

Case 5 is based on Case 3 but a 400 KVA cogenerator is added at bus 10. The bus voltages of this balanced case are shown in Tables III. Table IV is the voltage comparison table for Case 3 and Case 5. The voltage profiles are improved and the system loss is reduced from 0.0896 + j 0.2086 MVA to 0.0763 + j 0.1489 MVA, because the location of the cogenerator is much nearer to the loads than the source. The three-phase power flows of this case, not shown here, are also balanced. The power supplied from the cogenerator is

3*(0.01067 + j 0.008) * 10,000 KVA = 320 + j 240 KVA

This value is equal to the cogenerator's rating and is the expected value because a constant total P and Q model is used for the cogenerator.

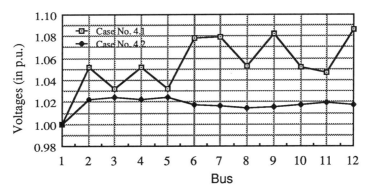

Figure 31 Voltage Comparison for Cases No. 4.1 and 4.2

Table III Final Converged Voltage Solutions for Case No. 5

Bus Number	Phase A		Phase B		Phase C	
	|V|	Ang.	|V|	Ang.	|V|	Ang.
1	1.000	0.000	1.000	-120.000	1.000	120.000
2	1.019	-1.612	1.019	-121.612	1.019	118.388
3	1.021	-0.792	1.021	-120.792	1.021	119.208
4	1.018	-1.638	1.018	-121.638	1.018	118.362
5	1.021	-0.803	1.021	-120.803	1.021	119.197
6	1.003	-3.730	1.003	-123.730	1.003	116.270
7	0.992	-3.952	0.992	-123.952	0.992	116.048
8	0.960	-3.523	0.960	-123.523	0.960	116.477
9	0.995	-3.983	0.995	-123.983	0.995	116.017
10	1.000	-3.048	1.000	-123.048	1.000	116.952
11	1.006	-2.535	1.006	-122.535	1.006	117.465
12	1.011	-3.195	1.011	-123.195	1.011	116.805

Table IV Voltage Comparison Table for Case No.3 and Case No. 5

Bus No.	Bus Name	Voltage Magnitude		
		Case No. 3	Case No. 5	Case No 5 - Case No.3
1	Swing Bus	1.000	1.000	0.000
2	HV Bus #02	1.016	1.019	+0.003
3	HV Bus #03	1.016	1.021	+0.005
4	HV Bus #04	1.016	1.018	+0.002
5	HV Bus #05	1.015	1.021	+0.006
6	LV Bus #06	0.993	1.003	+0.010
7	LV Bus #07	0.981	0.992	+0.011
8	LV Bus #08	0.942	0.960	+0.018
9	LV Bus #09	0.982	0.995	+0.013
10	LV Bus #10	0.972	1.000	+0.028
11	LV Bus #11	0.986	1.006	+0.020
12	LV Bus #12	1.008	1.011	+0.003

Case 6 is used to demonstrate the transformer models. It is based on Case 5 and uses various winding connections. Tables V, VI, and VII show the voltage magnitudes and angles of Phase A, Phase B, and Phase C , respectively. In this case, the voltage magnitudes of different connections are the same because the system is balanced. The angles of bus voltages of grounded wye-delta has expected -30 degree phase shift, and delta-grounded wye has +30 degree phase shift. This case verifies that the transformer model can achieve the phase shifting automatically.

TableV Voltage Profiles of Phase A

Bus No.	|V| (P.U.)	Angles (Degree)		
		G W-G W	G W-Delta	Delta-G W
1	1.000	0.000	0.000	30.000
2	1.019	-1.612	-31.612	28.388
3	1.021	-0.792	-30.792	29.208
4	1.018	-1.638	-31.638	28.362
5	1.021	-0.803	-30.803	29.197
6	1.003	-3.730	-63.730	56.270
7	0.992	-3.952	-63.951	56.049
8	0.960	-3.523	-63.523	56.477
9	0.995	-3.983	-63.983	56.017
10	1.000	-3.048	-63.047	56.953
11	1.006	-2.535	-62.535	57.465
12	1.011	-3.195	-63.195	56.805

Table VI Voltage Profiles of Phase B

Bus	|V|	Angles (Degree)		
No.	(P.U.)	G W-G W	G W-Delta	Delta-G W
1	1.000	-120.000	-120.000	-120.000
2	1.019	-121.612	-151.612	-91.612
3	1.021	-120.792	-150.792	-90.792
4	1.018	-121.638	-151.638	-91.638
5	1.021	-120.803	-150.803	-90.803
6	1.003	-123.730	176.270	-63.730
7	0.992	-123.952	176.049	-63.951
8	0.960	-123.523	176.477	-63.523
9	0.995	-123.983	176.017	-63.983
10	1.000	-123.048	176.953	-63.047
11	1.006	-122.535	177.465	-62.535
12	1.011	-123.195	176.805	-63.195

Table VII Voltage Profiles of Phase C

Bus	|V|	Angles (Degree)		
No.	(P.U.)	G W-G W	G W-Delta	Delta-G W
1	1.000	120.000	120.000	120.000
2	1.019	118.388	88.388	148.388
3	1.021	119.208	89.208	149.208
4	1.018	118.362	88.362	148.362
5	1.021	119.197	89.197	149.197
6	1.003	116.270	56.270	176.270
7	0.992	116.048	56.049	176.049
8	0.960	116.477	56.477	176.477
9	0.995	116.017	56.017	176.017
10	1.000	116.952	56.953	176.953
11	1.006	117.465	57.465	177.465
12	1.011	116.805	56.805	176.805

Case 7 is used to demonstrate the effects of the voltage dependent load model. Buses 7, 8, 9, 10, and 12 use a load window which is comprised of 19% heater and oven, 7% incandescent lighting, 4% clothes dryer, 11% three-phase air conditioning, 44% ductor heater, and 15% window unit air conditioning loads, shown in Figure 29.

The voltage profiles for this case are shown in Table VIII. Note that Table VIII is very similar to Table III, which shows the results of the constant P,Q load model.

The loads are voltage dependent, however. For example, the load at bus 7 is:

$3 * (0.01238 + j\, 0.01098) * 10$ MVA $= 371.4 + j\, 329.4$ KVA

which is slightly different (2.7 + j 1.1 KVA) from the input data (375 + j 331 KVA) because the converged voltage at that bus is 0.993 p.u. compared with 1.0 p.u. at the start.

The voltage magnitude of bus 12 is 1.011 p.u., and the load at bus 12 is:

3 * (0.01634 + j 0.02150) * 10 MVA = 490.2 + j 645.0 KVA

The difference between this value and the input data (or the results obtained from the system using the constant P and Q model) is:

(490.2 + j 645.0) - (480.0 + j 640.0) = 10.2 + 5.0 KVA

This difference is significant, illustrating that the use of a constant power load model can distort power flow results and give the user an inaccurate index of system performance. There is no doubt that, when used correctly, the more accurate load model will yield better results.

Table VIII Final Converged Voltage Solutions for Case No. 7

Bus Number	Phase A		Phase B		Phase C	
	|V|	Ang.	|V|	Ang.	|V|	Ang.
1	1.000	0.000	1.000	-120.000	1.000	120.000
2	1.019	28.392	1.019	-91.608	1.019	148.392
3	1.021	29.219	1.021	-90.781	1.021	149.219
4	1.019	28.366	1.019	-91.634	1.019	148.366
5	1.021	29.208	1.021	-90.792	1.021	149.208
6	1.003	56.305	1.003	-63.695	1.003	176.305
7	0.993	56.087	0.993	-63.913	0.993	176.087
8	0.962	56.565	0.962	-63.435	0.962	176.565
9	0.996	56.052	0.996	-63.948	0.996	176.052
10	1.000	56.988	1.000	-63.012	1.000	176.988
11	1.006	57.501	1.006	-62.499	1.006	177.501
12	1.011	56.788	1.011	-63.212	1.011	176.788

Case 8 is used to demonstrate the effects of an unbalanced system on the power flow results. Figure 28 is used for this case, but the loads at buses 7 and 9 are assumed to be unbalanced. The other loads are still assumed to be balanced. At bus 7, 50% of the total load is on phase A, 30% on phase B, and 20% on phase C. At bus 9, there is no load on phase A, 30% on phase B, and 70% load on phase C.

Two different transformer connections are used: delta-grounded wye (Case 8.1) and grounded wye-grounded wye (Case 8.2). The system voltage profiles for these two cases are shown in Table IX and X, respectively. Comparing the results of this case with the balanced case (Case 6) illustrates the need for non-trivial transformer model in the unbalanced case.

The bus voltages are invarient with respect to a change in the transformer windings in the balanced case. This is not evident in the unbalanced case.

Table IX Final Converged Voltage Solutions for Case No. 8.1

Bus Number	Phase A		Phase B		Phase C	
	|V|	Ang.	|V|	Ang.	|V|	Ang.
1	1.000	0.000	1.000	-120.000	1.000	120.000
2	1.016	28.565	1.023	-91.583	1.017	148.120
3	1.020	29.257	1.022	-90.773	1.021	149.106
4	1.015	28.547	1.023	-91.613	1.016	148.090
5	1.019	29.249	1.022	-90.785	1.020	149.094
6	1.020	56.635	1.009	-63.680	0.973	175.643
7	0.993	56.717	1.000	-64.516	0.978	175.758
8	0.962	56.776	0.966	-63.738	0.949	176.251
9	1.046	54.426	1.022	-62.202	0.911	175.669
10	1.005	57.145	1.003	-63.156	0.988	176.754
11	1.008	57.706	1.008	-62.710	0.998	177.296
12	1.012	57.037	1.013	-63.414	1.005	176.726

Table X Final Converged Voltage Solutions for Case No. 8.2

Bus Number	Phase A		Phase B		Phase C	
	|V|	Ang.	|V|	Ang.	|V|	Ang.
1	1.000	0.000	1.000	-120.000	1.000	120.000
2	1.026	-1.505	1.021	-121.497	1.009	118.098
3	1.022	-0.716	1.022	-120.831	1.019	119.136
4	1.026	-1.569	1.021	-121.499	1.007	118.087
5	1.021	-0.729	1.022	-120.841	1.018	119.126
6	1.026	-3.514	1.009	-123.341	0.967	115.440
7	0.999	-3.429	1.000	-124.198	0.972	115.573
8	0.964	-3.244	0.965	-123.623	0.947	116.151
9	1.051	-5.694	1.022	-121.908	0.905	115.486
10	1.006	-2.843	1.003	-123.130	0.987	116.714
11	1.009	-2.289	1.008	-122.678	0.998	117.257
12	1.019	-3.110	1.013	-123.051	0.999	116.504

Tables VIII and IX show the voltage profiles of balanced and unbalanced cases, respectively. For a balanced case, the magnitudes of the voltages of phases A, B, and C for each bus in the system are the same and the voltage angles show exactly 120 degrees difference between any two phases. This is not true for an unbalanced case.

In the GDAS program, each branch loss can be obtained, such as feeders, cables, and transformers. Additionally, total transformer core loss, copper loss, cable loss and total system losses can be obtained too.

B. SHORT CIRCUIT STUDY RESULTS

Assuming the fault occurs at bus 8, the following three cases are used to demonstrate the GDAS short circuit study program.

Case 1 : Single-Line-To-Ground fault

        (a) Fault on phase A

        (b) Fault on phase B

        (c) Fault on phase C

Case 2: Double-Line-To-Ground fault

        (a) Fault on phases A and B

        (b) Fault on phases B and C

        (c) Fault on phases C and A

Case 3: Three-Line-To-Ground fault

For SLG faults, the magnitude of total fault currents to ground, is the same for cases 1(a), 1(b), and 1(c), i.e. no matter the fault occurring on phase A, B, or C. The total fault current is the same in each case because the system is balanced. The three DLG faults exhibit this same, expected behavior.

Voltage profiles during the fault period of the SLG fault on phase A, DLG fault on phase A and B, and 3LG fault are shown in Figures 32,33, and 34, respectively.

Figure 32 shows that the SLG fault results in an increase in voltage magnitude on the other two phases, especially at the same bus. The amount of voltage rise depends upon the system's X/R ratio at the fault bus where the X and R are the real and imaginary part of driving point impedance. Referring to Figure 33, the DLG fault also demonstrates this phenomenon but not as severely as the SLG fault. These transient overvoltages can be harmful to system equipments such as insulated cables, and transformers.

The 3LG fault is a balanced fault, the voltage magnitude of the three phases are equal as shown in Figure 34. For comparison, the pre-fault voltage profile of the normal case is included in Figure 34. The voltage magnitudes of the primary buses (Bus 1 to 5) are only slightly different for these two cases, however, the voltages of the 208 V secondary buses (Bus 6 to 11) show significant change. The comparison of total fault currents for three different type (SLG, DLG, and 3LG) faults is shown in Figure 35.

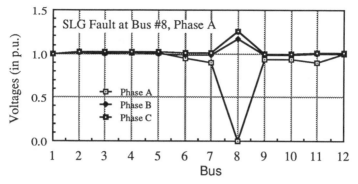

Figure 32 Voltage Profile of SLG Fault on Phase A

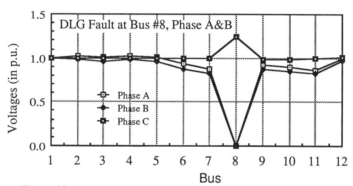

Figure 33 Voltage Profile of DLG Fault on Phase A and B

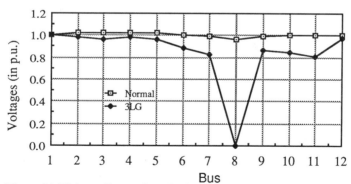

Figure 34 Voltage Comparison for 3LG Fault and Normal case

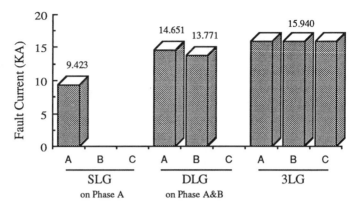

Figure 35 Comparison of Total Fault Currents for Three Different Type Faults

IV. DISCUSSION AND CONCLUSIONS

An advanced distribution analysis system capable of simulating very large electric distribution networks has been introduced. This system is a very detailed analysis and simulation program written for electrical distribution engineering. The methodologies applied in this program have been issued above. The optimal ordering bi-factorization Y_{Bus} method was used for distribution power flow analysis; and the iterative compensation method for short circuit analysis. The major features and capabilities of this system are listed below:

- (a) can simultaneously analyze multivoltage primary circuits supplying a large secondary network in a single computer run; electrical construction can be either radial or network with both primary and secondary at more than one voltage level;
- (b) can model cogenerators at both primary and secondary sides, and the cogenerators can be either induction or synchronous generators;
- (c) can model three phase transformers with various connections;
- (d) can model composition loads;
- (e) can analyze contingency cases on the combined primary-secondary models;

(f) can analyze short circuit currents on a distribution system that contains cogenerators;

(g) can calculate transformer back-feed current;

(h) can analyze system loss

(i) can analyze multibanks, spot networks, isolated networks;

(j) can calculate simple impedance, etc..

A program capable of these features and capabilities is very complex. Programming techniques such as structured programming, sparse matrix factorization, and object oriented representation are required. The mathematical models and solution techniques are very important, especially for a large-scale system. For a system combining a primary and secondary system, the number of significant digital of variables is also very important.

In highly complicated systems, it is almost impossible to analyze actual system conditions without tools with advanced capabilities. The GDAS program with these features would provide distribution engineers a powerful tool to do more accurate distribution simulation. With this powerful simulation tool, the quality of the distribution system could be improved, and the performance of the electrical power system will be more easily realized by the distribution engineers. Furthermore, based upon the results achieved by this research many advanced functions can be added.

REFERENCES

1. S. K. Chang, " Distribution Load Flow Automation," Ph. D. Dissertation, May 1982, The University of Texas at Arlington, Arlington, Texas.

2. N. Vemptati, R.R. Shoults, M.S. Chen and L. Schwobel, "Simplified Feeder Modeling for Load Flow Calculations," *IEEE Trans. on Power Systems*,Vol. PWRS-2, No.1, pp. 168-174, February 1987.

3. M. S. Chen and W. E. Dillon, "Power System Modeling," *Proc. IEEE*, Vol.62, No.7, PP.901-915, July 1974.

4. Energy Systems Research Center, The University of Texas at Arlington, "Determining Load Characteristics for Transient Performance," Final Report of EPRI Project RP 849-3, Vol. I, II, III, December 1978.

5. J. R. Carson, "Wave Propogation in Overhead Wires with Ground Return, " *Bell System Technical Journal*, Vol. 5, Oct. 1926, pp. 539-554.

6. E. Clark, *Circuit Analysis of A-C Power Systems*, Vol. 1, John Wiley and Sons, N.Y. 1943, pp. 363-433.

7. W. A. Lewis and G. D. Allen, J. C. Wang, "Cable Constants for Concentric Neutral Underground Distribution Cable on a Per Phase Basis," *IEEE Transactions on Power Systems and Apparatus*, Vol. PAS-97, Jan./Feb. 1978, pp.200-207.

8. S. D. Hu, *Cogeneration*, Reston, Virginia : Reston Publishing, 1985.

9. G. Polimeros, *Energy Cogeneration Handbook*, New York : Industrial Press, 1981.

10. S. A. Spiewak, *Cogeneration and Small Power Production Manual*, Lilburn, Georgia : The Fairmont Press, 1987.

11. M. S. Sarma, *Synchronous Machines*, New York : Gordon and Breach Science Publishers, 1979.

12. D. B. Brown and E. P. Hamilton III, *Electromechanical Energy Conversion*, New York : Macmillan Publishing, 1984.

13. W. E. Dillon, "Modeling and Analysis of an Electrically Propelled Transportation System," Ph. D. Dissertation, The University of Texas at Arlington, May 1972.

14. W. E. Dillon and M. S. Chen, "Transformer Modeling in Unbalanced Three-Phase Networks," *IEEE Summer Power Meeting*, Vancouver, Canada, July 1972.

15. J. Arrillaga and C. P. Arnold, *Computer Modelling of Electrical Power Systems*, New Zealand: John Wiley and Sons, 1983.

16. D. I. Sun, "Distribution System Loss Analysis and Optimal Planning," Ph. D. Dissertation, The University of Texas at Arlington, May 1980.

17. D. I. Sun, S. Abe, R. R. Shoults, M. S. Chen, P. Eichenberger, and D. Farris, "Calculation of Energy Losses in a Distribution System" *IEEE Transactions on Power Apparatus and Systems*, Vol. PAS-99, No.4, July/Aug 1980.

18. William F. Tinney, "Compensation Methods for Network Solutions by Optimally Ordered Triangular Factorization," *IEEE Transactions on Power*

Apparatus and Systems, Vol. PAS-91, No.1, PP. 123-127, January/February 1972.

19. L. Roy, "Generalized Polyphase Fault-Analysis Program: Calculation of Cross-Country Fault," *Proc. IEE*, Vol. 126, No. 10, October, 1979.

20. H. E. Brown, C. E. Person, L. K. Kirchmayer, and G. W. Stagg, "Digital Calculation of Three-phase Short Circuits by Matrix Method," *Trans. AIEE*, Part III, Vol. 79, pp. 1277-1282, February, 1961.

21. A. H. El-Abiad, "Digital Calculations of Line-to-Ground Short Circuits by Matrix Method," *Trans. AIEE*, Part III, Vol. 79, pp. 323-332, June, 1960.

POWER SYSTEM TRANSIENT STABILITY ASSESSMENT USING THE TRANSIENT ENERGY FUNCTION METHOD

A. A. FOUAD and V. VITTAL

Iowa State University
Ames, Iowa 50011

I. INTRODUCTION

I.A Power System Transient Stability

Power system transient stability deals with the power system's response to disturbances. The period of interest is the transient period before the new steady state conditions are reached. If the power system is subjected to a large disturbance, and it is able to reach an acceptable steady-state condition, it is (transiently) stable. A large disturbance is a sudden change in a power system parameter or operating condition such that linearization of the system equations, for the purpose of analysis, cannot be justified. Transient stability studies are made by electric utilities in North America (and in other parts of the world) to ensure the reliability of the bulk power supply.

The conventional, and still the standard, method to determine the transient stability of a power system, is to solve for the system variables and parameters for a given sequence of events. From this solution it is determined whether an acceptable steady state condition will be reached, hence whether the system will be stable. When direct methods are used, the value of a certain function (e.g., describing the system transient energy) is computed at the end of the disturbance period. This value is compared with a critical, or threshold, value of the function for transient stability assessment.

CONTROL AND DYNAMIC SYSTEMS, VOL. 43

115

Power system transient stability assessment using energy functions have been around for many years. Until recently the power system model used is the classical model (see Section II below). The well known equal area criterion [1], which has been in use since the 1920's, is applied to a one-machine-infinite-bus power system. Early investigations of energy criteria for multi-machine power systems were conducted in the Soviet Union (in the 1930s and 1940s) [2,3]. Some results appeared in the technical literature in the West in the 1940s and 1950s [4,5]. In the 1960s the application of Lyapunov's second method to the power system transient stability problem became the main focus of direct transient stability research [2,6]. This has continued until the late 1970s when the interest in energy functions was revived by the development of the new energy function proposed by Athay et al [7,8]. The transient energy function (TEF) discussed in this chapter is based on that function.

I.B Foundation of Direct Transient Stability Assessment

Whenever power system transient stability assessment is made by direct methods, two important ingredients are involved: 1) development or determination of a function which must satisfy certain criteria, e.g., sign definiteness (Lyapunov) or accounting for energy responsible for separation of some generators (TEF) etc., and 2) determination of the critical value of that function. The history of power system direct transient stability analysis is the documentation of the progress in these two areas. Furthermore, there are some very important practical implications that must be kept in mind: 1) accurate transient stability assessment requires accurate determination of both the function and its threshold value, and 2) to obtain answers to practical industry problems, an accurate formulation of the problem in terms of the attributes of the function or its threshold value is essential.

I.C Overview of Progress Areas

Since the 1970s considerable progress has been achieved in power system transient stability analysis using direct methods based on energy functions, especially the transient energy function (TEF) method discussed in this chapter. This progress has covered two broad fronts: (1) the development of TEF as an accurate and reliable 'tool' for transient stability assessment; and (2) the diversity of the type of applications analyzed, and hence the associated problems addressed, by this tool.

The present TEF is based on the energy function proposed by Athay, et al. [7,8] in the late 1970s. However, the development of the TEF method as a tool for stability analysis required improvements in two main areas: (1) identifying the system transient energy responsible for the separation of some (severely disturbed) generators from the rest of the system [9,10]; and (2) estimating the critical value of this transient energy for the particular disturbance under investigation.

Investigation of industry-related applications by direct methods of transient analysis have revealed a host of issues which needed to be successfully dealt with if these direct methods are to be seriously considered by the industry. Some of these relate to the type of disturbance investigated (e.g., simple faults, complex disturbance sequence, loss of generation, etc.). Other issues are related to the type of information required (e.g., whether loss of synchronism will occur, stability-limited loading of a plant or power flow over a key transmission circuit, maximum transient voltage dip at a certain bus, minimum apparent impedance at a certain relay, etc.). The size of the power network used in the investigation can be important in revealing some dynamic problems which may not appear in smaller networks, especially at higher system loading (e.g., occurrence of system splits involving large groups of generators known as the interarea mode). Computational difficulties can be encountered when large heavily-loaded power networks are analyzed [12-14]. Furthermore, some applications require better system models than the model commonly used in direct stability analysis [15-17]. Associated with all these problems, the analyst must also deal with the questions of: how fast will the solution be obtained, as well as how accurate and reliable should that solution be [13].

I.D The Context in Which Direct Stability Analysis is Made

Transient stability studies are made by the*electric utilities in North America to ascertain that criteria established by the North American Electric Reliability Council (NERC) are met [18]. These criteria state that when certain disturbances occur, no cascading outages in the interconnected power network must result. Different regions in North America establish their own disturbance criteria according to the conditions which prevail in their own bulk power supply.

To meet the NERC criteria for avoiding cascading outages, engineers in North American utilities conduct many transient stability studies: when new generation or transmission facilities are planned; and when these studies are warranted by the operating conditions in the system.

Because of heavier transmission loadings, greater interchange of power among neighboring systems, and because fewer transmission facilities are being built, it is becoming increasingly difficult to ascertain that NERC criteria will be met at all times. Therefore, transient stability studies are being conducted routinely to ensure that under various operating conditions cascading outages will not result when certain credible contingent events occur. The type and scope of these transient stability studies are varied. For several power systems in North America which have become transient stability-limited the number of such studies is large and is increasing [19].

The above trend represents a rather profound change in the electric utilities environment in North America. In addition to being of concern to the system planner, transient stability has also become an operating problem. Operations planning groups have been established in many electric utilities whose concern is to determine safe operating limits for critical plants or transmission elements. This trend also explains the increased interest in direct transient stability analysis by the electric utility industry, and by the research community. At the same time, it offers important constraints for the development of direct methods and their use.

The following establishes the context in which direct transient stability analysis methods are to be used by the electric utility industry. Direct methods must meet the following criteria: (1) they should deal with actual industry problems, i.e., investigate the type of disturbances of interest to the industry; (2) they must provide the kind of answers, and information, needed by the industry; (3) they should offer some advantage over conventional (e.g., time solution) methods; and (4) they must ensure that the effort required to adopt them is justified by the gains to be achieved by using them. These criteria must be kept in mind because in the past, well-meaning researchers have made claims about what direct methods can offer the power system analyst, which were difficult to justify.

II. THE MATHEMATICAL MODEL

II.A The Classical Power System Model

For the classical power system model (see Chapter 2 of reference [20]) the equations of motion of the synchronous generators, written with respect to an arbitrary synchronous frame, are given by

$$M_i \dot{\omega}_i = P_{mi} - P_{ei} \qquad\qquad (2.1)$$

$$\dot{\delta}_i = \omega_i \qquad i = 1, 2, ..., n$$

where n is the number of generators,

$$P_{ei} = \sum_{j=1}^{n} \left[C_{ij} \sin(\delta_i - \delta_j) + D_{ij} \cos(\delta_i - \delta_j) \right]$$

$$P_i = P_{mi} - E_i^2 G_{ii}$$

$$C_{ij} = E_i E_j B_{ij} \qquad D_{ij} = E_i E_j G_{ij}$$

and, for generator i,

P_{mi} mechanical power input

G_{ii} driving point conductance

E_i constant voltage behind the direct axis transient reactance

ω_i, δ_i generator rotor speed and angle respectively

M_i inertia constant

$B_{ij}(G_{ij})$ transfer susceptance (conductance) in the reduced bus admittance matrix.

In the above, the power network has been reduced to the internal generator nodes. The generators are represented by constant voltage sources. The loads which are represented by constant impedances are included in the resulting admittance matrix after elimination of all other network nodes. The resulting network is shown schematically in Figure 2.1.

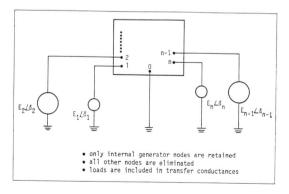

- only internal generator nodes are retained
- all other nodes are eliminated
- loads are included in transfer conductances

Figure 2.1. Classical model

Transformation of equations (2.1) into the center of inertia (COI) coordinates is done by defining the position of the center of inertia by the equations

$$\delta_o = \frac{1}{M_T} \sum_{i=1}^{n} M_i \delta_i \tag{2.2}$$

$$\dot{\delta}_o = \frac{1}{M_T} \sum_{i=1}^{n} M_i \dot{\delta}_i$$

where

$$M_T = \sum_{i=1}^{n} M_i$$

The COI motion is defined by the equations

$$M_T \dot{\omega}_o = \sum_{i=1}^{n} P_{mi} - P_{ei}$$

$$= \sum_{i=1}^{n} P_i - 2 \sum_{i=1}^{n-1} \sum_{j=i+1}^{n} D_{ij} \cos \delta_{ij} \overset{\Delta}{=} P_{COI} \tag{2.3}$$

$$\dot{\delta}_o = \omega_o \quad i = 1,...,n$$

We now define the generators' angles and speeds relative to the COI by

$$\theta_i = \delta_i - \delta_o \quad \tilde{\omega}_i = \dot{\delta}_i - \dot{\delta}_o \quad i = 1,2,...,n \tag{2.4}$$

The system equations of motion become

$$M_i \dot{\tilde{\omega}}_i = P_{mi} - P_{ei} - \frac{M_i}{M_T} P_{COI} \tag{2.5}$$

$$\dot{\theta}_i = \tilde{\omega}_i \quad i = 1,...,n$$

We note that the center of inertia variables satisfy the constraints

$$\sum_{i=1}^{n} M_i \theta_i = \sum_{i=1}^{n} M_i \tilde{\omega}_i = 0 \tag{2.6}$$

II.B Load Model

II.B.1 Classical model

Consider load bus k. Let the pre-disturbance conditions at bus k be: voltage $\overline{V}_k$, and the load power and reactive power be P_{Lk} and Q_{Lk} respectively. Then the equivalent load admittance is given by

$$P_{Lk} + j\,Q_{Lk} = V_k^2 (G_{Lk} - j\,B_{Lk}) \tag{2.7}$$

or

$$G_{Lk} = P_{Lk}\big/V_k^2 \,, \tag{2.8}$$

$$B_{Lk} = -Q_{Lk}\big/V_k^2$$

As stated in Section II.A, the load admittances are included in the network admittance matrix. A reduced network admittance matrix is then obtained by eliminating all the network nodes, including the load buses.

II.B.2 Nonlinear load model

If the load at bus k is represented as a constant current or a constant MVA load,

$$P_{Lk} + j\,Q_{Lk} = V_k\,\underline{/\psi_k}\,\,\overline{I}_{Lk}^{*} \tag{2.9}$$

or

$$\overline{I}_{Lk} = (P_{Lk} - j\,Q_{Lk})\big/V_k\,\underline{/-\psi_k} \tag{2.10}$$

When constant current and constant power loads are included, a special network procedure is used to incorporate these loads in the classical model solution. This procedure must meet the requirement that the current $\overline{I}_{Lk}$ needs to be appropriately preserved during the transient. When the network is reduced to the internal generator nodes, $\overline{I}_{Lk}$ is preserved by determining equivalent current injections at the internal generator nodes. A procedure is also developed to update the value of the current injection during the transient. This procedure is outlined in Section V of this chapter.

II.C Higher Order Generator Model with Excitation Control

II.C.1 Synchronous generator model

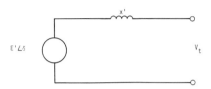

Figure 2.2. Synchronous generator model

When excitation control is of interest, the synchronous generator is assumed to be represented by the so–called two–axis model (see Section 4.15.3 of [20]). Figure 2.2 shows a schematic of the model; x' is the transient reactance (usually the average of the d-axis transient reactance x'_d and q-axis transient reactance x'_q). The voltage E' is the transient EMF or the voltage behind x'; its components E'_q and E'_d vary and have different time constants (see below). The angle δ is now the generator rotor angle (or torque angle).

For generator k, the equations describing the dynamic behavior of the generator EMFs are given by

$$\tau'_{dok}\,\dot{E}'_{qk} = E_{FDk} - E'_{qk} + (x_{dk} - x'_{dk})\,I_{dk} \tag{2.11}$$

$$\tau'_{qok}\,\dot{E}'_{dk} = -E'_{dk} - (x_{qk} - x'_{qk})\,I_{qk} \tag{2.12}$$

and the equations for the rotor motion, with respect to a synchronous frame, are given by

$$M_k\,\dot{\omega}_k = P_{mk} - E'_{dk}I_{dk} - E'_{qk}I_{qk} + (x'_{qk} - x'_{dk})\,I_{dk}I_{qk} \tag{2.13}$$

$$\dot{\delta} = \omega - 1 \qquad k=1,2,...,m.$$

where

$\qquad P_m =$ mechanical power

$\qquad m=$ no. of generators represented in detail

$\tau'_{do}, \tau'_{qo} =$ open circuit direct and quadrature axes time constants respectively

$E_{FD}=$ stator EMF corresponding to field voltage

$x_d, x'_d=$ direct axis synchronous and transient reactances

$x_q, x'_q=$ quadrature axis synchronous and transient reactances

$E'_d, E'_q=$ direct and quadrature axes stator EMFs corresponding to rotor flux components

$I_d, I_q=$ direct and quadrature axes stator currents

When the COI formulation is used, (2.17) becomes

$$M_k \overset{\cdot\cdot}{\omega}_k = P_{mk} - E'_{dk} I_{dk} - E'_{qk} I_{qk} + (x'_{qk} - x'_{dk}) I_{dk} I_{qk}$$

$$-\frac{M_k}{M_T} P_{COI} \tag{2.14}$$

where

$$P_{COI} = \sum_{j=1}^{n} P_{mj} - \sum_{k=1}^{m} [E'_{dk} I_{dk} + E'_{qk} I_{qk} - (x'_{qk} - x'_{dk}) I_{dk} I_{qk}]$$

$$-\sum_{\ell=m+1}^{n} E_\ell I_{q\ell} \tag{2.15}$$

where $\ell = m+1, ..., n$ represent the number of generators represented classically.

II.C.2 Generator currents–constant impedance loads

When the loads are represented by the classical model, the generator currents and voltages are represented by the following equations (see Chapter 9 of [1]).

$$I_{qk} = \sum_{j=1}^{m} [F_{G+B}(\theta_{kj}) E'_{qj} - F_{B-G}(\theta_{kj}) E'_{dj}] + \sum_{\ell=m+1}^{n} F_{G+B}(\theta_{k\ell}) E_\ell$$

$$I_{dk} = \sum_{j=1}^{m} [F_{B-G}(\theta_{kj}) E'_{qj} + F_{G+B}(\theta_{kj}) E'_{dj}] + \sum_{\ell=m+1}^{n} F_{B-G}(\theta_{k\ell}) E_\ell$$

$$k=1,2,...,m \tag{2.16}$$

$$I_{qi} = \sum_{j=1}^{m} F_{G+B}(\theta_{ij}) E'_{qj} + \sum_{\ell=m+1}^{n} F_{G+B}(\theta_{i\ell}) E_\ell - \sum_{j=1}^{m} F_{B-G}(\theta_{ij}) E'_{dj}$$

$$i=m+1,...,n \tag{2.17}$$

where

$$F_{G+B}(\theta_{ij}) = G_{ij} \cos\theta_{ij} + B_{ij} \sin\theta_{ij}$$

$$F_{B-G}(\theta_{ij}) = B_{ij} \cos\theta_{ij} - G_{ij} \sin\theta_{ij} \qquad (2.18)$$

II.C.3 Generator currents—nonlinear loads

To accommodate both constant impedance and nonlinear loads, the synchronous generator model shown in Figure 2.3 is used

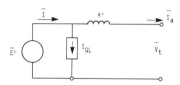

Figure 2.3. Synchronous generator model with current injection representing nonlinear load

where

$\overline{I} =$ current in the generator

$\overline{I}_a =$ current into the network

$=$ current for classical model portion of the network

$\overline{I}_{GL} =$ constant current or constant MVA component reflected at the generator node

In the following, the capital subscripts D and Q will refer to the network (i.e., synchronous) reference frame, while lower case symbols, e.g., q and d, will refer to the generator rotor reference frame. In addition, the terms $[(x'_q - x'_d)I_d I_k]$ in the power equations are neglected.

In the steady state, we can easily show that the generator electrical power P_e is given by

$$P_e = [E'_Q I_{aQ} + E'_D I_{aD}] + [E'_Q I_{GLQ} + E'_Q I_{GLD}]$$

$$= P'_e + E' I_{GL} \cos(\theta - \phi) \qquad (2.19)$$

where

$P'_e =$ electrical power corresponding to classical portion of the network, i.e., constant impedance loads

$E' \underline{/\theta} =$ generator EMF

$I_{GL} \underline{/\phi} =$ current injection at generator node due to nonlinear loads

In the transient state we can show that the generator i current, to the $d-$ and $q-$ frames, are given by

$$I_{qi} + j I_{di} = [I_{aqi} + j I_{adi}] + I_{GLi} \, \epsilon^{j \phi_i} \qquad (2.20)$$

Where phasor notation is used for convenience (as in Chapter 9 of [20], I_{aqi} and I_{adi} can be obtained from equations similar to (2.16) and (2.17).

Then the electrical power for generator i is given by

$$P_{ei} = P'_{ei} + [E'_{qi} I_{GLqi} + E'_{di} I_{GLdi}]$$

$$= P'_{ei} + P_{eGLi} \qquad (2.21)$$

and P_{COI} becomes

$$P_{COI} = P'_{COI} - \sum_{i=1}^{n} P_{eGLi} \qquad (2.22)$$

The generator equations will now become

$$\tau'_{dok} \dot{E}_{qi} = E_{FDi} - E'_{qi} + (x_{di} - x'_{di}) I_{adi}$$

$$\tau'_{qoi} \dot{E}_{di} = -E'_{di} - (x_{qi} - x'_{qi}) I_{aqi} \qquad (2.23)$$

$$M_i \dot{\omega}_i = P_{mi} - P_{ei} - \frac{M_i}{M_T} P_{COI}$$

$$= (P_{mi} - P'_{ei} - \frac{M_i}{M_T} P'_{COI}) - \left[P_{eGLi} + \frac{M_i}{M_T} \sum_{j=1}^{n} P_{eGLj} \right] \qquad (2.24)$$

II.C.4 Excitation system model

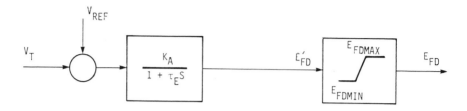

Figure 2.4. Exciter model

The exciter is assumed to be represented by an equivalent one gain, one time constant, and one limiter model, as shown in Figure 2.4. This model is used to standardize the treatment of incorporation of the effect of excitation control on the flux variation in the synchronous generator; and a higher order exciter model can be used if desired. Standard techniques are available in the literature to reduce a higher order system to an equivalent one gain, one time constant in the frequency range of interest (see for example Section 3.5 of [21]). By proper choice of the parameters K_A, τ_E; $E_{FD_{max}}$ and $E_{FD_{min}}$ in Figure 2.4 most modern excitation systems can be modeled.

The exciter equations for generator k are given by

$$\tau_{Ek}\dot{E'}_{FDk} = g\left(E'_{FDk}, V_{tk}, V_{REFk}\right)$$

$$= -E'_{FDk} + K_{Ak}\left(V_{REFk} - V_{tk}\right) \tag{2.25}$$

$E_{FD}, E'_{FD}=$ Stator emf corresponding to the field voltage,

$E_{FD}=$ Value after the limiter, and

$E'_{FD}=$ Value before the limiter.

$V_t=$ Generator terminal voltage.

$V_{REF}=$ Exciter reference voltage.

III. STABILITY ANALYSIS

In this section we shall introduce a few definitions which are commonly used in analytical work related to stability theory, and also present some results on the characterization of the stability boundary based on the controlling unstable equilibrium point (UEP). Many of the definitions, theorems, etc., that follow are stated for differential equations of the type

$$\dot{x}(t) = f(t, x(t)) \tag{3.1}$$

The system described by (3.1) is said to be autonomous if $f(t, x) \equiv f(x)$, i.e., independent of t.

III.A Stability Definitions and Theorems

Definition III.1

A point $x_o \, \epsilon R^n$ is called an equilibrium point for system (3.1) at time t_o if $f(t, x_o) \equiv 0$ for all $t \geqslant t_o$. An equilibrium point x_o of (3.1) is said to be an isolated equilibrium point if there exists some neighborhood N of x_o which does not contain any other equilibrium point of (3.1).

Definition III.2

The equilibrium $x = o$ of (3.1) is said to be stable in the sense of Lyapunov, or simply stable, if for every real number $\epsilon > o$ and initial time $t_o > o$ there exists a real number $\delta(\epsilon, t_o) > o$ such that for all initial conditions satisfying the inequality

$$\| x(t_o) \| = \| x_o \| < \delta$$

the motion satisfies $\| x(t) \| < \epsilon$ for all $t \geqslant t_o$. The symbol $\|\cdot\|$ stands for a norm.

Definition III.3

The equilibrium $x = o$ of (3.1) is asymptotically stable at time t_o, if

i. $x = o$ is stable at $t = t_o$

ii. for every $t_o \geqslant 0$ there exists an $\eta(t_o) > o$ such that

$$\lim_{t \to \infty} \| x(t) \| \to 0 \text{ whenever } \| x(t_o) \| < \eta$$

III.A.1 Lyapunov's method

The basic concept of Lyapunov's method can be summarized as follows:

If $E(x)$ is the system energy, and if $\dfrac{dE}{dt}$ is negative except at the equilibrium x_e, then $E(x)$ will continue to decrease until it assumes the value $E(x_e)$, or $E(o)$ when the equilibrium is at the origin.

In the mathematical treatment $E(x)$ is replaced with a scalar function $V(x)$ referred to as the Lyapunov function, and the invariance theory of ordinary differential equations is invoked to obtain the important stability results given in Theorem III.2 below.

We apply the results to autonomous system of the form

$$\dot{x} = f(x) \tag{3.2}$$

Here we assume that f and $\dfrac{\partial f}{\partial x_i}$, i=1,2,...,n, are continuous in a region D, which is a subset of R^n, denoted by $D \subset R^n$ (where D may be all of R^n) and we assume that $x = o$ is an isolated equilibrium

Definition III.4

A set Γ of points in R^n is invariant with respect to (3.2) if every solution of (3.2) starting in Γ remains in Γ for all time.

Theorem III.1

Let V be a continuously differentiable, real valued function defined on some domain $D \subset R^n$ containing the origin and assume that $\dot{V}_{(3.2)} \leqslant o$ on D. Assume that $V(o)=o$. For some real constant $k \geqslant o$, let H_k be the component of the set $S_k = \{x : V(x) \leqslant k\}$ which contains the origin. Suppose that H_k is a closed and bounded subset of D. Let $E = \{x \in D : \dot{V}_{(3.2)} = 0\}$. Let M be the largest invariant subset of E with respect of (3.2). Then every solution of (3.2) starting on H_k at $t=o$ approaches the set M as $t \to \infty$.

The above theorem by itself does not provide a stability result, but describes the qualitative behavior of the system. Using this theorem we can now establish the following stability result.

Theorem III.2

Assume that for system (3.2) there exists a continuously differentiable, real valued, positive definite function V defined on some set $D \subset R^n$ containing the origin. Assume that $\dot{V}_{(3.2)} \leqslant o$ on D. Suppose that the origin is the only invariant subset with respect to (3.2) of the set $E = \{x \in D : \dot{V}_{(3.2)} = o\}$. Then the equilibrium $x = o$ of (3.2) is asymptotically stable.

Definition III.5

The set of all $x(t_o) \in R^n$ such that $x(t, t_o, x(t_o)) \to o$ as $t \to \infty$ for some $t_o > o$ is called the domain of attraction of the equilibrium $x = o$ of (3.1).

III.B Analytical Justification for the Controlling UEP

Most of the results dealing with the TEF method applied to power systems have been derived largely based on physical reasoning supported by simulations. In [22] and [23], a complete topological characterization of the stability boundary of a stable equilibrium point is derived. We will now present some results from [22], where the domain of attraction of a multi-machine power system is characterized and the concept of the controlling UEP analytically justified.

Consider the stability boundary of the system described by (3.2). Some standard terminology is established first to describe the concept. The derivative of f at a point x^a is called the Jacobian matrix at x^a and is denoted by $J(x^s)$. An equilibrium point (Definition III.1) is hyperbolic if the Jacobian has no eigenvalues with zero real part. The stability region or domain of attraction (Definition III.5) of a stable equilibrium point x^s is denoted by $A(x^s)$. The stability boundary (boundary of the stability region) and the closure of $A(x^s)$ are denoted by $\partial A(x^s)$ and $\bar{A}(x^s)$. E denotes the set of all equilibrium points of (3.2). The type of the equilibrium point x^a is defined according to the number of eigenvalues of $J(x^a)$ with positive real part. If the Jacobian of the equilibrium point x^a has exactly one eigenvalue with positive real part, we call it a type-one equilibrium point.

Definition III.6

Let x^a be a hyperbolic equilibrium point of (3.2); its stable manifold $W^s(x^a)$ and unstable manifold $W^u(x^a)$ are defined as

$$W^s(x^a) = \left\{ x_0 : x(t, x_0) \to x^a \text{ as } t \to \infty \right\}$$

$$W^u(x^a) = \left\{ x_0 : x(t, x_0) \to x^a \text{ as } t \to \infty \right\}$$

The stable and unstable manifolds are invariant sets.

Definition III.7

Consider two manifolds A, B in R^n. We say that the intersection of A and B denoted by $A \cap B$, satisfies the transversality condition if (i) at every point of the intersection, $x \in (A \cap B)$, the tangent spaces of A and B at x span the tangent space of R^n at x

$$T_x(A) + T_x(B) = R^n \text{ for } x \in A \cap B$$

or (ii) they do not intersect at all.

Furthermore, the following assumptions concerning (3.2) are made.

(A1) All the equilibrium points on the stability boundary of (3.2) are hyperbolic.

(A2) The intersection of $W^s(x^a)$ and $W^u(x^b)$ satisfies the transversality condition, for all the equilibrium point x^a, x^b on the stability boundary.

(A3) There exists a continuously differentiable function $V: R^n \to R$ for (3.2) such that

   i. $\dot{V}_{(3.2)}(x) \leqslant o$ at $x \in E$;

   ii. If x is not an equilibrium point, then the set $\{t \in R: \dot{V}_{(3.2)}(x) = o\}$ has measure o in R; i.e., its length is zero.

   iii. $V(x)$ is bounded.

We shall refer to this function as the "energy function" for later reference. Based on these assumptions the following three theorems, characterizing the stability boundary of system (3.2), are proved in [22].

Theorem III.3

For the dynamical system (3.2) satisfying assumptions (A1)-(A3), x^i is an unstable equilibrium point on the stability boundary $\partial A(x^s)$ of a stable equilibrium point x^s if and only if $W^u(x^i) \cap A(x^s) \neq \phi$.

Theorem III.4

For the dynamical system (3.2) satisfying assumptions (A1)-(A3), let x^i, $i = 1, 2, ..., n$ be the unstable equilibrium points on the stability boundary $\partial A(x^s)$ of a stable equilibrium point x^s; then

$$\partial A(x^s) = \bigcup_{x^i \in E \cap \partial A} W^s(x^i)$$

i.e., the stability boundary ∂A is the union of all the stable manifolds of those unstable equilibrium points contained in the intersection of E and ∂A.

Theorem III.5

If the dynamical system (3.2) satisfies the assumption (A3), then on the stable manifold $W^s(x^i)$ of an equilibrium point x^i, the point at which the energy function achieves the minimum is the equilibrium point x^i itself, i.e.,

$$V(x^i) = \min_{x \in W^s(x^i)} V(x)$$

For the classical power system model developed in Chapter II, for the case with transfer conductances neglected, these theorems can be applied [22] and the concept of the region of stability can be rigorously justified. Furthermore, the concept of the controlling UEP can also be justified.

Consider the fault-on trajectory $x(t)$, starting from x_0 at the time of the fault, as shown in Figure 3.1. If UEP x^3 was selected to be the controlling UEP, then the set of points on the dotted curve denoted by $R(V_{x^3})$ would represent the boundary of stability ∂A. It is obvious that, for this case, the set $R(V_{x^3}) = \{x : V(x) < V_{x^3}\}$ is an underestimate of the "true" critical energy. Let us call the point x^e where the fault-on trajectory $x(t)$ crosses the stability boundary $\partial A(x^s)$, the exit point of the trajectory $x(t)$. The value of the energy function at the exit point is actually the true critical energy.

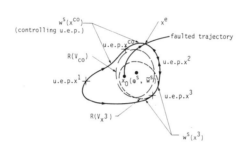

Figure 3.1. The concept of controlling UEP

Suppose that x^e belongs to the stable manifold $W^s(x^{co})$ of the UEP x^{co} on the stability boundary, and $V(x)$ assumes its minimum on the stable manifold $W^s(x^{co})$ at the equilibrium point x^{co} (Theorem III.5), the fault-on trajectory must pass through the constant energy surface $\{x : V(x) = V(x^{co})\}$ before it reaches the exit point x^e. Let us define $V_{co} = V(x^{co})$. Clearly, $V_{co} \geq V_{x^3}$; hence, V_{co} is a better estimate of the critical energy. For a given fault-on trajectory $x(t)$, if the exit point x^e lies on the stable manifold of x^{co}, then we call x^{co} the controlling UEP.

IV. THE TRANSIENT ENERGY FUNCTION METHOD APPLIED TO THE CLASSICAL POWER SYSTEM MODEL

In this section we present a general overview of the transient energy function (TEF) method applied to the classical power system model developed in Section II. The various steps involved in conducting the transient stability assessment procedure using the TEF method will be described. The concept of the controlling unstable equilibrium point will be developed. The notion of the mode of disturbance (MOD) will be introduced. Details regarding the numerical and computational aspects of the various steps in the TEF method will also be presented.

IV.A The Transient Energy Function

The transient energy function can be derived directly from the dynamic equations in the COI frame (equation (2.5)).

In these equations, parameters of the post-disturbance system are used. Multiplying the ith post-disturbance dynamic equation by $\dot{\theta}_i$, and integrating the sum of the equations using as a lower limit $t = t^s$ where $\omega(t^s) = 0$, and $\theta(t^s) = \theta^{s2}$ the post-disturbance stable equilibrium point, we obtain

$$V = \frac{1}{2} \sum_{i=1}^{n} M_i \tilde{\omega}_i^2 - \sum_{i=1}^{n} P_i(\theta_i - \theta_i^{s2})$$

$$- \sum_{i=1}^{n-1} \sum_{j=i+1}^{n} \left[C_{ij}(\cos\theta_{ij} - \cos\theta_{ij}^{s2}) - \int_{\theta_i^{s2}+\theta_j^{s2}}^{\theta_i+\theta_j} D_{ij}\cos\theta_{ij}d(\theta_i+\theta_j) \right]$$

$$(4.1)$$

The first term in the energy expression represents the total change in kinetic energy of all generator rotors relative to the COI. The second term represents the change in the position energy of all rotors relative to the COI. The third and fourth terms represent the change in stored magnetic energy and change in the dissipation energy of branch ij respectively.

It is to be noted that the term in the energy expression corresponding to the dissipation component consists of a path dependent integral. This cannot be evaluated unless the system trajectory is known. Uemura, et al. [24] proposed several approximations to this term; of these the linear angle trajectory assumption was used by Athay, et al. [8], and by other workers in the field since. The authors have examined the actual dissipation energy component evaluated along the trajectory from numerous simulations. The authors have found that for a first swing transient, this approximation is fairly accurate and has become standard practice in the TEF method to incorporate the effect of transfer conductances.

Let θ^a and θ^b represent the initial and final vectors of angular positions for the n-machines. The approximation to the integral term is then given by

$$I_{ij} = D_{ij} \frac{\theta_i^b - \theta_i^a + \theta_j^b - \theta_j^a}{\theta_{ij}^b - \theta_{ij}^a} (\sin\theta_{ij}^b - \sin\theta_{ij}^a) \tag{4.2}$$

In [9] and [10] a detailed analysis of the energy behavior along the time domain trajectory was conducted using a 17-generator equivalent of the network of the State of Iowa. The various components of transient energy developed earlier were carefully analyzed along the time domain trajectories. A detailed analysis of these components provided a significant characteristic of the energy behavior. It was observed that in all cases where the system was stable following the removal of the disturbance, the kinetic energy does not become zero, indicating that a certain amount of the total kinetic energy in the system is not absorbed. This clearly indicates that not all the transient kinetic energy created by the disturbance contributes to the instability of the system. Some of this kinetic energy is responsible for the inter-machine motion between the generators, and does not contribute to the separation of the severely disturbed generators from the rest of the system.

Based on the above observation, it is postulated that the transient kinetic energy which tends to separate the critical generators from the rest of the system is that associated with the gross motion of the critical generators, i.e., that of their inertial center with regard to the inertial center of the other generators. The kinetic energy associated with the gross motion of a group of k machines having angular speeds $\tilde{\omega}_1, \tilde{\omega}_2, ..., \tilde{\omega}_k$ is the same as the kinetic energy of their inertial center. The speed of the inertial center of that group and its kinetic energy are given by

$$\tilde{\omega}_{cr} = \sum_{i=1}^k M_i \tilde{\omega}_i \bigg/ \sum_{i=1}^k M_i \tag{4.3}$$

$$V_{KE_{cr}} = \frac{1}{2} \left| \sum_{i=1}^k M_i \right| (\tilde{\omega}_{cr})^2 \tag{4.4}$$

The gross motion of the two groups approximates that of a two-machine system. The kinetic energy causing the separation of the two groups is the same as that of an equivalent one-machine-infinite-bus system having inertia constant M_{eq} and angular speed ω_{eq} given by

$$M_{eq} = \frac{M_{cr} \cdot M_{sys}}{M_{cr} + M_{sys}}$$

$$\tilde{\omega}_{eq} = (\tilde{\omega}_{cr} - \tilde{\omega}_{sys}) \qquad (4.5)$$

where M_{cr}, $\tilde{\omega}_{cr}$, and M_{sys}, $\tilde{\omega}_{sys}$ are the inertia constants and angular speeds of the critical machines and the rest of the generators respectively. The corresponding kinetic energy is given by

$$V_{KEcorr} = \frac{1}{2} M_{eq} (\tilde{\omega}_{eq})^2 \qquad (4.6)$$

The kinetic energy term in (4.1) should be replaced by (4.6).

IV.B Transient Stability Assessment

Transient stability assessment using the TEF method is made only for the last transient. Thus, for a case where the disturbance scenario involves multiple transients or switchings, a stability assessment using the TEF method can only be done for the last transient or switching.

The stability assessment is made by comparing two values of the transient energy V. The value of V is computed when the last disturbance is removed, e.g., if the disturbance is a simple fault the value of V at fault clearing, V_{cl}.

The other value of V which to a great extent determines the accuracy of the stability assessment is the critical value of V, V_{cr} which is the potential energy at the controlling unstable equilibrium point (UEP), for the particular disturbance under investigation.

If for a fault disturbance $V_{cl} < V_{cr}$, the system is stable; and if $V_{cl} > V_{cr}$, the system is unstable. Alternatively the assessment can be made by computing the energy margin ΔV given by

$$\Delta V = V_{cr} - V_{cl} \qquad (4.7)$$

Substituting for V_{cr}, and V_{cl} from (4.1) with the kinetic energy corrected by (4.6) and using the linear angle path assumption for the dissipation terms between the conditions at the end of the disturbance and the controlling UEP, we have

$$\Delta V = -\frac{1}{2} M_{eq} \tilde{\omega}_{eq}^{cl\,2} - \sum_{i=1}^{n} P_i(\theta_i^u - \theta_i^{cl})$$

$$-\sum_{i=1}^{n-1} \sum_{j=i+1}^{n} [C_{ij}(\cos\theta_{ij}^u - \cos\theta_{ij}^{cl})$$

$$-D_{ij}\frac{\theta_i^u - \theta_i^{cl} + \theta_j^u - \theta_j^{cl}}{(\theta_{ij}^u - \theta_{ij}^{cl})} (\sin\theta_{ij}^u - \sin\theta_{ij}^{cl})] \tag{4.8}$$

where $(\theta^{cl}, \tilde{\omega}^{cl})$ are the conditions at the end of the disturbance and $(\theta^u, 0)$ represents the controlling UEP.

If ΔV is greater than zero the system is stable, and if ΔV is less than zero the system is unstable. Transient stability assessment using the energy margin has two advantages: (i) the energy margin can be obtained by one computation only by evaluating V between the conditions at fault clearing and those at the UEP; and (ii) a qualitative measure of the degree of stability (or instability) can be obtained if ΔV is normalized with respect to the corrected kinetic energy at the end of the disturbance [9]

$$\Delta V_n = \frac{\Delta V}{V_{KEcorr}} \tag{4.9}$$

In the transient stability assessment framework described above, it is to be noted that for disturbances involving faults (even for unbalanced faults) a balanced positive sequence network is assumed. For disturbances other than faults, the parameters describing the system at the end of the disturbance should be related to ΔV, for assessment to be made.

IV.C Concept of the Controlling Unstable Equilibrium Point (UEP)

The analytical concepts behind the controlling UEP and its importance in determining an accurate estimate of the region of stability were discussed earlier in Section III. In this section we present a technique to determine the correct controlling UEP based on the physical behavior of the power system.

The equilibrium points represent a set of generators' angles which satisfy equation (4.10)

$$f_i = P_{mi} - P_{ei} - \frac{M_i}{M_T} P_{COI} = 0 \qquad i=1,2,...,n \tag{4.10}$$

Of those possible solutions, one set represents a stable equilibrium, known as SEP, all others represent unstable equilibrium points (UEPs). For multi-machine power systems with n–generators, there are $(2^{n-1}-1)$ solutions possible. Each solution will give a certain value of potential energy. Thus, identifying the "correct" UEP for a given disturbance means identifying the correct energy level against which stability assessment is made.

Based on the evidence provided by [26,27,8,9,10] the authors of [11] carefully re-examined the physical behavior of a multi-machine power system subjected to a disturbance. They determined that during the ensuing transient, the more severely disturbed generators may or may not lose synchronism with the rest of the system, depending on whether the potential energy absorbing capacity of the network is adequate to convert the transient kinetic energy at the end of the disturbance into potential energy. Therefore, the controlling UEP must properly account for two aspects of the transient phenomena: (i) the effect of the disturbance on the various generators, and (ii) information on the energy absorbing capacity of the post–disturbance network. These two aspects are incorporated in the following criterion proposed in [11] for determining the controlling UEP.

"The post–disturbance trajectory approaches (if the disturbance is large enough) the controlling UEP. This is the UEP with the lowest normalized potential energy margin at the instant the disturbance is removed."

Stated mathematically

$$\Delta V_{PEn} = \frac{\Delta V_{PE}}{V_{KEcorr}} \qquad (4.11)$$

where

$$\Delta V_{PE} = V_{PE}^u - V_{PE}^{c\ell}$$

$$= - \sum_{i=1}^{n} P_i(\theta_i^u - \theta_i^{c\ell})$$

$$- \sum_{i=1}^{n-1} \sum_{j=i+1}^{n} \left[C_{ij}(\cos\theta_{ij}^u - \cos\theta_{ij}^{c\ell}) \right.$$

$$\left. - D_{ij} \left(\frac{\theta_i^u - \theta_i^{c\ell} + \theta_j^u - \theta_j^{c\ell}}{\theta_{ij}^u - \theta_{ij}^{c\ell}} \right) (\sin\theta_{ij}^u - \sin\theta_{ij}^{c\ell}) \right] \qquad (4.12)$$

Determining the advanced generators for the controlling UEP is referred to as determining the MOD. For a given UEP the components of the rotor positions of the advanced generators have angles greater than $90°$. Hence, if for a given UEP, machines j and k have their angles greater than $90°$ (the MOD for this UEP is referred to as "jk").

IV.D Main Steps in the TEF Procedure

IV.D.1 Initial computations

These calculations involve the pre-disturbance power flow data and the appropriate generator dynamic model data. The power flow data provides the level of generation, the network connectivity and parameters, and the magnitude and angles of the voltages at all buses. For the classical generator model, the generator data consists of the generator inertia (H), and the direct axis transient reactance (x'_d).

From the input power flow data and the synchronous generator data which are obtained on a common MVA base, the internal voltages of the generators are computed from the pre-disturbance generator terminal voltages and current as follows.

Define $\overline{I}_i = I_{i1} + j\,I_{i2}$,

then from the relation $P_i + j\,Q_i = \overline{V}_i\overline{I}_i{}^*$

we have $I_{i1} + j\,I_{i2} = (P_i - j\,Q_i)^i/\overline{V}_i.$
The internal EMF of a generator i is computed from

$$E_i\underline{/\delta'_i} = \overline{V}_i + j\,x'_{di}\overline{I}_i \tag{4.13}$$

The initial generator angle δ_{io} is then obtained by adding the pre-disturbance terminal angle α_i to δ'_i or

$$\delta_{io} = \delta'_i + \alpha_i \quad i=1,2,...,n \tag{4.14}$$

These angles are then converted into the center of inertia reference frame to obtain the internal voltage of each machine, given by $E_i\underline{/\theta_i^{s1}}, \quad i=1,2,...,n.$

IV.D.2 Construction of admittance matrices

The admittance matrices used in the TEF analysis are evaluated systematically using the input power flow data and generator data. The first step in this procedure consists of building the pre-disturbance network admittance matrix. This matrix is modified to obtain the other network matrices, which include the disturbed network admittance matrix and the post-disturbance network admittance matrix.

IV.D.3 Conditions at the end of the disturbance

Disturbances in general could be classified into two categories: fault and non-fault. Based on the type of the disturbance, the procedure to evaluate the conditions at the end of the disturbance and its effect on the system energy could be (vastly) different for the two categories.

Conditions at fault clearing

A constant acceleration assumption is used for fast computation of the end of fault conditions. The time period from the instant of application of the fault t_o to the instant of fault clearing t_{cl} is divided into (several) intervals, each of length Δt. The following procedure is then used to determine the conditions at t_{cl}.

1. Using the reduced faulted network admittance matrix, the accelerating power for each machine is computed using

$$P_{a_i}(t_o^+) = P_{m_i} - P_{e_i}(t_o^+), \quad i=1,2,...,n \tag{4.15}$$

2. Assuming a constant acceleration over the first time interval, the change in angular speed is given by

$$\Delta\omega_i(t_1) = \frac{P_{a_i}(t_o^+)\,\Delta t}{2H}\ p.u., \quad i=1,2,...,n \tag{4.16}$$

3. The angular change is given by

$$\Delta\delta_i(t_1) = (2\pi f \times \frac{180}{\pi})\frac{P_{a_i}(t_o^+)(\Delta t)^2}{4H}\ \text{degrees}\quad i=1,2,...,n \tag{4.17}$$

   and the angles and speeds at the end of the interval are given by

$$\delta_i(t_1) = \delta_i(t_o) + \Delta\delta_i(t_1)\quad i=1,2,...,n \tag{4.18}$$

$$\omega_i(t_1) = 0.0 + \Delta\omega_i(t_1)\quad i=1,2,...,n \tag{4.19}$$

4. Using the updated angles from step 3, the accelerating power of each machine at time t_1, $P_{a_i}(t_1)$ $i=1,2,...,n$ is evaluated as described in step 1.

5. The parameters at the end of the next interval are given by

$$\Delta\omega_i(t_2) = \frac{P_{a_i}(t_1)\Delta t,(\theta^u)}{2H} \quad p.u. \quad i=1,2,...,n \tag{4.20}$$

and

$$\Delta\delta_i(t_2) = 2\pi f \times \frac{180}{\pi} \frac{P_{a_i}(t_1)(\Delta t)^2}{4H} \quad i=1,2,...,n \tag{4.21}$$

6. The procedure is then repeated until the instant of fault clearing is reached.

Disturbances other than faults

For disturbances other than faults, it is essential to analyze the effect of the disturbance on the system transient energy. In some cases the disturbance itself is a sudden change in the potential energy (e.g., a sudden opening of a loaded transmission line).

From a stability assessment viewpoint, the situation boils down to determining the effect of the disturbance on the energy margin ΔV. In terms of the components of the energy margin, this relation can be expressed as

$$\Delta V = \Delta V_{KE} + \Delta V_{PE} \tag{4.22}$$

IV.D.4 Determination of the mode of disturbance (MOD)

As described in subsection IV.C, the determination of the mode of disturbance constitutes a very important step in the TEF procedure. The criterion developed in equation (4.11) requires the determination of the normalized potential energy margin ΔV_{PEn}. The controlling UEP would be that among the possible ones in the direction of the post-disturbance trajectory having the lowest ΔV_{PEn}.

Based on this criterion it is evident that an automated practical technique to determine the MOD would first have to identify the candidate modes which need to be tested. Furthermore, calculating ΔV_{PEn} requires knowledge of the UEP, i.e., θ^u, for all the candidate modes. Computationally, this can be an expensive and time-consuming task that may make the large scale application of the TEF method impractical. Keeping these issues in mind, an automated procedure has been developed. The details of this procedure are given in [11,13,14].

IV.D.5 Calculation of equilibrium points

The technique to calculate the equilibrium points is a general solution technique which can be used to evaluate both the post-disturbance SEP and the controlling UEP. The only difference between these two solutions is the starting point provided for the solution procedure.

The procedure to obtain the equilibrium points consists of determining a solution to a system of nonlinear algebraic equations by (4.10).

In general, there are two approaches to solving such a system of equations. They are:

1. The direct solution approach: In this case equation (4.10) is directly solved using an iterative technique. The Newton-Raphson method had been previously used. However, it has in some cases, exhibited divergent characteristics and has not been completely reliable [13,8]. In [28] a modified Newton-Raphson technique has been used with greater success.

2. Indirect solution approach: In this case, the solution technique is formulated as a nonlinear least squares minimization problem [8]. The objective function to be minimized is given by

$$F(\theta) = \sum_{i=1}^{n} f_i^2 (\theta) \qquad (4.23)$$

where f_i is given in equation (4.10).

In [13], a careful search of the various methods available to solve the minimization problem was made, keeping in consideration the system size dealt with and the stressed nature of the system. Based on this search, the corrected Gauss-Newton (CGN) method [29] was selected as a replacement of the DFP technique. The CGN method is a very robust and reliable technique; it however, requires a singular value decomposition of the Jacobian matrix at every iteration. This results in severe computational burden, as a result, the method has been replaced by the full Newton method [29].

V. MODELING IMPROVEMENTS IN THE TEF METHOD

V.A Introduction

In the previous sections, the basic elements of power system direct transient stability assessment using the TEF method were presented. In all this, a relatively simple (i.e., the classical) power system model has been used. In present day power networks this simple model is of limited validity, even in analysis of first swing transients. To overcome this shortcoming, some improvements in the modeling of power system components have been introduced in the TEF method.

These improvements are discussed in this section. The modeling improvements in the TEF method are those which would primarily improve the accuracy of the results during the first swing and are examined through the behavior of the transient energy responsible for the separation of one or more generators from the rest of the system. Therefore, in the TEF method model improvements are of interest only to the extent that they may give a better estimate of the relevant system transient energy during the first swing transient.

The power system components to be discussed in this section are: excitation control which involves both the synchronous generator and excitation system models; two-terminal HVDC model; and voltage dependent nonlinear load models.

V.B Modeling of Excitation Control Effects

Again we stress the fact that the model improvement is intended to give a better estimate of the transient energy in the first swing transient. For this purpose we use the two-axis model detailed in Section II.C.1. For the excitation system model, a simple one gain-one time constant-one limiter model described in Section II.C.4 is used. This model is used to standardize the treatment of incorporating excitation control in the TEF method. All other excitation system models are assumed to have been converted into this model format in the frequency range of interest using the procedure developed in [30].

The incorporation of the excitation control in the TEF method is accomplished by focusing on the energy in the first swing transient, which is primarily an inertial transient. In such a transient the generators' motion is determined by the interaction between the synchronizing torques and the inertias of the generators. The exciters primarily change the generators' internal EMF (mainly E'_q) and hence alter their synchronizing torques.

In obtaining the system transient energy in the first swing transient, the following assumptions are made:

1. The generator flux does not change abruptly, i.e., a reasonably well-behaved function can describe the flux variation in the first swing. It is also assumed that the flux varies smoothly between conditions at fault clearing and at the peak of the first swing.

2. The key parameters needed to compute the transient energy can be obtained from conditions at fault clearing, and conditions at the point where the system potential energy is maximum on the critically stable trajectory (also the peak of the critical angle swing for the severely disturbed generators). The last conditions will be referred to as the "peak point".

3. The transient energy in the first swing can be evaluated using a
   constant value of E'. This value is estimated from the two
   values at clearing and at the peak point (a simple average value
   has been used).

4. When the disturbance is removed the total transient energy is
   constant.

V.B.1 The transient energy function

The new transient energy function is obtained from the integral

$$V = \int \sum_{i=1}^{n} \left| M_i \dot{\tilde{\omega}}_i - P_{mi} + P_{ei} - \frac{M_i}{M_T} P_{COI} \right| \dot{\theta}_i \, dt \qquad (5.1)$$

In the above, the following expression is used for P_{ei} (for the generator
represented in detail)

$$P_{ei} = E'_{di} I_{di} + E'_{qi} I_{qi}$$

where I_d, I_q are given by (2.16); and the term $(x'_q - x'_d) I_d I_q$ is
neglected in P_{ei}.
The resulting expression for P_{ei} can be put in the form

$$P_{ei} = \alpha_{ii} G_{ii} + \sum_{\substack{j=1 \\ j \neq i}}^{n} \left\{ B_{ij}[\alpha_{ij}\sin\theta_{ij} + \beta_{ij}\cos\theta_{ij}] + G_{ij}[\alpha_{ij}\cos\theta_{ij} - \beta_{ij}\sin\theta_{ij}] \right\}$$

$$(5.2)$$

where

$$\alpha_{ij} = E'_{di} E'_{dj} + E'_{qi} E'_{qj}$$

$$\beta_{ij} = E'_{di} E'_{qj} - E'_{qi} E'_{dj} \qquad (5.3)$$

we note that $\alpha_{ij} = \alpha_{ji}$, $\beta_{ij} = -\beta_{ji}$
To obtain the new TEF, equation (5.1) is integrated between the
post-disturbance equilibrium condition ($\theta = \theta^s$, $\tilde{\omega} = 0$), and the system
state of interest ($\theta, \tilde{\omega}$). The transient energy function has two
components: kinetic energy and potential energy. They are given
below.

Kinetic Energy:
Strictly speaking the system kinetic energy is given by

$$V_{KE} = \frac{1}{2} \sum_{i=1}^{n} M_i \tilde{\omega}_i^2 \tag{5.4}$$

However, the kinetic energy responsible for the separation of the group of severely disturbed (critical) generators from the rest of the system is the corrected kinetic energy given by (4.7) repeated here for convenience

$$V_{KEcorr} = \frac{1}{2} M_{eq} \tilde{\omega}_{eq}^2 \tag{5.5}$$

where M_{eq}, ω_{eq} are the inertia constant and speed of the equivalent one-generator-infinite-bus system formed by the two groups of generators: the critical group and the rest of the system.

Potential Energy:

The potential energy, with respect to the stable equilibrium conditions, is given by

$$V_{PE} = \sum_{i=1}^{n} -\int_{\theta_i^s}^{\theta_i} (P_{mi} - \alpha_{ii} G_{ii}) d\theta_i + \sum_{i=1}^{n-1} \sum_{j=i+1}^{n} \int_{\theta_{ij}^s}^{\theta_{ij}} B_{ij} [\alpha_{ij} \sin\theta_{ij}$$

$$+ \beta_{ij} \cos\theta_{ij}] d\theta_{ij} + \sum_{i=1}^{n-1} \sum_{j=i+1}^{n} \int_{\theta_i^s+\theta_j^s}^{\theta_i+\theta_j} G_{ij} [\alpha_{ij} \cos\theta_{ij} - \beta_{ij} \sin\theta_{ij}] d(\theta_i + \theta_j)$$

$$\tag{5.6}$$

The above integrals are evaluated using the usual assumption of a linear trajectory approximation [15,16] and a constant value for each of E'_{di} and E'_{qi} (average of the values at fault clearing and at the peak point); the α and β parameters in (5.6) are thus treated as constant. For a generator modeled classically, $E'_{qi} = E_i$, and $E'_{di} = \beta_{ij} = 0$. The resulting potential energy expression is given by

$$V_{PE} = \sum_{i=1}^{n} -(P_{mi} - \alpha_{ii} G_{ii})(\theta_i - \theta_i^s) + \sum_{i=1}^{n-1} \sum_{j=i+1}^{n} \{B_{ij} \alpha_{ij} [-\cos\theta_{ij} + \cos\theta_{ij}^s]$$

$$+ B_{ij} \beta_{ij} [\sin\theta_{ij} - \sin\theta_{ij}^s]\} + \sum_{i=1}^{n-1} \sum_{j=i+1}^{n} \{G_{ij} \alpha_{ij} \frac{\theta_i + \theta_j - \theta_i^s - \theta_j^s}{\theta_{ij} - \theta_{ij}^s} [\sin\theta_{ij} - \sin\theta_{ij}^s]$$

$$+ G_{ij} \beta_{ij} \frac{\theta_i - \theta_j - \theta_i^s - \theta_j^s}{\theta_{ij} - \theta_{ij}^s} [\cos\theta_{ij} - \cos\theta_{ij}^s]\}$$

$$\tag{5.7}$$

V.B.2 Critical value of the transient energy and the energy margin

Transient stability assessment is made by computing the difference between the critical value of the transient energy $V_{critical}$, and the value of the transient energy at the end of the disturbance (i.e., at fault clearing). This energy difference is called the energy margin ΔV. The system is stable or unstable depending on whether $\Delta V \gtrless 0$.

For the classical power system model, the critical value of the transient energy is the value of the potential energy at the controlling unstable equilibrium point (UEP). This UEP represents a system equilibrium conditions, attainable only at infinite time, which separates stable and unstable system trajectories, and represents the limiting stable conditions.

Careful investigation of the system trajectories in the post-disturbance period was made by the authors and their co-workers [16]. These investigations indicated that at the peak of the stable trajectories, while the derivatives of the generator angles and speeds are zero, derivatives of other state variables (e.g., E'_q, E'_d, E_{FD}) are usually not zero. This is found to be true even in the critically stable system conditions. Thus, at the critically stable trajectory the peak point is characterized by the following: a) for a critical generator i, $\dot{\theta}_i^P = 0$ and $\dot{\omega} = 0$ and b) for the system V_{PE}^P is maximum. A procedure for determining the peak point has been developed and is given below.

The critical value of the transient energy used in the final transient stability assessment is the value of the potential energy (5.7) evaluated at the rotor angles θ^P, and using E'^{av}_q; E'^{av}_d given by

$$E'^{av}_q = \frac{1}{2}\left[E'^{cl}_q + E'^{p}_q\right]$$

$$E'^{av}_d = \frac{1}{2}\left[E'^{cl}_d + E'^{p}_d\right] \tag{5.8}$$

where E'^{p}_q, E'^{p}_d are the values at the peak point.

V.B.3 Determination of the Peak Point with MOD Procedure

MOD Determination

When some of the generators are modeled in detail, the following procedure is used [16]:

1. The candidate modes to be checked are identified by inspection (e.g., if the possible severely disturbed generators can be readily identified), or by using the automatic mode selection procedure.

2. The rotor angles at the (approximate) corner point are computed, for all the generators, using (5.9) below. For the generators represented in detail, values of E'_q and E'_d at fault clearing are used.

   For a generator j in the critical (advanced) generator group, the rotor angles at the corner point are given by

   $$\theta_j^{CRN} = \theta_j^s + (\pi - 2\theta_{eq}^s)\frac{M_{T-k}}{M_T}, \tag{5.9a}$$

   and for a generator j in the rest of the system

   $$\theta_j^{CRN} = \theta_j^s - (\pi - 2\theta_{eq}^s)\frac{M_k}{M_T} \tag{5.9b}$$

   where

   $$\theta_{eq}^s = \theta_k^s - \theta_{T-k}^s$$

3. For each of the candidate modes, the corrected kinetic energy at fault clearing and the potential energy margin (between the angles at the approximate corner point and the angles at fault clearing) are computed.

4. The normalized potential energy margin, $\Delta V_{PEn} = \Delta V_{PE}/V_{KEcorr}$, is computed for each candidate mode. The mode with the lowest ΔV_{PEn} is selected.

   NOTE: Usually only 2-3 modes need to be checked for the above steps.

5. For the selected mode, the "actual" corner point voltages and angles, for the generators represented in detail, the corresponding ray point, and the exact peak point are computed as outlined below.

Determination of the Peak Point

It is assumed that $E'_q(\theta)$ varies exponentially between conditions at fault clearing and those at the limiting conditions, or

$$E'_q(\theta) = E_q'^{lim} - (E_q'^{lim} - E_q'^{cl})e^{\frac{\theta - \theta^{cl}}{\Delta\theta}}$$

where $\Delta\theta$ is obtained from

$$\Delta\theta = (E'^{\ell im}_q - E'^{c\ell}_q) \Big/ \left.\left|\frac{dE'_q}{d\theta}\right|\right._{c\ell} \tag{5.10}$$

Experimentally, it was found that for the best estimate of $\Delta\theta$, $\left|\dfrac{dE'_q}{d\theta}\right|$ is to be evaluated just prior to the instant of fault clearing.

The proper choice of $E'^{\ell im}_q$ would be the value of E'_q at the UEP. However, an approximate value of $E'^{\ell im}_q$ is obtained at the so-called "corner point" [16,31]. For a known mode of disturbance, i.e., assuming that the most severely disturbed generators (whose rotor angles are to be advanced), the corner point determination involves determining the angles of all the generators, and estimating E'_q, E'_d, and E_{FD} for the generators equipped with exciters.

For a detailed generator i, the solution procedure is started with the angles computed by (5.9) and the pre-disturbance values of EMFs. The corner point solution is obtained by solving the following set of equations with the detailed generator angles adjusted to satisfy

$$\dot{E}'_{qi} = 0,$$

$$\dot{E}'_{di} = 0,$$

$$\dot{E}_{FDi} = 0, \text{ and}$$

$$\dot{\tilde{\omega}}_i = 0$$

$$i = 1, 2, ..., m \tag{5.11}$$

Conditions at the peak point are determined by the following two-step procedure. First, an approximate peak point, called the ray point, is obtained by maximizing the potential energy on the ray from the post fault stable equilibrium point to the corner point. A one-dimensional search is used

$$\frac{dV_{PE}}{d\alpha} \simeq \sum_{i=1}^{n} P_{acc,i}\Big[E'_q(\theta^{ray}), E'^{ray}_d, \theta^{ray}\Big] \cdot \Big[\theta^{CRN}_i - \theta^s_i\Big] = 0 \tag{5.12}$$

where

$$\theta^{ray} = \theta^s + \alpha\,(\theta^{CRN} - \theta^s)$$

E'_q is given in (5.10)

$$E_d^{'ray} \simeq \frac{1}{2} \left[E_d^{'s} + E_d^{'CRN} \right] = \text{const}$$

In the second step the exact peak point is determined by solving

$$P_{acc,i} \left[E'(\theta^p), \theta^p \right] = 0 \tag{5.13}$$

$$i = 1, 2, ..., m$$

In (5.13) the generators represented in detail are actually modeled classically. However, their internal EMF E' are adjusted according to the updated values of E'_q obtained by (5.10).

Again, it is to be emphasized that at the peak point $\dot{E}'_{qi}$, $\dot{E}'_{di}$, and $\dot{E}_{FDi}$ may not be equal to zero.

V.C Modeling of Nonlinear Loads

V.C.1 Introduction

Again, as we consider load models other than those represented by constant impedance, we must keep in mind that the TEF method is a "first swing" tool of analysis. Our interest will then be in the changes which can be made in load models to improve the accuracy of the results during the first swing transient.

The experience of the authors, supported by the experience of engineers from several utilities in North America, indicates that for the first swing transient the power system load can be adequately represented by a mixture of constant current, constant impedance and constant MVA components. This representation will be used when nonlinear representation is called for. Since the representation of the constant impedance loads had been addressed, discussion of the constant current and constant MVA load models is presented.

V.C.2 General approach [17]

If the load at bus k is represented as a constant current or a constant MVA load, then

$$P_{Lk} + jQ_{LK} = V_k \underline{/\psi_k}\, \overline{I}_{Lk}^*$$

or

$$\overline{I}_{Lk}=(P_{Lk}-jQ_{LK})\big/V_k\,\underline{/\psi_k}$$ (5.14)

where

$V_k=V_k\,\underline{/\psi_k}$ = voltage at bus k

P_{LK},Q_{Lk}= power and reactive power at bus k

The current $\overline{I}_{Lk}$ must be preserved during the transient. This is accomplished by determining equivalent current injections at the internal generator nodes. In [32] Pai, et al., use similar current injections at the internal generator nodes, with the assumption that the complex ratio of internal generator voltage to load bus voltage is constant. This assumption could be quite restrictive and is not invoked in the development presented here. The current injections are re-calculated at each iteration of the SEP and UEP solution procedure based on the updated values of the load bus voltages. Hence, these current injections reflect the nonlinear load components at the internal generator nodes.

V.C.3 Deterining I_{GL} for classical generator model

It is assumed that the constant impedance loads have been folded into the post disturbance Y_{BUS}. For the constant current and constant MVA portions of the load, the initial current components are evaluated by

$$I^{\circ}_{CI}=\left[\left(p_2P^{\circ}_L+jq_2Q^{\circ}_L\right)\big/V^{\circ}_L\right]^*$$

$$I^{\circ}_{CM}=\left[\left(p_3P^{\circ}_L+jq_3Q^{\circ}_L\right)\big/V^{\circ}_L\right]^*$$ (5.15)

where
P°_L,Q°_L = pre-disturbance MW and MVAR load

V°_L = pre-disturbance load bus voltage

p_2,q_2 multipliers for constant current portions of the load

p_3,q_3 multipliers for constant MVA portions of the load

A nonlinear load current vector I_1, for initial conditions, is formed by

$$I_1 = I_{CI} + I_{CM} \tag{5.16}$$

The current injections at the internal generator nodes are determined, for the classical generator model, by the following procedure:

1. A vector of currents injected in the generator buses is computed from

    $$I_2 = E \underline{/\theta} \, / \, j \, x'_d \tag{5.17}$$

    we note that the angle θ is used, indicating that for the generator rotor angles a COI reference frame is used.

2. The set of voltages $\begin{vmatrix} V_L \\ V_G \end{vmatrix}$ is solved for using the relation

    $$\begin{vmatrix} -I_1 \\ I_2 \end{vmatrix} = \begin{vmatrix} Y^A_{BUS} \end{vmatrix} \begin{vmatrix} V_L \\ V_G \end{vmatrix} \tag{5.18}$$

    where Y^A_{BUS} includes the load buses and the generator terminal buses, with the terms $\dfrac{1}{j \, x'_d}$ incorporated on the diagonal entries of the generator terminal buses.

3. The nonlinear load current components I_{CI} and I_{CM} are updated, according to the new V_L by

    $$I_{CI} = \left| \frac{P^\circ_{LP2} \dfrac{|V_{Lnew}|}{|V^\circ_L|} + j \, Q^\circ_{LQ2} \dfrac{|V_{Lnew}|}{|V^\circ_L|}}{V_{Lnew}} \right|^*$$

    $$I_{CM} = \left| \frac{P^\circ_{LP3} + j \, Q^\circ_{LQ3}}{V_{Lnew}} \right|^* \tag{5.19}$$

(5.18) and (5.19) are used iteratively until successive iterations give nearly the same magnitude of the current vector (i.e., within a given tolerance).

4. The set of current injections I_{GL} is obtained from the final values of $-I_1$ using

$$I_{GL} = \left| Y_{21}^B \right| \left| Y_{11}^B \right|^{-1} \left| -I_1 \right|$$ (5.20)

where Y_{BUS}^B is the same as Y_{BUS}^A augmented with the internal generator buses so that the voltage vector is

$$\left| \begin{matrix} V_L \\ E \end{matrix} \right|.$$

V.C.4 Incorporation of nonlinear loads in the TEF method

Incorporation of the nonlinear load models in the TEF method affects the following steps in the TEF procedure: i) the SEP and UEP solution, ii) conditions at fault clearing, iii) mode of disturbance evaluation, and iv) energy margin computation. These steps are modified as follows.

1. SEP and UEP solution:
   The electrical power for generator i becomes

$$P_{ei} = P'_{ei} + E_i I_{GLi} \cos(\theta_i - \phi_i)$$ (5.21)

   where

   $P'_{ei} = $ electrical power in classical network model

   $I_{GLi} \underline{/\phi_i} = $ complex current, representing nonlinear loads, reflected at generator i

   The SEP and UEP solutions are obtained iteratively from

$$P_{mi} - P_{ei} - \frac{M_i}{M_T} P_{COI} = 0 \quad i = 1, 2, ..., n$$ (5.22)

   where $P_{COI} = \sum_{i=1}^{n} (P_i - P_{ei})$

2. Conditions at fault clearing
   The injected currents, at the generator terminals, are obtained using the faulted Y_{BUS}. The constant acceleration assumption is also used during each time step in the fault-on period.

3. Mode of disturbance evaluation
   In the MOD, the effect of the nonlinear load model comes in the SEP solution, which affects the ray point solution.

4. Energy margin computation
   The expression for the energy margin is obtained using a linear trajectory approximation [17]:

$$\Delta V = -\frac{1}{2} M_{eq} \tilde{\omega}_{eq}^{cl\,2} - \sum_{i=1}^{n} P_i(\theta_i^u - \theta_i^{cl}) - \sum_{i=1}^{n-1} \sum_{j=i+1}^{n} \{ C_{ij}(\cos\theta_{ij}^u $$

$$-\cos\theta_{ij}^{cl}) - \left[\frac{\theta_i^u + \theta_j^u - \theta_i^{cl} - \theta_j^{cl}}{\theta_{ij}^{cl} - \theta_{ij}^{cl}} \right] D_{ij} \left[\sin\theta_{ij}^u - \sin\theta_{ij}^{cl} \right] \}$$

$$+ \sum_{i=1}^{n} \int_{\theta_i^{cl}}^{\theta_i} E_i I_{GLi} \cos(\theta_i - \phi_i)\, d\theta_i \qquad (5.23)$$

All the terms in (5.23) have been previously defined. The last term, which is path dependent, is also evaluated assuming a linear trajectory between θ^{cl} and θ^u. Several points are selected on this trajectory. At each point, the injected currents I_{GL} are determined using the procedure in the previous section and the values of θ.

V.D Incorporation of Two-Terminal HVDC in the Transient Energy Function Method

The process of incorporating the two-terminal HVDC in the TEF method involves the following steps:

1. Development of a simplified two-terminal HVDC model suited for first swing transient; and a method to determine the dc control mode (when this model is used).

2. Development of a procedure to interface the ac/dc systems. The procedure assumes that the ac network is modeled classically.

3. Incorporation of the HVDC model in the expression for the transient energy.

4. Introducing a procedure for transient stability assessment using the new TEF expression.

For a detailed description of steps 1 and 2, we refer the readers to references [33,15]. Steps 3 and 4 are described below.

V.E Transient Energy with the HVDC Model

The effect of the two-terminal HVDC on the expression for the transient energy will be only in the calculation of the potential energy. Since the dc converter buses are retained, new terms which relate to the converter bus voltages should be included.

The new transient energy function is developed as follows. For generator j, the accelerating power in the COI coordinate is given by

$$P_{acc,j}(COI) = P_{mj} - P_{ej} - \frac{M_j}{M_T} P_{COI} \tag{5.24}$$

where P_m, P_e are the mechanical and electrical power respectively,

$$P_{COI} = \sum_{j=1}^{NR} (P_{mj} - P_{ej}),$$

M = rotor inertia, $M_T = \sum_{j=1}^{NR} M_j$, and NR is the number of rotors.

For the classical power system model, the rotor accelerating power is given by

$$P_{acc,j} = P_j - \sum_{\substack{k \neq 1 \\ k=j}}^{NR} (C_{jk}\sin\theta_{jk} + D_{jk}\cos\theta_{jk})$$

$$- \sum_{\ell \in 2ND} (C_{j\ell}\sin\theta_{j\ell} + D_{j\ell}\cos\theta_{j\ell}) \tag{5.25}$$

where $\qquad\qquad P_j = P_{mj} - E_j^2 G_{jj}$

$$C_{jk} = E_j E_k B_{jk}, \ D_{jk} = E_j E_k G_{jk}$$

$$C_{j\ell} = E_j V_\ell B_{j\ell}, \ D_{j\ell} = E_j V_\ell G_{j\ell}$$

$$\theta_{jk} = \theta_j - \theta_k, \ \theta_{j\ell} = \theta_j - \theta_\ell$$

ND = number of two-terminal dc lines.

2ND: number of converters. (2ND = 2 x ND)

j,k are generator internal nodes.

ℓ is converter bus.

Using the stable equilibrium point (SEP) as reference, the energy function at any point along the trajectory (denoted as *) is defined as:

$$V\Big|_s^{\cdot} \overset{\Delta}{=} \sum_{j=1}^{NR} \frac{1}{2}M_j\tilde{\omega}_j^{\cdot 2} - \sum_{j=1}^{NR} P_j(\theta_j^{\cdot}-\theta_j^s)$$

$$+\int_s^{\cdot} \sum_{j=1}^{NR} \sum_{\substack{k\neq j \\ k=1}}^{NR} (C_{jk}\sin\theta_{jk}+D_{jk}\sin\theta_{jk})d\theta_j$$

$$+\int_s^{\cdot} \sum_{j=1}^{NR} \sum_{l\in 2ND} (C_{jl}\sin\theta_{jl}+D_{jl}\cos\theta_{jl})d\theta_j \tag{5.26}$$

We note that in equation (5.26), the last terms are the dc converter bus voltage-related magnetic and dissipative potential energy contributions.

Since some of the terms in (5.26) are path-dependent, a linear trajectory approximation is used to evaluate these terms.

The resulting TEF expression is given below.

$$V= \sum_{i=1}^{n} \frac{1}{2}M_i\tilde{\omega}_i^2 - \sum_{i=1}^{n} P_i(\theta_i-\theta_i^s) - \sum_{i=1}^{n-1} \sum_{j=i+1}^{n} C_{ij}(\cos\theta_{ij}-\cos\theta_{ij}^s)$$

$$+\sum_{i=1}^{n-1} \sum_{j=i+1}^{n} I_{ij}(\sin\theta_{ij}-\sin\theta_{ij}^s)$$

$$-\sum_{i=1}^{n} E_i(\theta_i-\theta_i^s) \sum_{l\in 2ND} \frac{B_{il}}{\theta_{il}-\theta_{il}^s}\Bigg[\Big[V_l\cos\theta_{il}-V_l^s\cos\theta_{il}^s\Big]$$

$$-\frac{V_l-V_l^s}{\theta_{il}-\theta_{il}^s}(\sin\theta_{il}-\sin\theta_{il}^s)\Big]$$

$$+\sum_{i=1}^{n} E_i(\theta_i-\theta_i^s) \sum_{l\in 2ND} \frac{G_{il}}{\theta_{il}-\theta_{il}^s}\Bigg[\Big[V_l\sin\theta_{il}-V_l^s\sin\theta_{il}^s\Big]$$

$$+\frac{V_l-V_l^s}{\theta_{il}-\theta_{il}^s}\Big[\cos\theta_{il}-\cos\theta_{il}^s\Big]\Bigg] \tag{5.27}$$

where superscript s refers to stable equilibrium condition

  n =   no. of generators in ac system

  ND = no. of HVDC terminals

2ND = no. of converters

subscript i,l refer to generator and converter respectively and

$$I_{ij}=D_{ij}\frac{\theta_i+\theta_j-\theta_i^s-\theta_j^s}{\theta_{ij}-\theta_{ij}^s}$$

Equilibrium conditions:

Stable equilibrium conditions (θ^s) and the unstable equilibrium conditions (θ^u) are obtained from

$$P_{mi}-P_{ei}-\frac{M_i}{M_T}P_{COI}=0$$

$$i=1,2,...,n-1$$

$$P_{el}^*-P_{el}=0 \qquad\qquad (5.28)$$

$$l\in 2ND$$

$$Q_{el}^* - Q_{el} = 0$$

where

$P_{el},Q_{el}=$ dc power and reactive power injections from the converter buses to the ac system calculated using bus voltage and the Y-matrix

$P_{el}^*,Q_{el}^*=$ dc power and reactive power injections calculated from the dc part load flow

V.E.1 Procedure for transient stability assessment

Transient stability assessment of an ac/dc system using TEF is accomplished by a six-step procedure similar to that of an ac system:

Step 1:   The generator rotor angles and speeds at fault clearing time are obtained by time simulation.

Step 2:   Starting with the pre-fault equilibrium point, solution for the post-fault stable equilibrium point (SEP) is obtained by a suitable solution technique, e.g., the Newton-Raphson method, with the same dc model as in Step 1.

Step 3: The same procedure as in the ac system TEF method is used to determine the mode of instability and to obtain the starting point for calculation of the UEP.

Step 4: Similar to step 2, the controlling UEP is solved by using the starting point obtained from Step 3.

Step 5: The kinetic energy at clearing time and the potential energy margin are computed.

Step 6: Transient stability assessment is made according to the value of the normalized potential energy margin.

VI. APPLICATIONS OF THE TEF METHOD

VI.A Introduction

The basic elements of transient stability assessment of a classically modeled power system, using the transient energy function method, have been described in Section IV. Stability information is obtained from the value of the transient energy margin ΔV and its normalized value ΔV_n.

In this section we will deal with what might be considered "simple" applications, i.e., when such complexities as multi-swing instability, complex disturbance scenarios, etc., are not of concern. Therefore, in this section we are concerned with first swing transient stability, i.e., whether the rotor angles of the severely disturbed generators reach a maximum and recover. The classical power system model is assumed.

VI.B Plant Mode vs. Inter-Area Mode Stability

In power system transient stability studies two conditions are encountered:

1. A very small group of generators (tightly coupled) swinging against the rest of the generators in the power system. The most common situation is when this group of generators are in the same plant, and the disturbance is electrically close to that plant. However, the following situations may be (and have often been) encountered: the group may include generators in neighboring plants, and when the generator most affected by a disturbance is not the one electrically closest to it.

We will call this type of transient "plant mode" transient stability. We note that such a study precludes the oscillations of the various generators in the group against each other (these oscillations are of interest primarily in the proper "tuning" of the automatic control equipment of these generators).

2. A large group of generators, not necessarily tightly coupled electrically, tending to split from the rest of the system. The boundary of separation may be large and may be electrically far from the disturbance. This is a relatively recent phenomenon which has been encountered in different parts of the North American interconnection. It has been associated with the heavier loading of the transmission network, especially under heavy power transfers between areas. We will call this type of transient "Inter-area Mode." When this phenomenon is encountered, the generators which appear to be most severely disturbed, initially experience "stable" oscillatory behavior. However, they may later on (typically 1-2 seconds later) join a larger group of generators in separating from the interconnected network.

VI.C Plant Mode Stability

As explained in the previous sections, in plant mode stability (or instability) a relatively small group of generators, electrically close to the disturbance location, are severely disturbed. When the disturbance is sufficiently large, some generators in this group separate from the rest of the generators in the system. This group of generators, called the critical group, is usually located at a plant or plants very close to the disturbance; and their number is small. This critical group is determined by the mode of disturbance (MOD) test. They form the generator grouping representing the minimum ratio of the network's potential energy absorbing capacity to the corrected kinetic energy associated with that group at the end of the disturbance.

In this section some of the issues commonly dealt with in plant mode stability are discussed.

VI.C.1 Degree of stability (or instability)

As previously explained, stability or instability at the end of the disturbance, are determined based on whether $\Delta V \gtrless o$. However, the question may arise as to how the severity of two disturbances, for the same power network and the same initial condition, may compare.

Early investigations revealed that the ratio of the transient energy margin to the corrected kinetic energy is indicative of the degree of stability or instability. This is called the *Normalized Energy Margin* ΔV_n.

$$\Delta V_n = \frac{\Delta V}{V_{KE \mid corr}} \tag{6.1}$$

Thus while for $\Delta V > o$ the system is stable and for $\Delta V < o$ the system is unstable, more information can be obtained from the value of ΔV_n. For example:

$$\Delta V_n \gtrless o, \text{ system is } \begin{cases} stable \\ unstable \end{cases} \tag{6.2a}$$

$$\Delta V_n^{(1)} > \Delta V_n^{(2)} > o, \qquad \text{system is more } \underline{\text{stable}} \text{ in scenario no. 1 than in scenario no. 2} \tag{6.2b}$$

and

$$\Delta V_n^{(1)} < \Delta V_n^{(2)} < o, \qquad \text{system is more unstable in scenario no. 1 than in scenario no. 2} \tag{6.2c}$$

To illustrate the significance of the normalized energy margin concept, the following data obtained from the 17–generator reduced Iowa network [34] is offered. Three-phase and single-line-to-ground faults at different locations on the 345-KV network, all cleared at 0.15 s, were analyzed using the TEF method. Values of the transient energy margin ΔV, and their normalized values ΔV_n were computed. The data, reproduced from Reference [34] is shown in Tables 6.1 and 6.2. The severity of the disturbance, as given by the values of ΔV_n, were found to be different for some cases from those given by the values of ΔV. For example, in Table 6.1, the fault at Cooper cleared by opening line 6-774 gives an energy margin $\Delta V = 5.2$, while the Raun fault cleared by opening line 3.72-482 gives $\Delta V = 9.159$. These values of ΔV may seem to indicate that the Raun fault is less severe than the Cooper fault. However, the values of ΔV_n of 1.332 and 1.647 respectively indicate that the opposite is true. Several cases in Tables 6.1 and 6.2 show that the fault severity as indicated by the relative values of ΔV_n may be different from that indicated by the values of ΔV.

To verify the relative severity of the faults as indicated by ΔV_n, time simulation results were obtained for the three-phase faults. These results also given in Table 6.1, confirm the validity of the relative severity of the disturbance as given by the values of ΔV_n, i.e., lower values of ΔV_n correspond to lower values of critical fault clearing times.

Table 6.1
Ranking of Three-Phase Faults
17-Generator System

Rank	$\dfrac{\Delta V}{\text{Corr. KE}}$	ΔV (pu)	Fault Location	Critical Clearing Time s Stable	Unstable
1	0.691	4.756	Raun, line 372–773	0.177	0.180
2	1.080	7.432	Raun, line 372–193	0.1923	0.1924
3	1.332	9.159	Raun, line 372–482	0.192	0.196
4	1.530	9.870	Raun, trans. 372–800	0.196	0.200
5	1.620	6.208	C.B. No. 3, line 436–771	0.200	0.204
6	1.647	5.200	Cooper, line 6–774	0.204	0.212
7	1.988	6.278	Cooper, line 6–16	0.212	0.216
8	2.059	6.502	Cooper, line 6–439	0.216	0.220
9	2.089	6.596	Cooper, line 6–393	---	---
10	3.283	12.579	C.B. No. 3, line 436–439	---	---

Table 6.2
Ranking of SLG Faults
17-Generator System

Rank	$\dfrac{\Delta V}{\text{Corrected KE}}$	ΔV (p.u.)	Disturbance Faulted Bus	Lines Removed
1	0.962	0.963	Cooper	6–774, 6–393
2	2.057	3.408	C.B. No. 3	436–771, 436–439
3	3.321	3.217	Cooper	6–439, 6–393
4	4.200	7.523	Raun	372–482, 372–193
5	4.319	7.879	Raun	372–193, 372–482
6	4.354	4.350	C.B. No. 3	436–439, 436–771
7	5.550	5.561	Cooper	6–774, 6–16
8	34.205	21.980	Ft. Calhoun	773–372, 773–775
9	54.379	27.097	Ft. Calhoun	773–779, 773–775

VI.C.2 Critical clearing time

For a given power system, a given initial operating condition, a given type of fault, and a given fault location, the critical fault-clearing time is "conceptually" very simple to determine. It corresponds to the fault clearing time at which the value of V given by

$$V = V_{cr} = V^u \tag{6.3}$$

This can be obtained (again, conceptually) by increasing the fault duration until the critical value of the transient energy V is reached. Alternatively, the critical value of V can be obtained by interpolation between two values slightly below and slightly above V_{cr}. The current TEF program, however, computes ΔV and ΔV_n. The critical fault clearing time is then given by

$$\Delta V = \Delta V_n = 0 \tag{6.4}$$

The recommended procedure is to obtain ΔV_n for two fault durations; by interpolation or extrapolation the fault clearing time corresponding to $\Delta V_n = 0$ is obtained. Because of the nonlinear nature of the relationship between fault clearing time and ΔV_n, it is important that at least one of the fault durations is selected to give $| \Delta V_n |$ approximately equal to zero.

VI.C.3 Power limits

This is typical of the type of practical answers sought by the analyst in a stability study performed in an operations planning environment. The power network condition is given, i.e., the load level, distribution of generation, transmission connectivity (which lines are not in service), etc. The breaker operating times are known (e.g., successful clearing at first relay zone, stuck breaker fault, etc.) The loading at a key power plant, or the power flow at a critical transmission interface (i.e., group of transmission lines connecting two areas of the network) may be transient stability-limited. The analyst then determines those limits and recommends that the system be operated at a safe margin below them.

To arrive at the desired answer using the TEF method, different stability runs are made until two cases are obtained; one for $\Delta V > 0$, and the other for $\Delta V < 0$. The power limit is obtained by interpolation, i.e., that corresponding to $\Delta V = 0$ (or $\Delta V_n = 0$).

To illustrate this procedure some data, obtained from reference [14], is given below. Tables 6.3 and 6.4 give the TEF results for the so–called Bruce 100–generator and the Nanticoke 50–generator networks. The power limits were obtained by interpolation, i.e., corresponding to $\Delta V = 0$.

Again, it should be pointed out that since an interpolation procedure is used, the data points should correspond to values of $| \Delta V |$ or $| \Delta V_n |$ close to zero, for the results to be considered accurate.

Table 6.3
TEF Results for Bruce - 100 Generator System

Case#	Bruce Output (MW)	Normalized Energy Margin	Comments
1	1500	0.150	Stable
2	1513	0.025	Stable
3	1514	–0.008	Unstable
4	1530	–0.150	Unstable

Stability Limit = 1514 MW

Table 6.4
TEF Results for Nanticoke - Normal System

Case#	Nanticoke Output (MW)	Normalized Energy Margin	Comments
1	3800	0.113	Stable
2	3840	0.001	Stable
3	3841	–0.002	Unstable
4	3900	–0.158	Unstable

Stability Limit = 3840.5 MW

VI.D The Inter-area Mode

Again as explained in Section VI.B, this phenomenon is associated with the tendency of a large group of generators to separate from the rest of the system. The separation boundary may occur away from the disturbance location. This phenomenon is associated with a stressed post disturbance transmission network due to heavy loading and/or large transmission impedances.

As with plant mode stability, analysis of stability limited conditions may take place in a variety of forms: stability classification with qualitative assessment of the degree of stability or instability; power limits at a plant or across a transmission interface; as well as other information on the generators which influence the dynamic system behavior. The analysis, however, is much more complex than with the case of plant mode instability.

VI.D.1 Stability classification

As previously stated, transient stability classification is based on the computed value of the normalized energy margin ΔV_n for the disturbance sequence under investigation. To obtain ΔV_n, the value of the corrected kinetic energy is required. An interesting aspect of the inter-area mode phenomenon is that the value of the corrected kinetic energy to be used is that corresponding to the group having the MAXIMUM $V_{KE\,|corr}$ for the advanced generators in the inter-area mode.

As an illustration of the above, for the stressed 50–generator system [13], the maximum $V_{KE\,|corr}$ corresponds to the two Nanticoke generators advanced (generator nos. 20 and 26). This value of $V_{KE\,|corr}$ is used to obtain the normalized energy margin ΔV_n. For a given system condition, i.e., a Nanticoke plant loading of 3800 MW, and a given disturbance, a value of $\Delta V_n = 0.001$ is obtained (indicating a critically stable condition). Time simulation results verify that the stability assessment using the TEF method is correct, although somewhat conservative (see Table 6.5 below).

VI.D.2 Power limits

Two examples, taken from reference [14], illustrate the use of the TEF method to obtain power limits for stability-limited conditions in a stressed power network.

In the same 50–generator test system discussed above, the stability-limited loading of the Nanticoke power plant (generators nos. 20,26) is obtained. The results are shown in Table 6.5.

Table 6.5
Stability Limited Plant Output
50–Generator System

No.	Output of M/Cs #20,26 in MW	Normalized Energy Margin	Remarks
1	3750	0.1936	Stable
2	3800	0.001	Critically Stable
3	3801	-0.002	Critically Unstable
4	3900	-0.3397	Unstable

The results of Table 6.5 show that the TEF method predicts a limit of 3800 MW which compares very well with the limit of 3935 MW predicted by time simulation.

In another test system comprising of 115 generators, while the MOD test for the stressed conditions analyzed showed a corrected V_{KE} based on 19 generators advanced, the UEP solution gave an inter-area mode which included 57 machines with advanced rotor angles [13,14]. The stability limits was obtained in terms of total generation of six machines located at a critical plant. The limit obtained from time simulation was 4400 MW. Figure 6.1 shows the result of the TEF analysis on this system. Corresponding to a zero margin the limit obtained is 4450 MW, which gives excellent comparison with time simulation results.

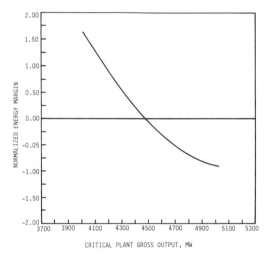

Figure 6.1. Results of TEF Analysis on 115 Generator System

VI.E Other Applications of the TEF Method

Some other applications of the TEF method have dealt with the following aspects:

i.   Loss of generation disturbance. We refer the reader to [35] for details.

ii. Determination of network conditions during transient [15].

iii. Determining operation of out–of–step impedance relays [15].

VI.F Energy Margin Sensitivity to Changes in System Conditions

VI.F.1 Introduction

In transient stability analysis of power systems the following question is often raised: given the stability results for a set of initial system conditions and disturbance scenario, how could these results be affected by a change in a certain key parameter or an initial system condition? The answer to this question is at the heart of many transient stability studies. To illustrate this point, let us assume that the analyst is seeking the transient stability limit for the loading of a particular power plant: i) by the conventional time simulation technique, and ii) by using the TEF method. If the former method is used, the analyst will select a convenient indicator of the severity of the disturbance, e.g., magnitude of the swing of a critical generator. Guided by this indicator, the analyst will conduct a series of transient stability runs until the limit is "bracketed" by two cases: a critically stable and a critically unstable case.

If the TEF method is used, we recall that an inherent advantage of the TEF method is the availability of a qualitative measure of the

If the TEF method is used, we recall that an inherent advantage of the TEF method is the availability of a qualitative measure of the degree of stability (or instability) in terms of the transient energy margin (ΔV). This was considered an important step toward reducing the number of transient stability runs needed to obtained transient stability information. It is likely therefore, that a fewer number of TEF runs will be needed to obtain the limit. The information, when displayed may look like that shown in Figure 6.2 where it is assumed that four TEF runs were made to determine the limit, which corresponds to the plant loading for $\Delta V = 0$.

The use of the sensitivity techniques in the TEF method is illustrated in Figure 6.2. If the TEF run is made for conditions shown as point a; and if the sensitivity of ΔV_a to change in P_a can be obtained with the desired accuracy, the limit can be determined quickly from only one stability run. This feature considerably increases the versatility of the TEF method.

Analytically derived sensitivity techniques for power system transient stability analysis are based on the techniques used with nonlinear systems, discussed in references [36-38]. For example, reference [38] proposed an analytical approach to energy margin sensitivity determination. In this approach, the transfer conductances were neglected and only self-clearing faults were considered. The technique discussed here is based on the authors' work discussed in [39].

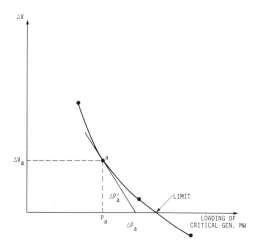

Figure 6.2. Determination of Stability-Limited Loading of a Plant
by Successive Use of the TEF Method

VI.F.2 Calculation of sensitivity factors

For the classical power system model, with n-machines, the transient energy margin is given by (4.9) repeated here for convenience

$$\Delta V = -\frac{1}{2} M_{eq} \tilde{\omega}_{eq}^{cl^2} - \sum_{i=1}^{n} P_i \left[\theta_i^u - \theta_i^{cl}\right] - \sum_{i=1}^{n-1} \sum_{j=i+1}^{n} \left[C_{ij}\left[\cos\theta_{ij}^u - \cos\theta_{ij}^{cl}\right]\right.$$
$$\left. - \frac{(\theta_i^u + \theta_j^u - \theta_i^{cl} - \theta_j^{cl})}{\theta_{ij}^u - \theta_{ij}^{cl}} D_{ij}\left[\sin\theta_{ij}^u - \sin\theta_{ij}^{cl}\right]\right] \tag{6.5}$$

where all the parameters are defined in Section IV.

In equation (6.5), the parameters P_i, C_{ij}, and D_{ij} are all evaluated with respect to the post disturbance network.

From equation (6.5), assuming that the controlling UEP does not change, it is seen that the energy margin ΔV is a function of several variables which can be expressed as

$$\Delta V = \Delta V\left[\theta^{cl}, \tilde{\omega}^{cl}, \theta^u, \left[P_i, i = 1, 2, ..., n\right],\right.$$
$$\left. \left[E_i E_j B_{ij}, E_i E_j G_{ij}, i, j = 1, 2, ..., n\right]\right] \tag{6.6}$$

VI.F.2.1 Functional dependence on system parameters

The effect of the change of a given system parameter α_k on the energy margin (ΔV) is obtained from the partial derivatives of (ΔV) with respect to the system parameters, since in general a change in a given parameter introduces other changes in the system conditions. The process is illustrated below (see [37,38] for a detailed treatment of the subject).

Let $\Delta V = \Delta V\,(\alpha_1, \alpha_2, ..., \alpha_N)$ where the α's are the relevant system parameters of interest. In general a change in ΔV is given by

$$\Delta(\Delta V) \simeq \sum_{k=1}^{N} \frac{\partial(\Delta V)}{\partial \alpha_k} \Delta \alpha_k + \frac{1}{2} \sum_{j=1}^{N} \sum_{k=1}^{N} \frac{\partial^2(\Delta V)}{\partial \alpha_j \partial \alpha_k} \Delta \alpha_j \Delta \alpha_k \qquad (6.7)$$

$$+ \text{ higher order terms}$$

The partial derivatives $\dfrac{\partial(\Delta V)}{\partial \alpha_k}$ must be strictly defined in terms of limit operations such as

$$\frac{\partial(\Delta V)}{\partial \alpha_k} = \lim_{\Delta \alpha_k \to 0} \frac{\Delta V(\alpha_k + \Delta \alpha_k) - \Delta V(\alpha_k)}{\Delta \alpha_k}$$

and

$$\frac{\partial(\Delta V)}{\partial \alpha_j\, \partial \alpha_k} = \lim_{\Delta \alpha_j \to 0} \lim_{\Delta \alpha_k \to 0} \frac{\Delta V(\alpha_{jk}) - \Delta V(\alpha_j) - \Delta V(\Delta \alpha_k) + \Delta V(\alpha_k)}{\Delta \alpha_j \Delta \alpha_k}$$

$$(6.8)$$

where

$$\Delta V(\Delta \alpha_{jk}) = \Delta V(\Delta \alpha_j,\, \Delta \alpha_k,\, \alpha_k) \qquad (6.9)$$

VI.F.3 First order sensitivity of ΔV

By using the first term in (6.7) only the first order sensitivity factor of the energy margin to a change in any parameter α_k, is given by the partial derivative of ΔV with respect to α_k. Applying the chain rule of differentiation, this can be obtained as

$$\frac{\partial(\Delta V)}{\partial \alpha_k} = -M_{eq}\, \tilde{\omega}_{eq}^{cl} \cdot \frac{\partial \left| \omega_{eq}^{cl} \right|}{\partial \alpha_k} - \sum_{i=1}^{n} \left[P_i \left(\frac{\partial \theta_i^u}{\partial \alpha_k} - \frac{\partial \theta_i^{cl}}{\partial \alpha_k} \right) \right.$$

$$+ \left[\theta_i^u - \theta_i^{cl} \right] \frac{\partial P_{mi}}{\partial \alpha_k} - \left[\theta_i^u - \theta_i^{cl} \right] \left[E_i^2 \frac{\partial G_{ii}}{\partial \theta_k} + 2E_i G_{ii} \frac{\partial E_i}{\partial \alpha_k} \right] \right]$$

$$- \sum_{i=1}^{n-1} \sum_{j=i+1}^{n} \left[\left(\cos\theta_{ij}^u - \cos\theta_{ij}^{cl} \right) \frac{\partial C_{ij}}{\partial \alpha_k} - B_{ij} \left(\sin\theta_{ij}^u - \sin\theta_{ij}^{cl} \right) \frac{\partial D_{ij}}{\partial \alpha_k} \right]$$

$$+ \sum_{i=1}^{n-1} \sum_{j=i+1}^{n} \left[\left(B_{ij} D_{ij} \cos\theta_{ij}^u + C_{ij} \sin\theta_{ij}^u \right) \frac{\partial \theta_i^u}{\partial \alpha_k} - \frac{\partial \theta_j^u}{\partial \alpha_k} \right]$$

$$- \left(B_{ij} D_{ij} \cos\theta_{ij}^{cl} + C_{ij} \sin\theta_{ij}^{cl} \right) \left[\frac{\partial \theta_i^{cl}}{\partial \alpha_k} - \frac{\partial \theta_j^u}{\partial \alpha_k} \right] \right]$$

$$+ \sum_{i=1}^{n-1} \sum_{j=i+1}^{n} D_{ij} \left(\sin\theta_{ij}^u - \sin\theta_{ij}^{cl} \right) \left[\gamma_{ij}^a \left(\frac{\partial \theta_i^u}{\partial \alpha_k} - \frac{\partial \theta_j^{cl}}{\partial \alpha_k} \right) + \gamma_{ij}^b \left(\frac{\partial \theta_j^u}{\partial \alpha_k} - \frac{\partial \theta_j^{cl}}{\partial \alpha_k} \right) \right]$$

$$(6.10)$$

where

$$\gamma_{ij}^a = \frac{-2 \left[\theta_j^u - \theta_j^{cl} \right]}{\left[\theta_{ij}^u - \theta_{ij}^{cl} \right]^2} , \quad \gamma_{ij}^b = \frac{2 \left[\theta_i^u - \theta_i^{cl} \right]}{\left[\theta_{ij}^u - \theta_{ij}^{cl} \right]^2} , \quad B_{ij} = \frac{\left[\theta_i^u + \theta_j^u - \theta_i^{cl} - \theta_j^{cl} \right]}{\left[\theta_{ij}^u - \theta_{ij}^{cl} \right]}$$

$$\frac{\partial C_{ij}}{\partial \alpha_k} = E_i E_j \frac{\partial B_{ij}}{\partial \alpha_k} + E_i B_{ij} \frac{\partial E_j}{\partial \alpha_k} + E_j B_{ij} \frac{\partial E_i}{\partial \alpha_k}$$

$$\frac{\partial D_{ij}}{\partial \alpha_k} = E_i E_j \frac{\partial G_{ij}}{\partial \alpha_k} + E_i G_{ij} \frac{\partial E_j}{\partial \alpha_k} + E_j G_{ij} \frac{\partial E_i}{\partial \alpha_k} \qquad (6.11)$$

In determining the energy margin sensitivity via (6.6), the sensitivity of a number of dependent variables to the parameter α_k also needs to be determined. These include:

$$\frac{\partial \theta_i^{cl}}{\partial \alpha_k}, \quad \frac{\partial \tilde{\omega}_i^{cl}}{\partial \alpha_k}, \quad \text{and} \quad \frac{\partial \theta_i^u}{\partial \alpha_k} \qquad i = 1, 2, ..., n$$

The sensitivity of the conditions at the end of the disturbance to the change in parameters α_k is obtained using dynamic sensitivity equations [37,38]. These are computed by considering the system dynamic equations [37] for the *disturbed* system and considering partial derivatives as follows:

$$M_i \frac{\partial \dot{\tilde{\omega}}_i}{\partial \alpha_k} = \frac{\partial P_{mi}}{\partial \alpha_k} - 2E_i \frac{\partial E_i}{\partial \alpha_k} G_{ii}^f - E_i^2 \frac{\partial G_{ii}^f}{\partial \alpha_k}$$

$$- \sum_{\substack{j=1 \\ j \neq i}}^{n} \left[C_{ij}^f \cos\theta_{ij} \left(\frac{\partial \theta_i}{\partial \alpha_k} - \frac{\partial \theta_j}{\partial \alpha_k} \right) - D_{ij}^f \sin\theta_{ij} \left(\frac{\partial \theta_i}{\partial \alpha_k} - \frac{\partial \theta_j}{\partial \alpha_k} \right) \right]$$

$$- \sum_{\substack{j=1 \\ j \neq i}}^{n} \left[\left(\frac{\partial E_i}{\partial \alpha_k} E_j + \frac{\partial E_j}{\partial \alpha_k} E_i \right) \left(B_{ij}^f \sin\theta_{ij} + G_{ij}^f \cos\theta_{ij} \right) \right]$$

$$- \sum_{\substack{j=1 \\ j \neq i}}^{n} \left[E_i E_j \sin\theta_{ij} \frac{\partial B_{ij}^f}{\partial \alpha_k} + E_k E_j \cos\theta_{ij} \frac{\partial G_{ij}^f}{\partial \alpha_k} \right]$$

$$- \frac{M_i}{M_T} \left[\sum_{j=1}^{n} \left(\frac{\partial P_{mj}}{\partial \alpha_k} - 2E_j \frac{\partial E_j}{\partial \alpha_k} G_{jj}^f - E_j^2 \frac{\partial G_{jj}^f}{\partial \alpha_k} \right) \right.$$

$$+ \sum_{i=1}^{n} \sum_{\substack{j=1 \\ j \neq i}}^{n} \left[D_{ij}^f \sin\theta_{ij} \left(\frac{\partial \theta_i}{\partial \alpha_k} - \frac{\partial \theta_j}{\partial \alpha_k} \right) \right.$$

$$\left. - \left(\frac{\partial E_i}{\partial \alpha_k} E_j + \frac{\partial E_j}{\partial \alpha_k} E_i \right) G_{ij}^f \cos\theta_{ij} - E_i E_j \cos\theta_{ij} \frac{\partial G_{ij}^f}{\partial \alpha_k} \right] \right]$$

$$\frac{\partial \dot{\theta}_i}{\partial \alpha_k} = \frac{\partial \tilde{\omega}_i}{\partial \alpha_k} \qquad i = 1, 2, \ldots, n-1 \qquad (6.12)$$

The above is a system of linear differential equations which can be integrated knowing the values of θ_i and $\tilde{\omega}_i$, i=1,...,n at each time step which are available from the procedure used to determined the conditions at the end of the disturbance in the base case. Thus,

$$\frac{\partial \theta_i^{cl}}{\partial \alpha_k}, \frac{\partial \tilde{\omega}_i^{cl}}{\partial \alpha_k} \qquad i = 1, 2, \ldots, n$$

can be determined.

The sensitivity of the controlling UEP to the change in parameters α_k is determined as follows. Consider the dynamic sensitivity equations corresponding to (6.12) for the post disturbance network. At the UEP, the derivatives of the variables will be zero resulting in a set of linear algebraic equations given by:

$$0 = \frac{\partial P_{mi}}{\partial \alpha_k} - 2E_i \frac{\partial E_i}{\partial \alpha_k} G_{ii} - E_i^2 \frac{\partial G_{ii}}{\partial \alpha_k}$$

$$- \sum_{\substack{j=1 \\ j \neq i}}^{n} \left[C_{ij} \cos\theta_{ij}^u \left(\frac{\partial \theta_i^u}{\partial \alpha_k} - \frac{\partial \theta_j^u}{\partial \alpha_k} \right) - D_{ij} \sin\theta_{ij}^u \left(\frac{\partial \theta_i^u}{\partial \alpha_k} - \frac{\partial \theta_j^u}{\partial \alpha_k} \right) \right]$$

$$- \sum_{\substack{j=1 \\ j \neq i}}^{n} \left[\left(\frac{\partial E_i}{\partial \alpha_k} E_j + \frac{\partial E_j}{\partial \alpha_k} E_i \right) \left(B_{ij}^f \sin\theta_{ij} + G_{ij}^f \cos\theta_{ij} \right) \right]$$

$$- \sum_{\substack{j=1 \\ j \neq i}}^{n} \left[\left(E_i E_j \sin\theta_{ij}^u \frac{\partial B_{ij}}{\partial \alpha_k} + E_i E_j \cos\theta_{ij}^u \frac{\partial G_{ij}}{\partial \alpha_k} \right) \right.$$

$$- \frac{M_i}{M_T} \left[\sum_{j=1}^{n} \left(\frac{\partial P_{mj}}{\partial \alpha_k} - 2E_j \frac{\partial E_j}{\partial \alpha_k} G_{jj} - E_j^2 \frac{\partial G_{jj}}{\partial \alpha_k} \right) \right.$$

$$+ \sum_{i=1}^{n} \sum_{\substack{j=1 \\ j \neq i}}^{n} \left[D_{ij} \sin\theta_{ij}^u \left(\frac{\partial \theta_i^u}{\partial \alpha_k} - \frac{\partial \theta_i^u}{\partial \alpha_k} \right) - \left(\frac{\partial E_i}{\partial \alpha_k} E_j + \frac{\partial E_j}{\partial \alpha_k} E_i \right) G_{ij} \cos\theta_{ij}^u \right.$$

$$\left. \left. - E_i E_j \cos\theta_{ij}^u \frac{\partial G_{ij}}{\partial \alpha_k} \right] \right] \qquad i=1,2,...,n-1 \qquad (6.13)$$

Equation (6.13) is solved knowing θ^u to obtain

$$\frac{\partial \theta_i^u}{\partial \alpha_k}, \quad i=1,2,...,n$$

It is to be noted that in the COI frame of reference only $(n-1)$ of the variables

$$\frac{\partial \theta_i^{cl}}{\partial \alpha_K}, \frac{\partial \theta_i^u}{\partial \alpha_k}, \text{ and } \frac{\partial \tilde{\omega}_i^{cl}}{\partial \alpha_k}, \text{ are independent.}$$

Equations similar to (6.13) are also solved with the predisturbance network parameters to obtain changes in the predisturbance operating point due to a change in parameter α_k. This is then used to obtain initial conditions for the system of equations (6.12).

The terms $\dfrac{\partial P_{mi}}{\partial \alpha_k}$ $i=1,2,...,n$ are calculated if the change in parameter α_k affects the mechanical power input of the machine.

$\dfrac{\partial E_i}{\partial \alpha_k}$ $i=1,2,...,n$ are evaluated by determining the effect of α_k on the predisturbance terminal voltage, V_i of each generator. Knowing the complex power generation at each generator bus and the transient reactance, the change in internal voltage E_i is determined.

To determine the terms $\dfrac{\partial B_{ij}}{\partial \alpha_k}$ and $\dfrac{\partial G_{in}}{\partial \alpha_k}$, a procedure based on the Householder method (see Section 4 of [9]) is used to estimate the changes in Y_{BUS} parameters due to changes in α_k. This completes the determination of all parameters required to evaluate the energy margin sensitivity.

There could be a change in more than one parameters, e.g., α_k, $k=1,2,...,m$. In this case the sensitivities to all the parameters are derived.

In arriving at the change in energy margin due to changes in parameters, the following expression is used

$$\Delta(\Delta V) = \sum_{k=1}^{m} \frac{\partial(\Delta V)}{\partial \alpha_k}(\Delta \alpha_k)$$

(6.14a)

In arriving at (6.14a) it is assumed that only first order changes of ΔV with respect to the parameters α_k are considered.

We note that the expression for $\dfrac{\partial(\Delta V)}{\partial \alpha_k}$ may include several terms. If ΔV is a function of the parameters $(x_1, x_2, ..., x_m)$ which are functions of α_k, then

$$\frac{\partial(\Delta V)}{\partial \alpha_k} = \frac{\partial(\Delta V)}{\partial x_1}\frac{dx_1}{d\alpha_k} + \frac{\partial(\Delta V)}{\partial x_2}\frac{dx_2}{d\alpha_k} + \cdots + \frac{\partial(\Delta V)}{\partial x_m}\frac{dx_m}{d\alpha_k}$$

(6.14b)

Higher order sensitivities can also be derived. In [40] a procedure to determine second order sensitivities is presented.

VI.G First Order Sensitivity Results

First order sensitivity factors derived in Section VI.F were applied to the 17-generator test system [9]. Reference [39] presented data for obtaining the stability limits for three types of system changes: 1) plant generation (assuming total system generation and load are constant), 2) load changes, and 3) network changes. In incorporating these changes, it is assumed that the magnitude of the changes are small, and that the advanced generators in the MOD are unchanged. The results obtained using the sensitivity techniques are compared with those obtained with repeated stability analysis using the TEF method. A summary of the results for the plant generation change is given below.

1. Plant generation change

The procedure was applied to determine for a given fault:

i. The amount of generation to be increased at the critical generator, before instability is reached, i.e., when the energy margin (ΔV) becomes zero. This data is presented in Table 6.1.a.

ii. The amount of generation to be reduced at the critical generators to make an unstable case stable, i.e., to make $\Delta V = 0$. This data is presented in Table 6.1.b.

The change in generation is implemented at the generators which are severely disturbed, i.e., the advanced generators in the MOD. These generators provide a greater effect on the energy margin sensitivity.

The results in Table 6.1 show that the first order analytical sensitivity technique gives an accurate estimate of the stability limits, i.e., comparable to those obtained by repetitive application of the TEF analysis. The largest error is the Ft. Calhoun case for the generation shift at generator #17 where the maximum error is less than 4%.

Table 6.1 Generation limits obtained with first order sensitivity
technique – 17-generator system

a. Generation change limits to absorb margin

Fault Location	t^{cl} sec	Generation Change Limit; First Order Sensitivity	Generation Change Limit; Repetitive TEF Method	% Error
Ft. Cal. (Bus #773) (Clear 773-339)	0.342	$\Delta P_{16} = 37.88$ MW	$\Delta P_{16} = 37.64$ MW	-0.63
		$\Delta P_{17} = 55.94$ MW	$\Delta P_{17} = 54.88$	-1.9
Cooper (Bus #6) (Clear 6-774)	0.191	$\Delta P_2 = 57.51$ MW	$\Delta P_2 = 56.69$ MW	-1.4
C. Bluffs (Bus 436) (Clear 436-771)	0.187	$\Delta P_{10} = 45.85$ MW	$\Delta P_{10} = 45.95$	0.22
		$\Delta P_{12} = 37.54$ MW	$\Delta P_{12} = 37.38$ MW	-0.42

b. Generation change limits to stabilize system

Fault Location	t^{cl} sec	Generation Change Limit; First Order Sensitivity	Generation Change Limit; Repetitive TEF Method	% Error
Ft. Cal. (Bus #773) (Clear 773-339)	0.357	$\Delta P_{16} = -8.379$ MW	$\Delta P_{16} = -8.18$ MW	-2.4
		$\Delta P_{17} = -12.68$ MW	$\Delta P_{17} = -12.21$	-3.8
Cooper (Bus #6) (Clear 6-774)	1.51	$\Delta P_2 = -55.56$ MW	$\Delta P_2 = -54.65$ MW	-1.66
C. Bluffs (Bus 436) (Clear 436-771)	0.209	$\Delta P_{10} = -45.44$ MW	$\Delta P_{10} = -44.69$	-1.7
		$\Delta P_{12} = -35.94$ MW	$\Delta P_{12} = -35.52$ MW	-1.2

VII. CURRENT DEVELOPMENTS

VII.A Introduction

Some of the recent development efforts in the TEF method have been to improve the technique as a tool of power system transient stability analysis. The majority of the development efforts, however, have focused on using the method in new applications. In contrast with the "simple" applications discussed in the previous section, the new applications tend to involve a higher degree of sophistication: either in the novelty of the application itself, or in combining two emerging technologies to deal with a given power system problem.

A brief review of the current development efforts is given in this section. They are grouped into the two broad categories discussed above: 1) development of the tool of analysis, and 2) the application of the tool.

VII.B Development of TEF as a Tool

A great deal of R & D work has been devoted to this area in the 1980's. Much of this work has been discussed in the previous sections. It includes: a) modeling improvements, e.g., higher order generator models, nonlinear loads, HVDC, etc., b) more useful information for analysis of transient system response, e.g., sensitivity to changes in system parameters and/or operating conditions, information on the network during the transient, etc., and c) dealing with issues related to reliability, accuracy and speed of solving realistic modern power networks. Current R & D efforts seem to be in the following areas.

VII.B.1 Consolidating recent improvements

While substantial improvements have been achieved in the above mentioned R & D efforts, they have mostly been separate efforts, each dealing with a specific issue or problem. A considerable amount of the current efforts, therefore, seem to be focused on consolidating and integrating the above developments into a coherent TEF program suited for use by industry engineers for analysis of realistic modern power networks. This program should be robust, reliable, accurate, and competitive in speed of solution. It should be capable of solving a power system of a size in the order of 200-300 generators, adequately modeled to deal with "first swing" transient issues associated with today's heavily loaded transmission. The direct stability analysis program called DIRECT, distributed by the Electric Power Research Institute (EPRI) reflects the latest achievements in these efforts [41].

VII.B.2 Parallel computation

Transient stability analysis of modern power systems usually requires numerous runs involving many scenarios: different initial network configurations and operating conditions, different disturbances, different post disturbance system configurations and so on. Especially when the stability analysis is performed to support operating functions, there is a strong motivation to perform the stability analysis quickly. In a recent investigation [42] the issue of how to expedite the transient stability analysis of a stream of runs by parallel computation to enhance the rate of processing the information and computation was explored. In this investigation power system transient stability analysis with the TEF method was conducted using the parallel computer architecture on the IBM 3090-600E super computer at Cornell University. In this work, the TEF program code was modified to suit the parallel environment, and the parallelism in the disturbance scenarios was exploited. Significant improvements in the speed of computations were reported.

VII.C Corrective Action to Prevent Instability

Power systems in North America are designed and operated to meet reliability criteria established by the North American Electric Reliability Council and the appropriate Regional Reliability Councils. When these criteria are adhered to, stability limits may be encountered in some systems. These limits may be in the form of a limit on the loading of a certain power plant, on the power flow on a group of transmission lines, on the voltage at a given bus, etc. When these conditions are recognized, engineers supporting system operations will have to plan actions such that the reliability criteria would not be violated. These actions can be in the form of: a) rescheduling the distribution of plant generations and transmission line flows, or if this solution is deemed undesirable (e.g., for economic reasons), b) apply discrete controllers to reduce the "shock" to the system when disturbances occur.

VII.C.1 Predisturbance action

Let us consider the rescheduling of generation and transmission flows such that the reliability criteria would be met if the contingencies causing potential instability were to occur. If this answer is deemed desirable, the engineers are still faced with the question of how to do this rescheduling most effectively.

The problem is formulated as follows. Given an initial operating condition and a designated contingency (or disturbance sequence) which would make the system unstable, the changes in the initial conditions which can be implemented to make the system stable (for the same contingency) are to be identified. In the TEF terminology, the contingency for the original system conditions results in a negative energy margin. The changes in the initial conditions sought are those which would make the energy margin zero or positive. The answer to this problem can be simple or exceedingly complex, depending on the type changes to be implemented. References [43–44] have dealt with this problem when limits on plant loadings as well as on flows on critical transmission circuits are included.

The problem is formulated mathematically as follows:

Given a vector of decision variables x, $x^T = [x_1, x_2, \ldots, x_n]$, and a vector of objective functions f, $f^T = [f_1(x), f_2(x), \ldots, f_k(x)]$ a solution x^* is sought such that $f(x^*) = opt.\ f(x)$.

Subject to constraints

$$\text{inequality:} \quad g_j(x) \geq 0 \quad j = 1,2,\ldots,m \qquad (7.1)$$

$$\text{equality:} \quad h_j(x) = 0 \quad j = 1,2,\ldots,p < n \qquad (7.2)$$

This is an optimization problem to which the traditional solution approach is to seek what is known as a number of Pareto Optimums, from which the final solution is selected.

Among the commonly used techniques to obtain Pareto optimal solution(s) of a vector function are which function scalarization techniques. The vector function $f(x)$ is transformed into a scalar function $f(x)$. The techniques most commonly used are:

a.   To use weighting coefficients, e.g.,

$$f(x) = \sum_{i=1}^{k} w_i f_i(x) \qquad (7.3)$$

The weighting coefficients assign different importance to the different objective functions.

b. Another method of scalarization is to use hierarchical optimization. In this method the criteria to be satisfied are ordered according to their importance, with the most important on top of the list, then the next important, and so on. Each objective function is minimized separately and in the order of importance.

VII.C.2 Emergency actions

If preventive action (e.g., rescheduling of generation) is too costly, emergency control actions are used to prevent cascading outages. the most common of these controls are discussed below.

a. Generation tripping

Generation tripping aims at reducing the excess transient energy, tending to separate a group of generators from the rest of the system. It is a rather drastic measure which is resorted to when the electrical power is produced at a level too high to maintain stability because of the fault on the limited capacity transmission corridor. When generation tripping is used there are several questions which need to be resolved. Among them are the following: 1) when to initiate the tripping, 2) how much generation to trip, and 3) whether it is to be used with other controls. It should be noted that the amount of generation to be tripped has to be in blocks of power equal to generation on specific units.

When generation tripping is used, the desired change in the energy margin is estimated by computing the change in θ^u and M_{eq}. These parameters in turn introducing changes in the kinetic energy at clearing as well as in the various components of the potential energy.

b. Load rejection

Assuming that the load is represented by constant impedance, load rejection is modeled by a sudden change in the impedance connected to some load bus, which in turn represents a sudden change in some elements of the Y-bus, i.e., in some of the G_{ii}, C_{ij}, and D_{ij} terms. This will result in a change in θ^u. This will cause a change in the potential energy which can be readily computed.

c. Braking resistor insertion

The effect of the insertion of a braking resistor is analogous to that of load rejection (but acts in the opposite direction). The change in (ΔV) is estimated by the same procedure as above: first as a change in the Y-matrix elements, then as a corresponding change in $\Delta \theta_i^u$ and (ΔV). The situation is more complicated, however, by the fact that the results will be affected by the size and location of the resistor, and by the time and duration of its insertion.

In reference [45] an application of the TEF method is used in a transient which involves generation tripping and braking resistor insertion.

VII.D Dynamic Security Assessment

As has been previously stated, it is getting increasingly difficult to study the combinations of outages and disturbances to ensure that NERC reliability criteria will be met at all times. For stability-limited power systems the number of scenarios to be analyzed for prevention of cascading outages would be very large. This is an important characteristic of the problem of the prevention of cascading outages: identifying the safe operating regimes requires analysis of a large number of combination of outages and processing and manipulating of (what amounts to) a massive amount of data. Therefore, dynamic security assessment has become one of the main concerns in several parts of the North American interconnected system.

VII.D.1 On-line derivation of stability limits

Reference [46] discusses the on-line derivation limits needed for an essentially radial portion of the Ontario Hydro (O.H.) network connecting the Bruce generating station to the rest of the O.H. network and the load and generation tripping scheme to maintain stability. The stability assessment load and generation rejection scheme is as follows. A manageably large number of post-contingency transmission configurations are selected. For each of these configurations, accurate stability limits are determined by detailed off-line studies for different pre-contingency generation levels. These are tabulated for on-line usage and are designed to cover the majority of the actual situations to be encountered.

The transient stability assessment scheme described in [46] is based on the equal area criterion of stability. At present the scheme is being adapted for use with direct energy-based methods including TEF. The adaptation involves a two-step process: i) stability assessment for current condition, and ii) use of energy margin sensitivity to obtain the stability limit.

VII.D.2 Use of expert system in dynamic security assessment

It has become apparent that, for stability-limited power networks, conventional stability analysis tools are presenting some problems to engineers supporting operating functions. Each stability study analyzes only one scenario (or disturbance sequence) at a time when many such studies are needed. The amount of the data required are too voluminous to be generated quickly, and needs good organization and classification for swift decision-making regarding security. Thus, it is sometimes felt that operations planning functions are supported by tools that are inadequate for the job at hand. What is needed are tools which would: i - allow quick determination of stability limits, ii - have the ability to compute the effect of changing system parameters and/or conditions on these limits, iii - and are suited for organization for group classification. Such tools would bring security-related analysis and decision-making closer to real time, and exploit the special knowledge about the power network which the system operators have. The new tools call for an imaginative use of a combination of the TEF method and expert systems to obtain a new framework for dynamic security assessment.

The TEF method attributes which are particularly valuable to the issue at hand are: the method's ability to express the degree of stability in terms of a quantifiable index (the transient energy margin ΔV); and its flexibility to obtain analytical sensitivity information, for the simulated contingencies, on how the energy margin is affected by changing system parameters or conditions.

The use of the expert systems technique offers the opportunity to capture the thinking of the power system operators and how they view power system security (and how they deal with it). In addition, the technique gives the analyst a good framework for organizing a large amount of data to provide information on: trends in system security, and in grouping of system dynamic behavior patterns for classification purposes for decision making.

Recent research efforts [47,48] have dealt with the above questions.

VII.E System Planning Applications

Use of the TEF method in applications involving planning of new facilities has been slow to come by. Among the applications explored in the past was the use of the TEF method to study the effects of various transmission reinforcement plans on the system transient stability (see Section 7.5.2 of [13]). The study assumed the addition of a number of HV transmission circuits to a certain area of the Ontario Hydro system. The affect of these additions on the stability limit of a certain power plant was investigated.

The following are two promising areas which are under investigation.

VII.E.1 Use of TEF as a screening tool

Although TEF is a tool for analysis of first swing transients, it is being investigated for use in system planning studies involving multi-swing transients. The idea is to identify, in a stream of runs involving many cases (typically hundreds), the critical cases to study in detail by conventional time domain stability programs. Certain attributes of the TEF method detect symptoms of stressed system condition (see Section IV) . Although these attributes are detected in the first swing, they are indicative of system conditions which may result in instability later on in the transient.

VII.E.2 Use of TEF in reliability computations

The aspect of power system reliability of interest here is the "ADEQUACY, which is the ability of the bulk power electric system to supply the aggregate electrical power and energy requirements of the consumers at all times, taking into account scheduled and unscheduled outages of system components." Probablistic indices are used as reliability measures, and reliability criteria are based on these indices.

One of the methods used to obtain reliability indices is the "state enumeration" method (see Chapter 4 of [49]) . In this method the states of the system under investigation are assumed to be known, along with the rates of the transitions between them. From this data the long-term probability, frequency, and mean duration of system failure are computed. For a stability-limited power network, reliability computation is proposed to be done by the state enumeration method, using TEF, as follows: a) by identifying the states with higher probability of instability or insecurity (e.g., greater sensitivity to changing system parameters, low critical energy, etc.), b) by speeding the computation of stability results for the numerous disturbance scenarios required, and c) by classification and grouping of states according to their energy margin characteristics.

REFERENCES

[1] Kimbark, E. W. *Power System Stability, Vol. 1.* New York: Wiley, 1948.

[2] Fouad, A. A. "Stability Theory - Criteria for Transient Stability." In *Proceedings of the Engineering Foundation Conference on System Engineering for Power: Status and Prospects.* NIT Publication No. Conf.-750867, August 1975.

[3] *Criteria of Stability of Electric Power Systems,* a report published by the all Union Institute of Scienctific and Technological Information and the Academy of Sciences of the USSR (in Russia). Electric Technology and Electric Power Series. Moscow, 1971.

[4] Magnusson, P. C. "Transient Energy Method of Calculating Stability." *AIEE Transactions* 66 (1947): 747-755.

[5] Aylett, P. D. "The Energy-Integral Criterion of Transient Stability Limits of Power Systems". In *Proceedings of Institution of Electrical Engineers (London)* 105C, no. 8 (Sept. 1958): 527-536.

[6] Pai, M. A. *Power System Stability - Analysis by the Direct Method of Lyapunov.* North Holland, 1981.

[7] Athay, T., R. Podmore, and S. Virmani. "A Practical Method for Direct Analysis of Transient Stability". *IEEE Transactions* PAS-98, no. 2 (1979): 573-584.

[8] Athay, T., V. R. Sherket, R. Podmore, S. Virmani, and C. Puech. "Transient Energy Stability Analysis". Systems Engineering for Power: Emergency Operating State Control - Section IV, U.S. Department of Energy Publication No. CONF-790904-PL.

[9] Fouad, A. A., et al. "Transient Stability Margin as a Tool for Dynamic Security Assessment". EPRI Report EL-1755, March 1981.

[10] Fouad, A. A., and S. E. Stanton. "Transient Stability of a Multimachine Power System. Part I: Investigation of the System Trajectory. Part II: Critical Transient Energy". *IEEE Transactions* PAS-100 (1981): 3408-3424.

[11] Fouad, A. A., V. Vittal, and T. Oh. "Critical Energy for Direct Transient Stability Assessment of Multimachine Power System". *IEEE Transactions* PAS-103 (Aug. 1984): 2199-2206.

[12] Fouad, A. A., et al. "Direct Transient Stability Analysis Using Energy Functions: Application to Large Power Networks". *IEEE Transactions on Power Systems* PWRS-2 (Feb. 1987): 37-44.

[13] Carvalho, V. F., M. A. El-Kady, E. Vaahedi, P. Kundur, C. K. Tang, G. Rogers, J. Libaque, D. Wong, A. A. Fouad, V. Vittal, S. Rajagopal. "Demonstration of Large Scale Direct Analysis of Power System Transient Stability". Electric Power Research Institute Report EL-4980, Dec. 1986.

[14] Vittal, V., et al. "Transient Stability Analysis of Stressed Power Systems Using the Transient Energy Function Method". *IEEE Transactions on Power Systems* PWRS-2 (Feb. 1988): 239-244.

[15] Fouad, A. A., et al. "Extending Applications of the Transient Energy Function Method". EPRI Report EL-5215, Sept. 1987.

[16] Fouad, A. A., V. Vittal, et al. "Direct Transient Stability Assessment with Excitation Control". *IEEE Transactions on Power Systems* 4 (Feb. 1989): 75-82.

[17] Vittal, V., N. Bhatia, A. A. Fouad, G. Maria, and H. Zein El-Din. "Incorporation of Nonlinear Load Models in the Transient Energy Function Method". *IEEE Transactions on Power Systems* 4, no. 3 (Aug. 1989): 1031-1036.

[18] Symposium on Reliability Criteria for System Dynamic Performance. IEEE/PES Publication No. 77CH1221-PW.

[19] Fouad, A. A. "Dynamic Security Assessment Practices in North America". An IEEE Committee Report, Chairman. *IEEE Transactions on Power Systems* 3, no. 3 (Aug. 1988): 1310-1321.

[20] Anderson, P. M., and A. A. Fouad. *Power System Control and Stability*. Ames, Ia.: Iowa State University Press, 1977.

[21] Brown, R. G. *Introduction to Random Signal Analysis and Kalman Filtering*. John Wiley & Sons, 1983.

[22] Chiang, H. D., F. F. Wu, and P. P. Varaiya. "Foundations of the Direct Methods for Power System Transient Stability Analysis". *IEEE Transactions Circuits Systems* CAS–34 (Feb. 1987): 160–173.

[23] Zaborszky, J., G. Huang, B. Zheng, and T-C Leung. "On the Phase-Portrait of a Class of Large Nonlinear Dynamic Systems Such as the Power System". *IEEE Transactions Automatic Control* 32 (Jan. 1988): 4–15.

[24] Uemura, K., J. Matsuki, I. Yamada, and T. Tsuji. "Approximation of an Energy Function in Transient Stability Analysis of Power Systems". *Electrical Engineering in Japan* 92, no. 4 (1972): 96–100.

[25] El-Abiad, A. H., and K. Nagappan. "Transient Stability Regions of Multimachine Power Systems". *IEEE Transactions* PAS–85 (Feb. 1966): 169–179.

[26] Gupta, C. L., and A. H. El-Abiad. "Determination of the Closest Unstable Equilibrium State for Lyapunov's Method in Transient Stability Studies". *IEEE Transactions* PAS–95 (Sept./Oct. 1976): 1699–1712.

[27] Ribbens-Pavella, M., P. G. Murthy, J. L. Horward. "The Acceleration Approach to Practical Stability Domain Estimation in Power Systems". In *Proceedings of the 20th IEEE Conference on Decision and Control*, 1981: 471–477.

[28] Sasson, A. M., et al. "Improved Newton's Load Flow Through a Minimization Technique". *IEEE Transactions* PAS–90, no. 5 (1971): 1974–1981.

[29] Gill, P. E., and W. Murray. "Algorithms for the Solution of the Nonlinear Least-Squares Problem". *Am. J. Numer. Anal.* 15(5) (Oct. 1978): 977–992.

[30] Vittal, V., and A. A. Fouad. "A Noise Equivalent Bandwidth Approach to Obtain Reduced Order Models for Power System Excitation Control". *Electrical Machines and Power Systems* 17, no. 1 (1989).

[31] Ni, Y-X., C. G. Shin, and A. A. Fouad. "TEF Method Solution with Exciter: Numerical Technique Used". In *Proceedings of the 1987 North American Power Symposium*, held in Edmonton, Alberta, October 1987.

[32] Pai, M. A., et al. "Transient Stability Analysis of Multimachine AC/DC Power Systems Via Energy Function Method". *IEEE Transactions* PAS–100, no. 12 (Dec. 1981): 5027–5035.

[33] Ni, Y-X., and A. A. Fouad. "A Simplified Two Terminal HVDC Model and its Use in Direct Transient Stability Assessment". *IEEE Transactions on Power Systems* PWRS-2, no. 4 (Nov. 1987): 1006-1013.

[34] Fouad, A. A., et al. "Contingency Analysis Using the Transient Energy Margin Technique". *IEEE Transactions* PAS-101, no. 4 (Apr. 1982): 757-766.

[35] Fouad, A. A., Vittal, V., Oh, T., and Raine, J. C. "Investigation of Loss of Generation Disturbances in the Florida Power & Light Co. Network by the Transient Energy Function Method". *IEEE Transactions on Power Systems* PWRS-1, no. 3 (Aug. 1986): 60-66.

[36] Pai, M. A., et al. "Direct Method of Stability Analysis in Dynamic Security Assessment". Paper no. 1.1/A4, IFAC World Congress, Budapest, July 1984.

[37] Tomovic, R. *Sensitivity Analysis of Dynamical Systems*. New York: McGraw-Hill, 1963.

[38] Kokotovic, P. V., and R. S. Rutman. "Sensitivity of Automatic Control Systems (Survey)". *Automation and Remote Control* 26 (1965): 727-749.

[39] Vittal, V., et al. "Derivation of Stability Limits Using Analytical Sensitivity of the Transient Energy Margin". *IEEE Transactions on Power Systems* 4, no. 4 (Nov. 1989): 1363-1372.

[40] Vittal, V., R. D'souza, and A. A. Fouad. "Analytical Sensitivity of Transient Energy Margin Including Second Order". In *Proceedings of the 10th PSCC*, Graz, Austria, August 1990: 481-486.

[41] "DIRECT" Version 1.2, User's Manual, RP2206-1, EPRI.

[42] Vittal, V., G. M. Prabhu, and S. L. Lim. "A Parallel Compuer Implementation of Power System Transient Stability Assessment Using the Transient Energy Function Method". Paper no. 90WM147-9 PWRS. Atlanta, Georgia, 1990 Winter Meeting of the Power Enginering Society of IEEE, February 4-9, 1990.

[43] Osyczka, Andrzej. *Multicriterion Optimization in Engineering*. New York: Ellis Horwood Ltd. (Div. of John Wiley), 1984.

[44] Ignizio, James P. "Generalized Goal Programming - An Overview". *Computer and Ops. Research*. Pergamon Press, 10, no. 4 (1983): 277-289.

[45] Fouad, A. A., et al. "Calculation of Generation-Shedding Requirements of the B.C. Hydro System Using Transient Energy Functions". *IEEE Transactions PWRS-1*, no. 2 (May 1986): 17-24.

[46] Findlay, J. A., G. A. Maria, and V. R. Wong. "Stability Limits for the Bruce LGR Scheme". in *Proceedings 8th Power Systems Computation Conference*. Helsinki, Finland, 1984: 1063-1069.

[47] El-Kady, M. A., A. A. Fouad, C. C. Liu. "Knowledge-Based System for Direct Stability Analysis". Final Report, EPRI EL-6796, Apr. 1990.

[48] El-Kady, M. A., A. A. Fouad, C. C. Liu, and S. Venkataraman. "Use of Expert Systems in Dynamic Security Assessment of Power Systems". In *Proceedings of the 10th Power System Computer Conference*, Graz, Austria, Aug. 1989: 913-920.

[49] Endrenyi, J. *Reliability Modeling in Electric Power Systems*. John Wiley & Sons, 1978.

DYNAMIC STABILITY ANALYSIS AND CONTROL OF POWER SYSTEMS

Dr. Young-Hwan Moon

Dr. Wei-Jen Lee

Dr. Mo-Shing Chen

The Energy Systems Research Center

The University of Texas at Arlington

CHAPTER 1

INTRODUCTION

1.1 Background

Every electric power system is changing every moment. Not only generation system but also transmission system are changing to enable a system to be operated optimally in security and efficiency. Due to the changes in both generation and transmission system, a power system may have very different dynamic characteristics. A change in the system loading, generation schedule, network-interconnections, or type of circuit protection may also give completely different results in a stability study for the same disturbance. Thus, even in a system normally operated without any problems, serious situations can emerge during disturbances although most of them could be eliminated during the system planning stage.

In order to maintain the integrity of the reliable power system should continue to provide electricity to consumers in all situations. Certainly, a system should be designed to withstand the specified contingencies with finite probability of occurring. The stability study can be separated into two distinct problems, the transient (short-term) problem and the dynamic (long-term) problem. In the transient stability analysis, the performance of the power system when subjected to sudden severe impacts is studied. The concern is whether the system is able to maintain synchronism during and following these disturbances. The period of interest is relatively short (a few seconds), with the first swing being of primary importance. The stability depends strongly upon the magnitude and location of the disturbance and to a lesser extent upon the initial state or operating condition of the system.

CONTROL AND DYNAMIC SYSTEMS, VOL. 43

In contrast to transient stability, dynamic stability tends to be a property of the state of the system. Dynamic stability indicate the ability of all machines in a system to adjust to small load changes or impacts. The concern is whether the system has growing oscillation phenomena and their damping. As modern power systems became equipped with more fast control systems and complicated, dynamic stability became more important. The stability strongly depends on the initial state or operating condition of the system. Without question, both of transient stability and dynamic stability must be secured for successful planning and operation of the system.

The procedures in dealing with the dynamic stability problem are (a) to reproduce or predict the potential oscillation phenomena of the system through computer simulation, (b) to identify the mode of oscillation and examine the most effective control location, (c) to adjust system control parameters or design compensation equipment to improve system damping, (d) to install these equipment at appropriate locations and perform fine tuning. A combination of time and frequency domain approach is recommended in this chapter. Since frequency domain analysis (eigenvalue analysis) is limited by the dimension of the matrix, it is difficult to handle huge system with detailed models. Normally, a simplified equivalent system is used to investigate the potential oscillation modes, and eigenvetors are used as sensitivity factors to determine the most effective control location for certain modes of oscillation. A detailed time domain analysis can be used to verify whether the fundamental characteristics of the system are maintained in the equivalent circuit. If any auxiliary control is necessary to compensate some oscillation frequencies and improve the system damping, frequency domain analysis determine the frequency to be compensated and the installation locations of control equipment while time domain analysis reveal the parameters of control equipment and verifies the results.

CHAPTER 2

SYNCHRONOUS MACHINE MODELING

2.1 The Three Phase Synchronous Machine

A synchronous machine has two principal parts; the stator and the rotor as shown in Figure 2.1. The stator is a stationary part which is a hollow laminated ferromagnetic cylinder with longitudinal slots on the inside surface. Placed in these slots are coils which are interconnected to form three separate windings. The stator serves two basic functions: a mounting structure for the stator windings and a low reluctance return path for the rotor magnetic field.

The rotor is a solid ferromagnetic cylindrical structure which rotates inside the hollow stator. Located on the rotor is a field winding which is supplied with dc current controlled by the exciter. The very high magnetomotive force (mmf) produced by this current in the field winding combines with the mmf produced by the current in the stator winding. The resultant flux across the air gap between the stator and rotor generates voltage in the coils of stator winding and provides the electromagnetic torque between the stator and rotor.

The stator and rotor are designed so that for a constant rotor speed a sinusoidal voltage is generated in each stator winding. These three voltages are identical in magnitude and frequency and are separated 120° in phase.

Besides the stator and field windings, synchronous machines usually have damper windings which consist of short-circuited copper bars through the pole face. The purpose of the damper windings is to reduce the mechanical oscillations of the rotor and to keep the rotor velocity as close to its synchronous speed as possible.

The synchronous machine is a challenging device to model. The main difficulties in modeling the three phase synchronous machine are:

- Stationary three windings in the stator and the rotating field winding and damper windings in the rotor;
- The presence of ferromagnetic material with saturation effects;
- The governor-turbine dynamics
- The voltage control system (the exciter)

The fundamental equations in modeling a synchronous machine are derived following the next steps:

(a) Obtain the self and mutual inductances of stator which are functions of the rotor position θ.

(b) Obtain the self and mutual inductances of rotor using the d-q axis.

(c) Obtain the mutual inductances between the stator and the rotor.

(d) Using Park's transformation, transform all stator quantities into the new variables based on the rotor side.

(e) Establish the flux linkage equations.

(f) Establish the voltage equations.

(g) Obtain the swing equation

(h) Obtain the normalized fundamental equations using per unit conversion.

The complexity level of a synchronous machine model varies depending on the purpose and its required accuracy. In this chapter, the model with two damper coils in the rotor will be used for detailed simulation of a synchronous machine.

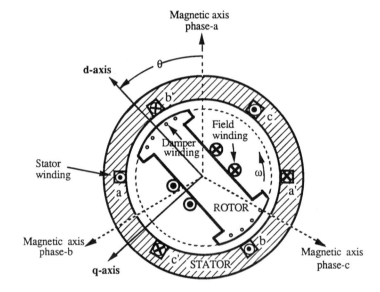

Figure 2.1 The Cross Section of an Elementary Three-Phase Synchronous Machine with a Salient Two-Pole Rotor

2.2 Machine Inductances

Three stator windings and one rotor winding have their own self inductances and mutual inductances between them are existing between them. Most of the self inductances L_{ii} and mutual inductances M_{ij} in the machine modeling are changing with the rotor angle θ with respect to a fixed reference position for which phase-a is chosen in Figure 2.1 for convenience.

2.2.1 Stator Self-Inductances

$$L_{aa} = L_1 + L_2 \cos(2\theta) \quad [H]$$
$$L_{bb} = L_1 + L_2 \cos2(\theta - 2\pi/3) \quad [H] \tag{2.1}$$
$$L_{cc} = L_1 + L_2 \cos2(\theta + 2\pi/3) \quad [H]$$

where $L_1 > L_2$ and L_1 and L_2 are constants. L_1 is the average self-inductance of each phase winding and L_2 is the maximum value of inductance variation of each phase winding. While $L_2 \neq 0$ in salient rotor machine, $L_2 = 0$ in round rotor machine.

2.2.2 Stator Mutual-Inductances

Since a positive current in a stator winding will give a negative flux linkage component in the two remaining stator windings, the stator mutual inductances have negative values. The phase-to-phase mutual inductances are functions of θ and are symmetric.

$$L_{ab} = L_{ba} = -L_3 - L_2 \cos2(\theta + \pi/6) \quad [H]$$
$$L_{bc} = L_{cb} = -L_3 - L_2 \cos2(\theta - \pi/2) \quad [H] \tag{2.2}$$
$$L_{ca} = L_{ac} = -L_3 - L_2 \cos2(\theta + 5\pi/6) \quad [H]$$

where $|L_3| > L_2$.

2.2.3 Rotor Self-Inductances

Since the circuit of magnetic flux induced by field current does not change with rotor position, the self-inductance of field winding, L_{fdfd}, does not vary with θ. Other rotor self-inductances are also constant for the same reason.

d-axis: Field winding, L_{fdfd} (constant)   [H]

Damper winding, L_{kdkd} (constant)   [H]

q-axis: Damper winding 1, L_{gqgq} (constant)   [H]   (2.3)

Damper winding 2, L_{kqkq} (constant)   [H]

2.2.4   Rotor Mutual-Inductances

The mutual inductance between two windings in the same axis (d-axis or q-axis) is constant and does not change with θ. The coupling coefficient between the d and q axis is zero, and all pairs of windings with 90° displacement have zero mutual inductance.

d-axis: Field-winding to kd-winding: $L_{fdkd} = L_{kdfd} = M_{fdkd}$ [H]

q-axis: gq-winding to kq-winding: $L_{gqkq} = L_{kqgq} = M_{gqkq}$ [H]   (2.4)

Others: $L_{fdgq} = L_{gqfd} = 0$ [H]

$L_{fdkq} = L_{kqfd} = 0$ [H]

$L_{kdgq} = L_{gqkd} = 0$ [H]

$L_{kdkq} = L_{kqkd} = 0$ [H]

2.2.5   Stator-to-Rotor Mutual-Inductances

The mutual inductances between stator and rotor windings are functions of the rotor angle θ. They are also symmetric and they have their maximum values when the magnetic axes of stator and rotor coincide (θ=0).

(1) Phase Winding-to-Field Winding

$L_{afd} = L_{fda} = M_{afd}\cos\theta$ [H]

$L_{bfd} = L_{fdb} = M_{afd}\cos(\theta-2\pi/3)$ [H]   (2.5)

$L_{cfd} = L_{fdc} = M_{afd}\cos(\theta+2\pi/3)$ [H]

(2) Phase winding-to-Damper winding, kd

$L_{akd} = L_{kda} = M_{akd}\cos\theta$ [H]

$L_{bkd} = L_{kdb} = M_{akd}\cos(\theta-2\pi/3)$ [H]   (2.6)

$L_{ckd} = L_{kdc} = M_{akd}\cos(\theta+2\pi/3)$ [H]

(3) Phase winding-to-Damper winding, gq

$L_{akd} = L_{kda} = M_{akd}\sin\theta$ [H]

$L_{bkd} = L_{kdb} = M_{akd}\sin(\theta-2\pi/3)$ [H]   (2.7)

$L_{ckd} = L_{kdc} = M_{akd}\sin(\theta+2\pi/3)$ [H]

(4) Phase winding-to-Damper winding, fq

$L_{akq} = L_{kqa} = M_{akq}\sin\theta$ [H]

$L_{bkq} = L_{kqb} = M_{akq}\sin(\theta-2\pi/3)$ [H]   (2.8)

$L_{ckq} = L_{kqc} = M_{akq}\sin(\theta+2\pi/3)$ [H]

2.3 Park's Transformation

In order to simplify the mathematical description of the synchronous machine, a certain transformation of variables is performed. The transformation is usually called Park's transformation. It defines a new set of stator variables such as currents, flux linkages, and flux linkages in terms of the actual widing variables. The new quantities are obtained by projecting the actual variables on three fictitious axis; d-axis, q-axis, and a stationary axis. We define the d-axis of the rotor at some instant of time to be at angle θ with respect to a fixed reference position, as shown in Figure 2.1. Projecting the stator currents i_a, i_b, and i_c along the d-, q-, and stationary axes, we get the relations

$$
\begin{aligned}
i_d &= \frac{2}{3} \{ i_a\cos\theta + i_b\cos(\theta - 2\pi/3) + i_c\cos(\theta + 2\pi/3) \} \\
i_q &= \frac{2}{3} \{ -i_a\sin\theta - i_b\sin(\theta - 2\pi/3) - i_c\sin(\theta + 2\pi/3) \} \\
i_0 &= \frac{1}{3} \{ i_a + i_b + i_c \}
\end{aligned}
\tag{2.9}
$$

where the stator currents i_a, i_b, and i_c are the currents leaving the generator terminals.
The above equation can be rewritten in matrix form as follows;

$$
\begin{bmatrix} i_d \\ i_q \\ i_0 \end{bmatrix} = \frac{2}{3}
\begin{bmatrix}
\cos\theta & \cos(\theta - 2\pi/3) & \cos(\theta + 2\pi/3) \\
-\sin\theta & -\sin(\theta - 2\pi/3) & -\sin(\theta + 2\pi/3) \\
1/2 & 1/2 & 1/2
\end{bmatrix}
\begin{bmatrix} i_a \\ i_b \\ i_c \end{bmatrix}
$$

$$
= [P] \begin{bmatrix} i_a \\ i_b \\ i_c \end{bmatrix}
\tag{2.10}
$$

or

$$
\begin{bmatrix} i_a \\ i_b \\ i_c \end{bmatrix} =
\begin{bmatrix}
\cos\theta & -\sin\theta & 1 \\
\cos(\theta - 2\pi/3) & -\sin(\theta - 2\pi/3) & 1 \\
\cos(\theta + 2\pi/3) & -\sin(\theta + 2\pi/3) & 1
\end{bmatrix}
\begin{bmatrix} i_d \\ i_q \\ i_0 \end{bmatrix}
$$

$$
= [P]^{-1} \begin{bmatrix} i_d \\ i_q \\ i_0 \end{bmatrix}
\tag{2.11}
$$

where Park's transformation, [P], is defined as:

$$[P] = \frac{2}{3} \begin{bmatrix} \cos\theta & \cos(\theta-2\pi/3) & \cos(\theta+2\pi/3) \\ -\sin\theta & -\sin(\theta-2\pi/3) & -\sin(\theta+2\pi/3) \\ 1/2 & 1/2 & 1/2 \end{bmatrix} \qquad (2.12)$$

Then, the inverse of [P] will be

$$[P]^{-1} = \begin{bmatrix} \cos\theta & -\sin\theta & 1 \\ \cos(\theta-2\pi/3) & -\sin(\theta-2\pi/3) & 1 \\ \cos(\theta+2\pi/3) & -\sin(\theta+2\pi/3) & 1 \end{bmatrix} \qquad (2.13)$$

The Park's transformation is to convert all stator quantities of a-b-c phases to new variables of d-q-0 axes. It would be more convenient, however, to expand the transformation to include all rotor quantities which are components of d-q-0 axes. The new transformation [T] would have the form:

$$[T] \equiv \begin{bmatrix} [P] & [0] \\ [0] & [U_4] \end{bmatrix} \qquad (2.14)$$

where [P] is Park's transformation and $[U_4]$ is the 4x4 unit matrix.

During the transformation using [T], rotor quantities remain the same as follows:
For currents,

$$\begin{bmatrix} i_d \\ i_q \\ i_0 \\ i_{fd} \\ i_{kd} \\ i_{gq} \\ i_{kq} \end{bmatrix} = \begin{bmatrix} [P] & [0] \\ [0] & [U_4] \end{bmatrix} \begin{bmatrix} i_a \\ i_b \\ i_c \\ i_{fd} \\ i_{kd} \\ i_{gq} \\ i_{kq} \end{bmatrix} \qquad \text{or simply} \qquad i_{dr} = [T]\, i \quad (2.15)$$

for flux linkages,

$$\begin{bmatrix} \psi_d \\ \psi_q \\ \psi_0 \\ \psi_{fd} \\ \psi_{kd} \\ \psi_{gq} \\ \psi_{kq} \end{bmatrix} = \begin{bmatrix} [P] & [0] \\ [0] & [U_4] \end{bmatrix} \begin{bmatrix} \psi_a \\ \psi_b \\ \psi_c \\ \psi_{fd} \\ \psi_{kd} \\ \psi_{gq} \\ \psi_{kq} \end{bmatrix} \qquad \text{or simply} \qquad \psi_{dr} = [T]\, \psi \ (2.17)$$

and finally for voltages,

$$\begin{bmatrix} v_d \\ v_q \\ v_0 \\ v_{fd} \\ v_{kd} \\ v_{gq} \\ v_{kq} \end{bmatrix} = \begin{bmatrix} [P] & [0] \\ [0] & [U_4] \end{bmatrix} \begin{bmatrix} v_a \\ v_b \\ v_c \\ v_{fd} \\ v_{kd} \\ v_{gq} \\ v_{kq} \end{bmatrix} \qquad \text{or simply} \qquad \mathbf{v}_{dr} = [T]\,\mathbf{v} \quad (2.16)$$

2.4 Magnetic Flux Linkage Equations

$$\begin{bmatrix} \Psi_a \\ \Psi_b \\ \Psi_c \\ \Psi_{fd} \\ \Psi_{kd} \\ \Psi_{gq} \\ \Psi_{kq} \end{bmatrix} = \begin{bmatrix} -L_{aa} & -L_{ab} & -L_{ac} & L_{afd} & L_{akd} & L_{agq} & L_{akq} \\ -L_{ba} & -L_{bb} & -L_{bb} & L_{bfd} & L_{bkd} & L_{bgq} & L_{bkq} \\ -L_{ca} & -L_{cb} & -L_{cc} & L_{cfd} & L_{ckd} & L_{cgq} & L_{ckq} \\ -L_{fda} & -L_{fdb} & -L_{fdc} & L_{fdfd} & L_{fdfd} & L_{fdgq} & L_{fdkq} \\ -L_{kda} & -L_{kdb} & -L_{kdc} & L_{kdfd} & L_{kdkd} & L_{kdgq} & L_{kdkq} \\ -L_{gqa} & -L_{gqb} & -L_{gqc} & L_{gqfd} & L_{gqkd} & L_{gqgq} & L_{gqkq} \\ -L_{kqa} & -L_{kqb} & -L_{kqc} & L_{kqfd} & L_{kqkd} & L_{kqgq} & L_{kqkq} \end{bmatrix} \begin{bmatrix} i_a \\ i_b \\ i_c \\ i_{fd} \\ i_{kd} \\ i_{gq} \\ i_{kq} \end{bmatrix} \quad (2.18)$$

or

$$\Psi = [L]\,\mathbf{i} \equiv \begin{bmatrix} [L_{SS}] & [L_{SR}] \\ [L_{RS}] & [L_{RR}] \end{bmatrix} \mathbf{i} \qquad (2.19)$$

where $[L_{SS}]$ = stator-stator inductances

$[L_{SR}], [L_{RS}]$ = stator-rotor inductances

$[L_{RR}]$ = rotor-rotor inductances

By premultiplying $[T]$ defined in (2.14) to (2.19),

$$[T]\,\Psi = [T]\,[L]\,\mathbf{i} \qquad (2.20)$$

The right side of (2.20) can be rewritten as

$$[T]\,[L]\,\mathbf{i} = [T]\,[L]\,\{[T]^{-1}\,[T]\,\mathbf{i}\} = [T]\,[L]\,[T]^{-1}\,\{[T]\,\mathbf{i}\} = [T]\,[L]\,[T]^{-1}\,\mathbf{i}_{dr} \quad (2.21)$$

From (2.17) and (2.21)

$$\psi_{dr} = [T][L][T]^{-1} i_{dr} \tag{2.22}$$

i.e.,

$$
\begin{bmatrix} \psi_d \\ \psi_q \\ \psi_0 \\ \psi_{fd} \\ \psi_{kd} \\ \psi_{gq} \\ \psi_{kq} \end{bmatrix}
=
\begin{bmatrix} [P] & [0] \\ [0] & [U_4] \end{bmatrix}
\begin{bmatrix} [L_{SS}] & [L_{SR}] \\ [L_{RS}] & [L_{RR}] \end{bmatrix}
\begin{bmatrix} [P] & [0] \\ [0] & [U_4] \end{bmatrix}^{-1}
\begin{bmatrix} i_d \\ i_q \\ i_0 \\ i_{fd} \\ i_{kd} \\ i_{gq} \\ i_{kq} \end{bmatrix}
$$

$$
=
\begin{bmatrix} [P][L_{SS}] & [P][L_{SR}] \\ [L_{RS}] & [L_{RR}] \end{bmatrix}
\begin{bmatrix} [P]^{-1} & 0 \\ 0 & [U_4] \end{bmatrix}
\begin{bmatrix} i_d \\ i_q \\ i_0 \\ i_{fd} \\ i_{kd} \\ i_{gq} \\ i_{kq} \end{bmatrix}
$$

$$
=
\begin{bmatrix} [P][L_{SS}][P]^{-1} & [P][L_{SR}] \\ [L_{RS}][P]^{-1} & [L_{RR}] \end{bmatrix}
\begin{bmatrix} i_d \\ i_q \\ i_0 \\ i_{fd} \\ i_{kd} \\ i_{gq} \\ i_{kq} \end{bmatrix}
\tag{2.23}
$$

If we define $[L_{dr}]$ as

$$
[L_{dr}] \equiv
\begin{bmatrix} [P][L_{SS}][P]^{-1} & [P][L_{SR}] \\ [L_{RS}][P]^{-1} & [L_{RR}] \end{bmatrix}
\tag{2.24}
$$

Performing the operation in (2.24), $[L_{dr}]$ can be computed as

$$
[L_{dr}] =
\left[
\begin{array}{ccc:cccc}
-L_d & 0 & 0 & M_{afd} & M_{akd} & 0 & 0 \\
0 & -L_q & 0 & 0 & 0 & M_{agq} & M_{akq} \\
0 & 0 & -L_0 & 0 & 0 & 0 & 0 \\
\hdashline
-(3/2)M_{afd} & 0 & 0 & L_{fdfd} & M_{fdkd} & 0 & 0 \\
-(3/2)M_{akd} & 0 & 0 & M_{kdfd} & L_{kdkd} & 0 & 0 \\
0 & -(3/2)M_{agq} & 0 & 0 & 0 & L_{gqgq} & M_{gqkq} \\
0 & -(3/2)M_{akq} & 0 & 0 & 0 & M_{kqgq} & L_{kqkq}
\end{array}
\right]
\tag{2.25}
$$

where the new constants are defined as follows;

$$L_d = L_1 + L_3 + \frac{3}{2} L_2 \quad [H]$$

$$L_q = L_1 + L_3 - \frac{3}{2} L_2 \quad [H] \tag{2.26}$$

$$L_0 = L_1 - 2 L_3 \quad [H]$$

Consequently, after applying the [T] transformation, the flux linkage equation (2.19) becomes

$$
\begin{bmatrix} \Psi_d \\ \Psi_q \\ \Psi_0 \\ \Psi_{fd} \\ \Psi_{kd} \\ \Psi_{gq} \\ \Psi_{kq} \end{bmatrix}
=
\begin{bmatrix}
-L_d & 0 & 0 & M_{afd} & M_{akd} & 0 & 0 \\
0 & -L_q & 0 & 0 & 0 & M_{agq} & M_{akq} \\
0 & 0 & -L_0 & 0 & 0 & 0 & 0 \\
-(3/2)M_{afd} & 0 & 0 & L_{fdfd} & M_{fdkd} & 0 & 0 \\
-(3/2)M_{akd} & 0 & 0 & M_{kdfd} & L_{kdkd} & 0 & 0 \\
0 & -(3/2)M_{agq} & 0 & 0 & 0 & L_{gqgq} & M_{gqkq} \\
0 & -(3/2)M_{akq} & 0 & 0 & 0 & M_{kqgq} & L_{kqkq}
\end{bmatrix}
\begin{bmatrix} i_d \\ i_q \\ i_0 \\ i_{fd} \\ i_{kd} \\ i_{gq} \\ i_{kq} \end{bmatrix}
\tag{2.27}
$$

or simply, $\Psi_{dr} = [L_{dr}] \, i_{dr}$ (2.28)

where $\Psi_{dr} = [\Psi_d, \Psi_q, \Psi_0, \Psi_{fd}, \Psi_{kd}, \Psi_{gq}, \Psi_{kq}]^t$

and $i_{dr} = [i_d, i_q, i_0, i_{fd}, i_{kd}, i_{gq}, i_{kq}]^t$.

2.5 Voltage Equations

In order to relate the flux linkage to the voltage which results from its time variation, Figure 2.2 establishes the conventions regarding the sign of the of the induced voltage with respect to the assumed current direction and the phase of flux linkage. From Figure 2.2, the instantaneous terminal voltages may be described by the following equations:

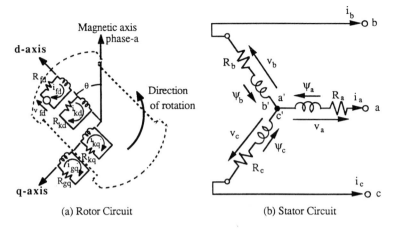

(a) Rotor Circuit   (b) Stator Circuit

Figure 2.2 Rotor circuit and Stator Circuit

$$
\begin{bmatrix} v_a \\ v_b \\ v_c \\ v_{fd} \\ v_{kd} \\ v_{gq} \\ v_{kq} \end{bmatrix} = \frac{d}{dt} \begin{bmatrix} \psi_a \\ \psi_b \\ \psi_c \\ \psi_{fd} \\ \psi_{kd} \\ \psi_{gq} \\ \psi_{kq} \end{bmatrix} - \begin{bmatrix} R_a & 0 & 0 & 0 & 0 & 0 & 0 \\ 0 & R_b & 0 & 0 & 0 & 0 & 0 \\ 0 & 0 & R_c & 0 & 0 & 0 & 0 \\ 0 & 0 & 0 & -R_{fd} & 0 & 0 & 0 \\ 0 & 0 & 0 & 0 & -R_{kd} & 0 & 0 \\ 0 & 0 & 0 & 0 & 0 & -R_{gq} & 0 \\ 0 & 0 & 0 & 0 & 0 & 0 & -R_{kq} \end{bmatrix} \begin{bmatrix} i_a \\ i_b \\ i_c \\ i_{fd} \\ i_{kd} \\ i_{gq} \\ i_{kq} \end{bmatrix}
$$

$$(2.29)$$

or

$$ v = \frac{d}{dt} \psi - [R] \, i \tag{2.30} $$

Multiplying [T] on both sides of the equation (2.30),

$$ [T] \, v = [T] \frac{d}{dt} \psi - [T][R] \, i \tag{2.31} $$

From (2.16), $[T] \, v = v_{dr}$ (2.32)

Since [R] is a diagonal matrix, the second term of (2.31) becomes

$$
\begin{aligned}
[T][R] \, i &= [T][R] \left\{ [T]^{-1} [T] \, i \right\} \\
&= [R][T] \, i \\
&= [R] \, i_{dr}
\end{aligned}
\tag{2.33}
$$

The first term (2.31) will be expressed as follows;

From (2.17),

$$
\begin{aligned}
[T] \frac{d}{dt} \psi &= [T] \frac{d}{dt} \left\{ [T]^{-1} \psi_{dr} \right\} \\
&= [T] \left\{ \frac{d}{dt} [T]^{-1} \right\} \psi_{dr} + [T][T]^{-1} \frac{d}{dt} \left\{ \psi_{dr} \right\} \\
&= [T] \frac{d}{dt} \left\{ [T]^{-1} \right\} \psi_{dr} + \frac{d}{dt} \psi_{dr}
\end{aligned}
\tag{2.34}
$$

where

$$
\begin{aligned}
\frac{d}{dt} [T]^{-1} &= \frac{d}{dt} \begin{bmatrix} [P] & [0] \\ [0] & [U_4] \end{bmatrix}^{-1} \\
&= \frac{d}{dt} \begin{bmatrix} [P]^{-1} & [0] \\ [0] & [U_4] \end{bmatrix}
\end{aligned}
\tag{2.35}
$$

From (2.13) and (2.35),

$$
\frac{d}{dt} [T]^{-1} = \frac{d}{dt}
\begin{bmatrix}
\cos\theta & -\sin\theta & 1 & 0 & 0 & 0 & 0 \\
\cos(\theta-2\pi/3) & -\sin(\theta-2\pi/3) & 1 & 0 & 0 & 0 & 0 \\
\cos(\theta+2\pi/3) & -\sin(\theta+2\pi/3) & 1 & 0 & 0 & 0 & 0 \\
0 & 0 & 0 & 1 & 0 & 0 & 0 \\
0 & 0 & 0 & 0 & 1 & 0 & 0 \\
0 & 0 & 0 & 0 & 0 & 1 & 0 \\
0 & 0 & 0 & 0 & 0 & 0 & 1
\end{bmatrix}
$$

$$
=
\begin{bmatrix}
-\sin\theta \frac{d\theta}{dt} & -\cos\theta \frac{d\theta}{dt} & 0 & 0 & 0 & 0 & 0 \\
-\sin(\theta-2\pi/3)\frac{d\theta}{dt} & -\cos(\theta-2\pi/3)\frac{d\theta}{dt} & 0 & 0 & 0 & 0 & 0 \\
-\sin(\theta+2\pi/3)\frac{d\theta}{dt} & -\cos(\theta+2\pi/3)\frac{d\theta}{dt} & 0 & 0 & 0 & 0 & 0 \\
0 & 0 & 0 & 0 & 0 & 0 & 0 \\
0 & 0 & 0 & 0 & 0 & 0 & 0 \\
0 & 0 & 0 & 0 & 0 & 0 & 0 \\
0 & 0 & 0 & 0 & 0 & 0 & 0
\end{bmatrix}
\tag{2.36}
$$

The matrix in the right side of (2.36) can be modified as the multiplication of the next two matrices;

$$
\begin{bmatrix}
\cos\theta & \sin\theta & 1 & 0 & 0 & 0 & 0 \\
\cos(\theta-2\pi/3) & \sin(\theta-2\pi/3) & 1 & 0 & 0 & 0 & 0 \\
\cos(\theta+2\pi/3) & \sin(\theta+2\pi/3) & 1 & 0 & 0 & 0 & 0 \\
0 & 0 & 0 & 1 & 0 & 0 & 0 \\
0 & 0 & 0 & 0 & 1 & 0 & 0 \\
0 & 0 & 0 & 0 & 0 & 1 & 0 \\
0 & 0 & 0 & 0 & 0 & 0 & 1
\end{bmatrix}
\begin{bmatrix}
0 & -d\theta/dt & 0 & 0 & 0 & 0 & 0 \\
d\theta/dt & 0 & 0 & 0 & 0 & 0 & 0 \\
0 & 0 & 0 & 0 & 0 & 0 & 0 \\
0 & 0 & 0 & 0 & 0 & 0 & 0 \\
0 & 0 & 0 & 0 & 0 & 0 & 0 \\
0 & 0 & 0 & 0 & 0 & 0 & 0 \\
0 & 0 & 0 & 0 & 0 & 0 & 0
\end{bmatrix}
$$

(2.37)

Thus, the equation (2.36) can be expressed as

$$\frac{d}{dt}[T]^{-1} = [T]^{-1}\frac{d}{dt}[\Theta]$$

(2.38)

where

$$
\frac{d}{dt}[\Theta] \equiv
\begin{bmatrix}
0 & -d\theta/dt & 0 & 0 & 0 & 0 & 0 \\
d\theta/dt & 0 & 0 & 0 & 0 & 0 & 0 \\
0 & 0 & 0 & 0 & 0 & 0 & 0 \\
0 & 0 & 0 & 0 & 0 & 0 & 0 \\
0 & 0 & 0 & 0 & 0 & 0 & 0 \\
0 & 0 & 0 & 0 & 0 & 0 & 0 \\
0 & 0 & 0 & 0 & 0 & 0 & 0
\end{bmatrix}
$$

(2.39)

Therefore, equation (2.34) becomes

$$
\begin{aligned}
[T]\left(\frac{d}{dt}\psi\right) &= [T][T]^{-1}\left(\frac{d}{dt}[\Theta]\right)\psi_{dr} + \frac{d}{dt}\psi_{dr} \\
&= \left(\frac{d}{dt}[\Theta]\right)\psi_{dr} + \frac{d}{dt}\psi_{dr}
\end{aligned}
$$

(2.40)

Finally, (2.31) can be rewritten as follows:

$$v_{dr} = \left(\frac{d}{dt}[\Theta]\right)\psi_{dr} + \frac{d}{dt}\psi_{dr} - [R]\,i_{dr}$$

(2.41)

or

$$v_d = \frac{d}{dt} \psi_d - \psi_q \frac{d\theta}{dt} - R_a i_d$$

$$v_q = \frac{d}{dt} \psi_q + \psi_d \frac{d\theta}{dt} - R_a i_q$$

$$v_0 = \frac{d}{dt} \psi_0 - R_a i_0$$

$$v_{fd} = \frac{d}{dt} \psi_{fd} + R_{fd} i_{fd}$$

$$v_{kd} = \frac{d}{dt} \psi_{kd} + R_{kd} i_{kd} \qquad (2.42)$$

$$v_{gq} = \frac{d}{dt} \psi_{gq} + R_{gq} i_{gq}$$

$$v_{kq} = \frac{d}{dt} \psi_{kq} + R_{kq} i_{kq}$$

where $R_a = R_b = R_c$ are assumed in the stator circuit.

2.6 Swing Equation

The swing equation governing the motion of the rotor of a synchronous machine is based on the elementary principle of dynamics and, in the MKS unit system, it can be written as follows;

$$J \frac{d^2\theta_m}{dt^2} = T_a = T_m - T_e \qquad [\text{N·m}] \qquad (2.43)$$

where
$\quad$ J : the moment of inertia of all rotating masses [kg·m²]

$\quad$ θ_m : the mechanical angle of the rotor with respect to a fixed reference [rad]

$\quad$ t : time [s]

$\quad$ T_a : the accelerating torque [N·m]

$\quad$ T_m : the mechanical torque supplied by the prime mover [N·m]

$\quad$ T_e : electrical torque [N·m]

In a synchronous generator, a positive T_m accelerates the shaft, whereas a positive T_e is a decelerating torque. It is convenient to measure the rotor angular position θ with respect to a reference axis which rotates at synchronous angular velocity ω_{m0} [rad/s]. Then,

$$\theta_m = \omega_{m0} t + \delta_m \qquad [\text{rad}] \qquad (2.44)$$

where δ_m is the rotor mechanical angle with respect to the synchronously rotating reference axis. Since $d^2\theta_m/dt^2 = d^2\delta_m/dt^2$ from (2.44), (2.43) can be rewritten as

$$J \frac{d^2\delta_m}{dt^2} = T_a = T_m - T_e \qquad [\text{N·m}] \tag{2.45}$$

Multiplying both sides of (2.45) by ω_m which is the rotor angular velocity in mechanical radians per second, another form can be obtained as

$$J \omega_m \frac{d^2\delta_m}{dt^2} = P_a = P_m - P_e \qquad [\text{W}] \tag{2.46}$$

where P_m is the shaft power input to the machine, P_e is the electrical power crossing the air-gap, and P_a is the accelerating power which accounts for any unbalance between those two quantities. The angular momentum of the rotor $J\omega_m$ is called the inertia constant and is denoted as M. It is related with the kinetic energy of the rotating masses W_k, which is $(1/2)\ J\omega_m^2$ [J] as follows;

$$M \equiv J\omega_m = 2\ W_k/\omega_m \qquad [\text{J·s}] \tag{2.47}$$

Thus, the swing equation can be written as

$$M \frac{d^2\delta_m}{dt^2} = M \frac{d}{dt}\omega_m = P_m - P_e \qquad [\text{W}] \tag{2.48}$$

where $\omega_m = d\delta_m/dt$. In practice, ω_m does not differ significantly from the synchronous speed ω_{m0} when the machine is stable, thus from (2.47),

$$M \cong J\omega_{m0} \tag{2.49}$$

In machine data supplied for stability studies, usually the H constant is used as machine inertia and it is defined as

$$H = \frac{\text{stored kinetic energy in MJ at synchronous speed}}{\text{machine rating in MVA}} \tag{2.50}$$

and

$$H \equiv \frac{W_k}{S_{B3}} = \frac{\frac{1}{2}\ J\ \omega_{m0}^2}{S_{B3}} \cong \frac{\frac{1}{2}\ M\ \omega_{m0}}{S_{B3}} \qquad [\text{s}] \tag{2.51}$$

where S_{B3} is the three-phase rating of the machine in MVA.

From (2.51),

$$J = \frac{2H}{\omega_{m0}^2}\ S_{B3} = \frac{2H}{\omega_{m0}}\ T_0 \tag{2.52}$$

where T_0 [N·m] is base torque equal to the rated torque at rated speed and is obtained as

$$T_0 = S_{B3}/\omega_{m0} \tag{2.53}$$

Substituting (2.53) for T_0 (2.52),

$$\frac{2H}{\omega_{m0}} \frac{d^2 \delta_m}{dt^2} = \frac{T_a}{T_0} = \frac{T_m - T_e}{T_0} \tag{2.54}$$

In order to relate the machine inertial performance to the electrical network, it would be more useful to write the swing equation (2.54) in terms of electrical angle. The electrical torque angle δ_e means the angle between the field MMF and the resultant MMF in the air gap, both rotating at synchronous speed. The relation between the electrical angle δ_e and the rotor mechanical angle δ_m is given as

$$\delta_e = (p/2) \delta_m \tag{2.55}$$

where p is the number of poles. Similarly, the electrical angular velocity can be obtained as

$$\omega_e = (p/2) \omega_m \tag{2.56}$$

For simplicity, δ and ω will be used for the electrical angle and its velocity without any subscripts from now on.

Then, (2.54) becomes

$$\frac{2H}{\omega_0} \frac{d^2 \delta}{dt^2} = \overline{T}_a = \overline{T}_m - \overline{T}_e \quad \text{[pu]} \tag{2.57}$$

where $\overline{T}_a$, $\overline{T}_m$, and $\overline{T}_e$ are in per unit, and H and t are in seconds.

The swing equation (2.57) is valid if

(1) both δ and ω_0 have consistent units which may be mechanical or electrical degrees or radians
(2) H and the torque base T_0 for $\overline{T}_a$, $\overline{T}_m$, and $\overline{T}_e$ have the same MVA base (S_{B3}) in (2.51) and (2.52)

Considering that the angular speed ω is not quite different from ω_0, the p.u. accelerating power is nearly equal to the accelerating torque T_a. Thus, another form of the swing equation (2.57) can be obtained as

$$\frac{2H}{\omega_0} \frac{d^2\delta}{dt^2} \cong \overline{P}_a = \overline{P}_m - \overline{P}_e \quad [pu] \tag{2.58}$$

2.7 Per Unit Conversion

In order to convert the derived equations to more convenient forms to use, it is necessary to normalize the quantities involved in those equations to base quantities and to express them in per unit of bases. The synchronous machine model considered here has stator a-b-c circuits, a field circuit, a d-axis damper circuit, and two q-axis damper circuits. In the simulation, most of those circuits are coupled each other with the mutual inductances. The base quantities for each circuit should be chosen to simplify numerical values and to provide convenient simulation of equations. There are four general principles regarding the selection of base quantities:

A. For each circuit, all the electrical quantities can expressed by three base quantities which should involve all three dimensions of voltage, current, and time.

B. All the circuits must have the same base power because of the conservation of energy.

C. For two mutually coupled circuits, the equivalence of per unit mutual inductances will ascertain the equivalence of the two base powers. In other words, The reciprocity of normalized mutual inductances is one of the constraints to define new bases.

D. For two mutual coupling circuits, any per unit quantity must have the same physical value regardless of the referred circuit used for evaluation and normalization.

The base quantities in each circuit will have the notations in Table 2.1;

Table 2.1 List of Base Quantities in Per Unit System

Quantity	Unit	Stator	Rotor Circuit			
			Field	kd	gq	kq
3-Phase MVA	MVA	P_0	P_0	P_0	P_0	P_0
Phase Voltage (peak)	kV	v_{a0}	v_{fd0}	v_{kd0}	v_{gq0}	v_{kq0}
Current (peak)	kA	i_{a0}	i_{fd0}	i_{kd0}	i_{gq0}	i_{kq0}
Resistance	Ω	R_{a0}	R_{fd0}	R_{kd0}	R_{gq0}	R_{kq0}
Inductance	H	L_{a0}	L_{fd0}	L_{kd0}	L_{gq0}	L_{kq0}
Flux Linkage	kWb turn	ψ_{a0}	ψ_{fd0}	ψ_{kd0}	ψ_{gq0}	ψ_{kq0}
Angular Speed	rad/s	ω_0	ω_0	ω_0	ω_0	ω_0
Time	s	t_0	t_0	t_0	t_0	t_0

2.7.1 Choosing Bases for Stator Quantities

The variables v_d, v_q, i_d, i_q, ψ_d, and ψ_q are stator quantities since they relate to the a-b-c phase quantities directly through Park's transformation. Power, voltage, and rotor speed are chosen as the stator base quantities involving all three dimensions (v, i, and t):

P_0 ≡ base three phase MVA $\qquad$ [MVA]

v_{a0} ≡ stator base phase voltage (peak) $\quad$ [kV]

ω_0 ≡ base rotor speed $\qquad$ [elec rad/s]

where the subscript "$_0$" to indicate "base" quantity.

By combining these quantities, other dependent base quantities in the stator can be computed as follows:

i_{a0} = stator base current (peak) = $P_0/(3/2)v_{a0}$ $\qquad$ [kA]

R_{a0} = stator base resistance $\quad$ = v_{a0}/i_{a0} $\qquad$ [Ω] $\qquad$ (2.59)

L_{a0} = stator base inductance $\quad$ = R_{a0}/ω_0 $\qquad$ [H]

ψ_{a0} = stator base flux linkage $\;$ = $L_{a0} \cdot i_{a0}$= v_{a0}/ω_0 $\quad$ [kWb turn]

2.7.2 Choosing Bases for Rotor Quantities

From (2.27),the flux linkage equations for stator and field circuits are as follows;

$$\psi_d = -L_d\, i_d + M_{afd}\, i_{fd} + M_{akd}\, i_{kd}$$
$$\psi_{fd} = -\frac{3}{2} M_{afd}\, i_d + L_{fdfd}\, i_{fd} + M_{fdkd}\, i_{kd} \tag{2.60}$$

Assuming ψ_{a0} and ψ_{fd0} as the bases for the stator and field circuits respectively, (2.60) can be converted to per unit values as follows;

$$\left[\frac{\psi_d}{\psi_{a0}}\right] = -\left[\frac{L_d}{L_{a0}}\right]\left[\frac{i_d}{i_{a0}}\right] + \left[\frac{M_{afd}\, i_{fd0}}{\psi_{a0}}\quad 1\right]\left[\frac{i_{fd}}{i_{fd0}}\right] + \left[\frac{M_{akd}\, i_{kd0}}{\psi_{a0}}\quad 1\right]\left[\frac{i_{kd}}{i_{kd0}}\right]$$

$$\left[\frac{\psi_{fd}}{\psi_{fd0}}\right] = -\left[\frac{\frac{3}{2} M_{afd}}{\psi_{fd0}}\, i_{a0}\quad 1\right]\left[\frac{i_d}{i_{a0}}\right] + \left[\frac{L_{fdfd}}{L_{fd0}}\right]\left[\frac{i_{fd}}{i_{fd0}}\right] + \left[\frac{M_{fdkd}\, i_{kd0}}{\psi_{fd0}}\quad 1\right]\left[\frac{i_{kd}}{i_{kd0}}\right] \tag{2.61}$$

where $\psi_{a0} = L_{a0}\cdot i_{a0}$ and $\psi_{fd0} = L_{fd0}\cdot i_{fd0}$.

Applying the principle-C in section 2.7, the per unit mutual inductances between the stator and the field circuit should be equal in the two equations in order to be reciprocal;

$$\left[\frac{M_{afd}\, i_{fd0}}{\psi_{a0}}\quad 1\right] = \left[\frac{\frac{3}{2} M_{afd}}{\psi_{fd0}}\, i_{a0}\quad 1\right] \equiv \overline{M}_{afd} \tag{2.62}$$

where $\overline{M}_{afd}$ is per unit mutual inductance.

Multiplying ω_0 on both sides of (2.62) and rearranging,

$$\psi_{fd0}\, i_{fd0}\, \omega_0 = \frac{3}{2}\, \psi_{a0}\, i_{a0}\, \omega_0 \tag{2.63}$$

Since $v_{a0} = \psi_{a0}\, \omega_0$ and $v_{fd0} = \psi_{fd0}\, \omega_0$,

$$v_{fd0}\, i_{fd0} = \frac{3}{2}\, v_{a0}\, i_{a0} = P_0 \quad [\text{MVA}] \tag{2.64}$$

which gives one constraint to choose base quantities.
In addition, (2.62) gives the base of M_{afd} ;

$$\overline{M}_{afd} = \frac{M_{afd}}{\dfrac{\psi_{a0}}{i_{fd0}}} = \frac{M_{afd}}{\dfrac{L_{a0}\, i_{a0}}{i_{fd0}}} \equiv \frac{M_{afd}}{M_{afd0}} \tag{2.65}$$

Thus, $M_{afd0} = \dfrac{L_{a0} \, i_{a0}}{i_{fd0}}$ (2.66)

For the principle-D mentioned in section 2.7, "L_{ad} base reciprocal per unit system" has been used in the simulation of synchronous machine to overcome unknown machine data which can not be measured. According to the L_{ad} reciprocal system, all per unit mutual inductances between stator and rotor are assumed to be equal to one another in each d and q-axis as follows;

$$\text{d-axis}: \quad \overline{L}_{ad} = \overline{M}_{afd} = \overline{M}_{akd}$$
$$\text{q-axis}: \quad \overline{L}_{aq} = \overline{M}_{agq} = \overline{M}_{akq}$$

(2.67)

where $\overline{L}_{ad} \equiv \overline{L}_d - \overline{L}_l$ and $\overline{L}_{aq} \equiv \overline{L}_q - \overline{L}_l$ as in (2.68), and $\overline{L}_l$. is the per unit stator inductances.

With this per unit system, each self inductance of the stator or rotor circuits is assumed to be the sum of mutual and leakage inductances as in Figure 2.3;

$$\text{d-axis:} \quad \overline{L}_d \equiv \overline{L}_{ad} + \overline{L}_l \qquad \text{q-axis:} \quad \overline{L}_q \equiv \overline{L}_{aq} + \overline{L}_l$$

$$\overline{L}_{fdfd} \equiv \overline{M}_{afd} + \overline{l}_{fd} \qquad\qquad \overline{L}_{gqgq} \equiv \overline{M}_{agq} + \overline{l}_{gq}$$

$$\overline{L}_{kdkd} \equiv \overline{M}_{akd} + \overline{l}_{kd} \qquad\qquad \overline{L}_{kqkq} \equiv \overline{M}_{akq} + \overline{l}_{kq}$$

(2.68)

where the l_{fd}, l_{kd}, l_{gq}, and l_{kq} are the leakage inductances for each circuit. In Figure 2.3, it would be notable that no leakage flux exists linking two circuits in the d-axis or q-axis.

Now, L_{ad} is a stator quantity and its pu value can be obtained as

$$\overline{L}_{ad} = \dfrac{L_{ad}}{L_{a0}} = \dfrac{L_{ad}}{\psi_{a0}} \, i_{a0}$$

(2.69)

From (2.65), the pu of M_{afd} is

$$\overline{M}_{afd} = \dfrac{M_{afd}}{\psi_{a0}} \, i_{fd0}$$

(2.70)

Since $\overline{L}_{ad} = \overline{M}_{afd}$ from (2.67),

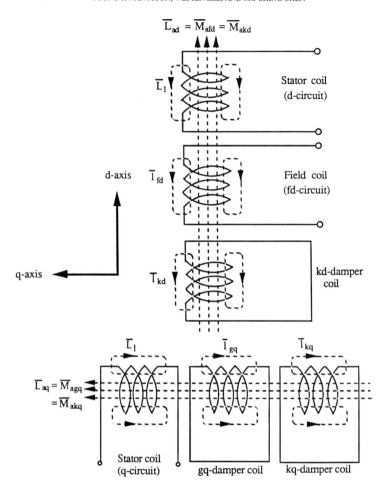

Figure 2.3 Mutual Inductances in the L_{ad} Based Reciprocal Per Unit System

$$\frac{L_{ad}}{\psi_{a0}} i_{a0} = \frac{M_{afd}}{\psi_{a0}} i_{fd0} \qquad (2.71)$$

or
$$i_{fd0} = \frac{L_{ad}}{M_{afd}} i_{a0} \qquad [kA] \qquad (2.72)$$

which gives another constraint to determine the base quantities.

In summary, the base quantities for the field circuit, which are i_{fd0}, v_{fd0}, and M_{afd0}, will be calculated from the following three equations;

From (2.72),
$$i_{fd0} = \frac{L_{ad}}{M_{afd}} i_{a0} \qquad [kA]$$

From (2.64),
$$v_{fd0} = \frac{P_0}{i_{fd0}} \qquad [MVA] \qquad (2.73)$$

From (2.66),
$$M_{afd0} = \frac{L_{a0} i_{a0}}{i_{fd0}} \qquad (2.74)$$

The base quantities in kd,gq,and kq damper circuits can be chosen similarly. Applying the same procedure, the following bases for the kd-circuit can be obtained:

$$i_{kd0} = \frac{L_{ad}}{M_{akd}} i_{a0} \qquad [kA]$$

$$v_{kd0} = \frac{P_0}{i_{kd0}} \qquad [MVA] \qquad (2.75)$$

$$M_{akd0} = \frac{L_{a0} i_{a0}}{i_{kd0}} \qquad [H]$$

Similarly, the bases for gq-circuit are

$$i_{gq0} = \frac{L_{aq}}{M_{agq}} i_{a0} \qquad [kA]$$

$$v_{gq0} = \frac{P_0}{i_{gq0}} \qquad [MVA] \qquad (2.76)$$

$$M_{agq0} = \frac{L_{a0} i_{a0}}{i_{gq0}} \qquad [H]$$

Finally, the bases for kq-circuit are

$$i_{kq0} = \frac{L_{aq}}{M_{akq}} i_{a0} \qquad [kA]$$

$$v_{kq0} = \frac{P_0}{i_{kq0}} \qquad [MVA] \qquad\qquad (2.77)$$

$$M_{akq0} = \frac{L_{a0} i_{a0}}{i_{kq0}} \qquad [H]$$

2.7.3 Normalized Fundamental Equations

2.7.3.1 The Normalized Flux-Linkage Equations

After normalizing, we can rewrite (3.27) in a matrix form as follows:

$$
\begin{bmatrix} \overline{\psi}_d \\ \overline{\psi}_q \\ \overline{\psi}_0 \\ \overline{\psi}_{fd} \\ \overline{\psi}_{kd} \\ \overline{\psi}_{gq} \\ \overline{\psi}_{kq} \end{bmatrix}
=
\begin{bmatrix}
-\overline{L}_d & 0 & 0 & \overline{M}_{afd} & \overline{M}_{akd} & 0 & 0 \\
0 & -\overline{L}_q & 0 & 0 & 0 & \overline{M}_{agq} & \overline{M}_{akq} \\
0 & 0 & -\overline{L}_0 & 0 & 0 & 0 & 0 \\
-\overline{M}_{afd} & 0 & 0 & \overline{L}_{fdfd} & \overline{M}_{fdkd} & 0 & 0 \\
-\overline{M}_{akd} & 0 & 0 & \overline{M}_{kdfd} & \overline{L}_{kdkd} & 0 & 0 \\
0 & -\overline{M}_{agq} & 0 & 0 & 0 & \overline{L}_{gqgq} & \overline{M}_{gqkq} \\
0 & -\overline{M}_{akq} & 0 & 0 & 0 & \overline{M}_{kqgq} & \overline{L}_{kqkq}
\end{bmatrix}
\begin{bmatrix} \overline{i}_d \\ \overline{i}_q \\ \overline{i}_0 \\ \overline{i}_{fd} \\ \overline{i}_{kd} \\ \overline{i}_{gq} \\ \overline{i}_{kq} \end{bmatrix}
\qquad (2.78)
$$

where $\overline{M}_{fdkd} = \overline{M}_{kdfd}$ and $\overline{M}_{fqkq} = \overline{M}_{kqfq}$.

Equation (2.78) can be rewritten as

$$\overline{\psi}_d = -\overline{L}_d \overline{i}_d + \overline{M}_{afd} \overline{i}_{fd} + \overline{M}_{akd} \overline{i}_{kd}$$

$$\overline{\psi}_q = -\overline{L}_q \overline{i}_q + \overline{M}_{agq} \overline{i}_{gq} + \overline{M}_{akq} \overline{i}_{kq} \qquad (2.79)$$

$$\overline{\psi}_0 = -\overline{L}_0 \overline{i}_0$$

$$\overline{\psi}_{fd} = -\overline{M}_{afd} \overline{i}_d + \overline{L}_{fdfd} \overline{i}_{fd} + \overline{M}_{fdkd} \overline{i}_{kd}$$

$$\overline{\psi}_{kd} = -\overline{M}_{akd} \overline{i}_d + \overline{M}_{fdkd} \overline{i}_{fd} + \overline{L}_{kdkd} \overline{i}_{kd}$$

$$\overline{\psi}_{gq} = -\overline{M}_{agq} \overline{i}_q + \overline{L}_{gqgq} \overline{i}_{gq} + \overline{M}_{gqkq} \overline{i}_{kq} \qquad (2.80)$$

$$\overline{\psi}_{kq} = -\overline{M}_{akq} \overline{i}_q + \overline{M}_{gqkq} \overline{i}_{gq} + \overline{L}_{kqkq} \overline{i}_{kq}$$

2.7.3.2   The Normalized Voltage Equations

Since $v_{a0} = \omega_0\, \psi_{a0} = R_{a0}\, i_{a0}$ from (2.59), the stator voltage equation in (2.42) can be normalized as

$$\frac{v_d}{v_{a0}} = \frac{v_d}{(\omega_0\, \psi_{a0})} = \frac{d}{d(\omega_0 t)}\frac{\psi_d}{\psi_{a0}} - \frac{\psi_q}{\psi_{a0}}\frac{d\theta}{d(\omega_0 t)} - \frac{R_a}{R_{a0}}\frac{i_d}{i_{a0}} \tag{2.81}$$

or can be expressed in per unit form;

$$\overline{v}_d = \frac{d}{d\overline{t}}\overline{\psi}_d - \overline{\psi}_q\frac{d\theta}{d\overline{t}} - \overline{R_a}\,\overline{i}_d \tag{2.82}$$

where $\overline{t} \equiv$ per unit time $= \omega_0\, t$

With the same procedure, the normalized voltage equations in (2.42) are obtained;

$$\overline{v}_d = \frac{d}{d\overline{t}}\overline{\psi}_d - \overline{\psi}_q\frac{d\theta}{d\overline{t}} - \overline{R}_a\,\overline{i}_d$$

$$\overline{v}_q = \frac{d}{d\overline{t}}\overline{\psi}_q + \overline{\psi}_d\frac{d\theta}{d\overline{t}} - \overline{R}_a\,\overline{i}_q \tag{2.83}$$

$$\overline{v}_0 = \frac{d}{d\overline{t}}\overline{\psi}_0 - \overline{R}_a\,\overline{i}_0$$

$$\overline{v}_{fd} = \frac{d}{d\overline{t}}\overline{\psi}_{fd} + \overline{R}_{fd}\,\overline{i}_{fd}$$

$$\overline{v}_{kd} = \frac{d}{d\overline{t}}\overline{\psi}_{kd} + \overline{R}_{kd}\,\overline{i}_{kd}$$

$$\overline{v}_{gq} = \frac{d}{d\overline{t}}\overline{\psi}_{gq} + \overline{R}_{gq}\,\overline{i}_{gq} \tag{2.84}$$

$$\overline{v}_{kq} = \frac{d}{d\overline{t}}\overline{\psi}_{kq} + \overline{R}_{kq}\,\overline{i}_{kq}$$

2.7.3.3   The Normalized Swing Equation

The equation (2.57) is not completely normalized because the angular speed ω and the time are given in MKS units. The swing equation in (2.57) can be written as

$$\frac{2H}{\omega_0}\frac{d\omega}{dt} = \overline{T}_m - \overline{T}_e \quad \text{[pu]} \tag{2.85}$$

Defining per unit time $\overline{t}$ and per unit angular velocity $\overline{\omega}$ as

$$\bar{t} \equiv \omega_0 t \qquad \bar{\omega} \equiv \frac{\omega}{\omega_0} \qquad (2.86)$$

and by substituting them into (2.85), the normalized swing equation is given as

$$2 H \omega_0 \frac{d\bar{\omega}}{d\bar{t}} = \bar{T}_m - \bar{T}_e \quad [pu] \qquad (2.87)$$

or

$$2 \bar{H} \frac{d\bar{\omega}}{d\bar{t}} = \bar{T}_m - \bar{T}_e \quad [pu] \qquad (2.88)$$

where $\bar{H} \equiv \omega_0 H$

Using power quantities, the swing equation becomes

$$2 \bar{H} \bar{\omega} \frac{d\bar{\omega}}{d\bar{t}} = \bar{P}_m - \bar{P}_e \quad [pu] \qquad (2.89)$$

2.8 Machine Constants

In the fundamental equations (2.79), (2.80), (2.83), and (2.84), the zero sequence circuit is completely uncoupled from the other circuits and can be treated separately. The flux linkage equations and the voltage equations except the zero sequence set are expressed with seventeen machine constants:

Five resistances - $R_a, R_{fd}, R_{kd}, R_{gq}, R_{kq}$ (2.90)

Twelve inductances - $L_d, L_q, L_{fdfd}, L_{kdkd}, L_{gqgq}, L_{kqkq}$

$M_{afd}, M_{akd}, M_{agq}, M_{akq}, M_{fdkd}, M_{gqkq}$ (2.91)

Unfortunately, only eleven data related with the seventeen constants can be obtained from tests:

One resistance - R_a

Six inductances - $L_d, L_d', L_d'', L_q, L_q', L_q''$

Four time constants - $T_{do}', T_{do}'', T_{qo}', T_{qo}''$

which are defined as

L_d ≡ d-axis synchronous inductance with both of the fd-circuit and the kd-circuit open

L_d' ≡ d-axis transient inductance with the fd-circuit short and the kd-circuit open.

L_d'' ≡ d-axis subtransient inductance with both of the fd-circuit and the kd-circuit short.

T_{do}' ≡ The time constant of the field current i_{fd} with both of the d-circuit and the kd-circuit open (d-axis transient open circuit time constant).

T_{do}'' ≡ The time constant of the damper current i_{kd} with the d-circuit open and the kd-circuit short (d-axis subtransient open circuit time constant).

Similarly, L_q, L_q', L_q'', T_{qo}', and T_{qo}'' can be defined. In the following sections, the relations between the required data for the machine model and the available data will be explained as well as the assumptions to overcome the unknown data.

2.8.1   Inductances

In this section, the relation between L_d' and L_d'' and the data in (2.91) will be explained. In the d-axis, the d-circuit, the fd-circuit, and the kd-circuit circuit have a common magnetic flux together as in Figure 2.3. . Assuming $R_{fd} \approx 0$, the following flux equations can be obtained from (2.79) and (2.80) by shorting the fd-circuit and opening the kd-winding:

$$\psi_d = -L_d\, i_d + M_{afd}\, i_{fd}$$
$$\psi_{fd} = -M_{afd}\, i_d + L_{fdfd}\, i_{fd} = 0 \tag{2.92}$$

Then, the d-axis transient inductance L_d' is obtained as

$$L_d' \equiv -\frac{\psi_d}{i_d} = L_d - \frac{M_{afd}^2}{L_{fdfd}} \tag{2.93}$$

Similarly, assuming $R_{fd} \approx R_{kd} \approx 0$, the next flux equations are obtained by shorting both of the fd-circuit and the kd-circuit:

$$\psi_d = -L_d\, i_d + M_{afd}\, i_{fd} + M_{akd}\, i_{kd}$$
$$\psi_{fd} = -M_{afd}\, i_d + L_{fdfd}\, i_{fd} + M_{fdkd}\, i_{kd} = 0$$
$$\psi_{kd} = -M_{akd}\, i_d + L_{fdkd}\, i_{fd} + L_{kdkd}\, i_{kd} = 0 \tag{2.94}$$

Then, the d-axis subtransient inductance L_d'' is obtained as

$$L_d'' \equiv -\frac{\psi_d}{i_d} = L_d - \frac{M_{akd}^2}{L_{kdkd}} - \frac{(L_{kdkd}\, M_{afd} - M_{akd}\, M_{fdkd})^2}{L_{kdkd}\left(L_{fdfd}\, L_{kdkd} - M_{fdkd}^2\right)} \tag{2.95}$$

With the same procedure, L_q' and L_q'' in the q-axis can be obtained to get the similar form. Generally, a synchronous machine has very small M_{agq}, $L_q' \approx L_q$ is a good approximation.

2.8.2 Time Constants

In general, the field circuit of a synchronous machine has a large inductance L_{fdfd} to excite the rotor while it has a small resistance R_{fd} to reduce loss. On the contrary, the kd-damper coil representing the eddy current effect in the rotor has a relatively small inductance L_{kdkd}. Therefore, the time constant (L/R) of the fd-circuit is much longer than that of the kd-circuit. In other words, the transient phenomenon diminishes much faster in the kd-circuit than in the fd-circuit.

T_{do}', as defined at the beginning of this section, is the time constant of i_{fd} with both of the d-circuit and the fd-circuit open. Thus

$$T_{do}' = L_{fdfd}/R_{fd} \qquad (2.96)$$

T_{do}'' is also defined at the beginning of this section as the time constant of i_{kd} with the d-circuit open and the fd-circuit short and assuming $R_{fd} \approx 0$. Thus, from (2.80) and (2.84)

$$\psi_{fd} = L_{fdfd}\, i_{fd} + M_{fdkd}\, i_{kd} = 0 \qquad (2.97)$$

$$v_{kd} = R_{kd}\, i_{kd} + L_{kdkd} \frac{d}{dt}(i_{kd}) + M_{fdkd} \frac{d}{dt}(i_{fd}) \qquad (2.98)$$

From (2.97) and (2.98),

$$T_{do}'' = \left(L_{kdkd} - \frac{M_{fdkd}^2}{L_{fdfd}} \right) / R_{kd} \qquad (2.99)$$

Besides T_{do}' and T_{do}'', T_d' and T_d'' are also defined and used in the synchronous machine modeling as follows:

$T_d' \equiv$ The time constant of the field current i_{fd} with the d-circuit short and the kd-circuit open (d-axis transient short circuit time constant)

$T_d'' \equiv$ The time constant of the damper current i_{kd} with both of the d-circuit and the kd-circuit short (d-axis subtransient short circuit time constant)

According to the definition of T_d', the following equations are obtained:

$$i_{kd} = 0$$

$$\psi_d = -L_d i_d + M_{afd} i_{fd} = 0$$

$$v_{fd} = R_{fd} i_{fd} + \frac{d}{dt}(- M_{afd} i_d + L_{fdfd} i_{fd}) \qquad (2.100)$$

From (2.100), T_d' is calculated as

$$T_d' = \frac{1}{R_{fd}} \frac{L_d L_{fdfd} - M_{afd}^2}{L_d} = T_{do}' \frac{L_d'}{L_d} \qquad (2.101)$$

where L_d' and T_{do}' are obtained in (2.93) and (2.96) respectively.

Regarding T_d'', the following equations are obtained ignoring the resistances in the d and fd circuits:

$$\psi_d = -L_d i_d + M_{afd} i_{fd} + M_{akd} i_{kd} = 0$$

$$\psi_{fd} = - M_{afd} i_d + L_{fdfd} i_{fd} + M_{fdkd} i_{kd} = 0$$

$$v_{kd} = R_{kd} i_{kd} + \frac{d}{dt}(- M_{akd} i_d + M_{fdkd} i_{fd} + L_{kdkd} i_{kd}) \qquad (2.102)$$

From (2.100), T_d'' is calculated as

$$T_d'' = T_{do}'' \frac{L_d''}{L_d'} \qquad (2.103)$$

where L_d'' and T_{do}'' are obtained in (2.95) and (2.99) respectively.

Consequently, T_d' and T_d'' can be calculated with T_{do}', T_{do}'', L_d, L_d', and L_d'' as in (2.101) and (2.103).

2.9   Park's Equations using Machine Constants and the Equivalent Circuit

In order to overcome the deficient machine data for the model considered here, several assumptions have been used with acceptable reasons. "L_{ad} base reciprocal per unit system" discussed in the section 3.6 is one of such assumptions which is necessary for building the Park's equation using machine constants.

The leakage flux in the d-circuit mainly attributed to the leakage at the coil edge can be measured. Setting the stator current very high to saturate the magnetic circuit, the incremental flux linkage due to the higher current corresponds to the leakage flux L_l. With the assumed three winding transformer in Figure 2.3, the equivalent circuit of the Park's equations ,which is seen from the stator, is shown in Figure 2.4. where r_f and l_f are the resistance and the inductance of the fd-circuit seen from d-winding, and r_{kd} and l_{kd} are those of the kd-circuit.

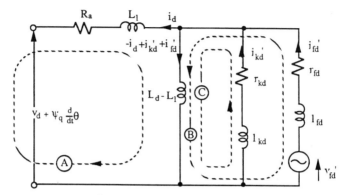

Figure 2.4 d-axis Equivalent Circuit

As defined in (2.67), L_{ad} is the inductance of the three winding transformer with no leakage flux;

$$L_{ad} \equiv L_d - L_l \qquad (2.104)$$

where L_l is the leakage inductance in the d-axis stator coil.

Similarly, in the q-axis,

$$L_{aq} \equiv L_q - L_l \qquad (2.105)$$

L_d', as defined in (2.93), is the inductance seen from d-circuit with the fd-circuit short and the kd-circuit open. Thus, from the equivalent circuit in Figure 2.4,

$$L_d' = L_l + \frac{(L_d - L_l)\, l_{fd}}{(L_d - L_l) + l_{fd}} \qquad (2.106)$$

From (2.106)

$$l_{fd} = \frac{(L_d' - L_l)(L_d - L_l)}{L_d - L_d'} \qquad (2.107)$$

L_d'', as defined in (2.95), is the inductance seen from d-circuit with both of the fd-circuit and the kd-circuit short. Again from the equivalent circuit in Figure 2. ,

$$L_d'' = L_l + \cfrac{1}{\cfrac{1}{L_d - L_l} + \cfrac{1}{l_{fd}} + \cfrac{1}{l_{kd}}} \tag{2.108}$$

From (2.108)

$$l_{kd} = \frac{(L_d'' - L_l)(L_d' - L_l)}{L_d' - L_d''} \tag{2.109}$$

T_{do}', as defined in (2.96), is the time constant of i_{fd} with both of the d-circuit and the kd-circuit open. Again from the equivalent circuit in Figure 2.4,

$$T_{do}' = \frac{1}{r_{fd}}(L_d - L_l + l_{fd}) \tag{2.110}$$

From (2.107) and (2.110)

$$r_{fd} = \frac{1}{T_{do}'} \frac{(L_d - L_l)^2}{L_d - L_d'} \tag{2.111}$$

T_{do}'', as defined in (2.99), is the time constant of i_{kd} with the d-circuit open and the fd-circuit short. Once again from the equivalent circuit in Figure 2.4,

$$T_{do}'' = \frac{1}{r_{kd}}\left\{l_{kd} + \frac{(L_d - L_l)l_{fd}}{(L_d - L_l) + l_{fd}}\right\} \tag{2.112}$$

From (2.107), (2.109), and (2.112),

$$r_{kd} = \frac{1}{T_{do}''} \frac{(L_d' - L_l)^2}{L_d' - L_d''} \tag{2.113}$$

In Figure 2.4, i_{fd}' and i_{kd}' are the equivalent current of the fd-circuit and the kd-circuit respectively, and v_{fd}' is the exciter voltage, all of which are seen from the d-circuit. The voltage-current equation for the loop A in Figure 2.4 is

$$\begin{aligned}
v_d + \psi_q \frac{d\theta}{dt} &= -R_a i_d - \frac{d}{dt}(L_l i_d) + \frac{d}{dt}(L_d - L_l)(-i_d + i_{fd}' + i_{kd}') \\
&= -R_a i_d - \frac{d}{dt}(L_d i_d) + \frac{d}{dt}(L_d - L_l) i_{fd}' + \frac{d}{dt}(L_d - L_l) i_{kd}'
\end{aligned} \tag{2.114}$$

On the other hand, by substituting the first equation of (2.79) for ψ_d in the first equation of (2.83),

$$v_d + \psi_q \frac{d\theta}{dt} = -R_a i_d - \frac{d}{dt}(L_d i_d) + \frac{d}{dt}M_{afd} i_{fd} + \frac{d}{dt}M_{akd} i_{kd} \qquad (2.115)$$

Comparing (2.114) with (2.115),

$$i_{fd}'(t) = \frac{M_{afd}}{L_d - L_l} i_{fd}(t)$$
$$i_{kd}'(t) = \frac{M_{akd}}{L_d - L_l} i_{kd}(t) \qquad (2.116)$$

Using the same procedure, the similar equations for the q-axis can be obtained:

$$i_{gq}'(t) = \frac{M_{agq}}{L_q - L_l} i_{gq}(t)$$
$$i_{kq}'(t) = \frac{M_{akq}}{L_q - L_l} i_{kq}(t) \qquad (2.117)$$

For the loop B in Figure 2.4,

$$v_{fd}' = r_{fd} i_{fd}' + \frac{d}{dt}(l_{fd} i_{fd}') + \frac{d}{dt}(L_d - L_l)(-i_d + i_{fd}' + i_{kd}') \qquad (2.118)$$

Substituting (2.111) and (2.113) for r_{fd} and l_{fd} in (2.118),

$$v_{fd}' = \frac{(L_d - L_l)^2}{(L_d - L_d')T_{do}'} \frac{M_{afd}}{(L_d - L_l)} i_{fd} + \frac{d}{dt}\left\{ \frac{(L_d' - L_l)(L_d - L_l)}{(L_d - L_d')} + (L_d - L_l) \right\} \frac{M_{afd}}{(L_d - L_l)} i_{fd}$$
$$- \frac{d}{dt}(L_d - L_l) i_d + \frac{d}{dt}(L_d - L_l) \frac{M_{akd}}{(L_d - L_l)} i_{kd} \qquad (2.119)$$

Since $T_{do}' = L_{fdfd}/R_{fd}$ and $L_d - L_d' = M_{afd}^2/L_{fdfd}$ from (2.96) and (2.93), both sides of (2.119) are multiplied by $M_{afd}/(L_d - L_l)$ to give

$$\frac{M_{afd}}{L_d - L_l} v_{fd}' = R_{fd} i_{fd} + \frac{d}{dt}L_{fdfd} i_{fd} - \frac{d}{dt}M_{afd} i_d + \frac{d}{dt}\frac{M_{akd} M_{afd}}{L_d - L_l} i_{kd} \qquad (2.120)$$

On the other hand, by substituting the first equation in (2.80) for $\psi_{fd}(t)$ in the first equation of (2.84),

$$v_{fd}(t) = R_{fd} i_{fd} + \frac{d}{dt}L_{fdfd} i_{fd} - \frac{d}{dt}M_{afd} i_d + \frac{d}{dt}M_{fdkd} i_{kd} \qquad (2.121)$$

Comparing (2.120) with (2.121),

$$v_{fd}'(t) = \frac{L_d - L_l}{M_{afd}} v_{fd}(t) \tag{2.122}$$

and

$$M_{fdkd} = \frac{M_{akd} M_{afd}}{L_d - L_l} \tag{2.123}$$

For the q-axis,

$$M_{gqkq} = \frac{M_{akq} M_{agq}}{L_d - L_l} \tag{2.124}$$

Finally, the equivalent circuit for the Park's equations in (2.83) and (2.84) is obtained as shown in Figure 2.5, where only measurable machine constants are used.

2.10   Modified Park's Equations

For the system analysis, it is not necessary to calculate all the four currents - $i_{fd}(t)$, $i_{kd}(t)$, $i_{gq}(t)$, and $i_{kq}(t)$. For example, in order to get the flux linkage ψ_d in the d-axis, which is expressed as

$$\psi_d = - L_d i_d + M_{afd} i_{fd} + M_{akd} i_{kd} \tag{2.125}$$

it is sufficient to know the two values of $M_{afd} i_{fd}$ and $M_{akd} i_{kd}$ instead of the values M_{afd}, i_{fd}, M_{akd}, and i_{kd} separately. To apply this idea, the following four variables are defined as

$$e_{q1}(t) \equiv \omega_0 M_{afd} i_{fd}(t)$$
$$e_{q2}(t) \equiv \omega_0 M_{akd} i_{kd}(t)$$
$$e_{d1}(t) \equiv \omega_0 M_{agq} i_{gq}(t)$$
$$e_{d2}(t) \equiv \omega_0 M_{akq} i_{kq}(t)$$

$$\tag{2.126}$$

all of which are fictitious voltage proportional to the rotor currents. Using these new variables, it is not necessary to get M_{afd}, M_{akd}, M_{agq}, and M_{akq} as far as the d- and q-circuits are concerned. The ω_0 multiplied on the right side of (2.126) is only a constant coefficient to express the new variables as voltages for convenience, and it does not matter to use any value for it. Each of the variables in (2.126) has either a subscript d or q indicating that the induced voltages proportional to $i_{fd}(t)$ and $i_{kd}(t)$ are the speed voltages in the q-axis direction while those proportional to $i_{gq}(t)$ and $i_{kq}(t)$ are in the d-axis direction.

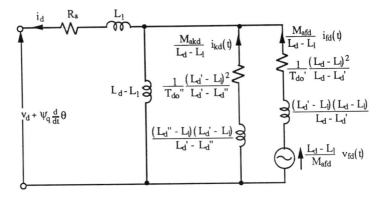

(a) d-axis Equivalent Circuit

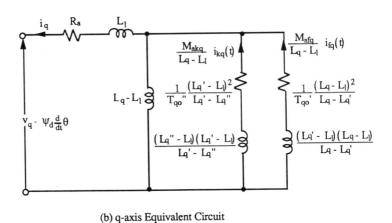

(b) q-axis Equivalent Circuit

Figure 2.5 Equivalent Circuits of a Synchronous Machine

In addition to the voltages proportional to the rotor currents, more variables related with the fluxes ψ_{fd}, ψ_{kd}, ψ_{gq}, and ψ_{kq} should be defined in order to accomplish the modified Park's equations:

$$e_q{}'(t) \equiv \frac{\omega_0 M_{afd}}{L_{fdfd}} \psi_{fd}(t)$$

$$e_q{}''(t) \equiv \frac{\omega_0 M_{akd}}{L_{kdkd}} \psi_{kd}(t)$$

$$e_d{}'(t) \equiv \frac{\omega_0 M_{agq}}{L_{gqgq}} \psi_{gq}(t)$$ (2.127)

$$e_d{}''(t) \equiv \frac{\omega_0 M_{akq}}{L_{kqkq}} \psi_{kq}(t)$$

Consequently, eight new variables are introduced as

$$e_{q1}, e_{q2}, e_{d1}, e_{d2}, e_q{}', e_q{}'', e_d{}', \text{ and } e_d{}''$$

substituting eight variables

$$i_{fd}(t), i_{kd}(t), i_{gq}(t), i_{kq}(t), \psi_{fd}, \psi_{kd}, \psi_{gq}, \text{ and } \psi_{kq}$$

and eight unknown machine data

$$L_{fdfd}, L_{kdkd}, L_{gqgq}, L_{kqkq}, M_{afd}, M_{akd}, M_{agq}, L_{akq}$$

Finally, the modified Park's equation corresponding to the Figure 2.4 using the assumptions in the previous section can be described as follows:

(1) Stator Winding Flux Linkages

$$\omega_0 \psi_d(t) = - \omega_0 L_d i_d(t) + e_{q1}(t) + e_{q2}(t)$$
$$\omega_0 \psi_q(t) = - \omega_0 L_q i_q(t) - e_{d1}(t) - e_{d2}(t)$$ (2.128)
$$\omega_0 \psi_0(t) = - \omega_0 L_0 i_0(t)$$

(2) Rotor Winding Flux Linkages

$$e_q'(t) = -\omega_0 (L_d - L_d') i_d(t) + e_{q1}(t) + \frac{L_d - L_d'}{L_d - L_l} e_{q2}(t)$$

$$e_d'(t) = \omega_0 (L_q - L_q') i_q(t) + e_{d1}(t) + \frac{L_q - L_q'}{L_q - L_q} e_{d2}(t)$$

$$e_q''(t) = -\frac{\omega_0 (L_d - L_l)}{K_d} i_d(t) + \frac{1}{K_d} e_{q1}(t) + e_{q2}(t)$$

$$e_d''(t) = \frac{\omega_0 (L_q - L_l)}{K_q} i_q(t) + \frac{1}{K_q} e_{d1}(t) + e_{d2}(t)$$

(2.129)

where

$$K_d = 1 + \frac{(L_d' - L_l)(L_d'' - L_l)}{(L_d' - L_d'')(L_d - L_l)}$$

$$K_q = 1 + \frac{(L_q' - L_l)(L_q'' - L_l)}{(L_q' - L_q'')(L_q - L_l)}$$

(2.130)

(3) Stator Winding Voltage

$$v_d(t) = \frac{d}{dt} \psi_d(t) - \psi_q \frac{d}{dt}\theta(t) - R_a i_d(t)$$

$$v_q(t) = \frac{d}{dt} \psi_q(t) + \psi_d \frac{d}{dt}\theta(t) - R_a i_q(t)$$

$$v_0(t) = \frac{d}{dt} \psi_0(t) \qquad\qquad - R_a i_d(t)$$

(2.131)

(4) Rotor Winding Voltage

$$\frac{N_{fd}}{N_a} \frac{\omega_0 (L_d - L_l)}{R_{fd}} v_{fd}(t) = T_{do}' \frac{d}{dt} e_q'(t) + e_{q1}(t)$$

$$T_{qo}' \frac{d}{dt} e_d'(t) + e_{d1}(t) = 0$$

$$T_{do}'' \frac{d}{dt} e_q''(t) + \frac{(L_d' - L_l)^2}{K_d (L_d' - L_d'')(L_d - L_l)} e_{q2}(t) = 0$$

$$T_{qo}'' \frac{d}{dt} e_d''(t) + \frac{(L_q' - L_l)^2}{K_q (L_q' - L_q'')(L_q - L_l)} e_{d2}(t) = 0$$

(2.132)

where N_a and N_{fd} are the number of turns of the stator coil and the field coil respectively.

CHAPTER 3

POWER SYSTEM STABILITY ANALYSIS

3.1 Fundamental Equations for Stability Analysis

In this section, fundamental equations used in the stability analysis are developed. The fundamental equations consist of the state differential equations for the generator internal states and the algebraic equations for the system network states.

3.1.1 Differential Equations for Generator Internal States

(a) The modified swing equation from (2.88) including a damping power coefficient D:

$$\frac{d\delta(t)}{dt} = \omega(t) - 1 \tag{3.1}$$

$$\frac{d\omega(t)}{dt} = \frac{1}{2H}[\, T_m(t) - T_e(t) - D\,\{\omega(t) - 1\}\,] \tag{3.2}$$

where all the units in (3.1) are in per unit.

(b) The modified rotor flux linkage equations from (2.132):

(e_{q1}, e_{q2}, e_{d1}, and e_{d2} are substituted by $e_q'(t)$, $e_d'(t)$, $e_q''(t)$, and $e_d''(t)$ using (2.129).)

- fd-winding on d-axis

$$\frac{de_q'(t)}{dt} = \frac{1}{T_{do}'}[e_{fd}(t) - e_{q1}(t)]$$

$$= \frac{1}{T_{do}'}\left[\, e_{fd}(t) + \frac{(L_d - L_d')(L_d' - L_d'')}{(L_d' - L_l)^2}K_d\, e_q''(t) - \left\{1 + \frac{(L_d - L_d')(L_d' - L_d'')}{(L_d' - L_l)^2}\right\}e_q'(t)\right.$$

$$\left. - \omega(t)\frac{(L_d - L_d')(L_d' - L_d'')}{L_d' - L_l}i_d(t)\,\right] \tag{3.3}$$

where $e_{fd}(t) \equiv \dfrac{N_{fd}}{N_a}\dfrac{\omega_0\,(L_d - L_l)}{R_{fd}}$ [p.u.]

- kd-winding on d-axis

$$\frac{de_q''(t)}{dt} = \frac{1}{T_{do}''} \frac{(L_d' - L_l)^2}{K_d(L_d' - L_d'')(L_d - L_l)} e_{q2}(t)$$

$$= -\frac{1}{T_{do}'' K_d} \left[K_d e_q''(t) - e_q'(t) + \omega(t)(L_d' - L_l) i_d(t) \right] \tag{3.4}$$

- gq-winding on q-axis

$$\frac{de_d'(t)}{dt} = -\frac{1}{T_{qo}'} e_{d1}(t)$$

$$= \frac{1}{T_{qo}'} \left[-\frac{(L_q - L_q')(L_q' - L_q'')}{(L_q' - L_l)^2} K_q e_d''(t) + \left\{ 1 + \frac{(L_q - L_q')(L_q' - L_q'')}{(L_q' - L_l)^2} \right\} e_d'(t) \right.$$

$$\left. - \omega(t)\frac{(L_q - L_q')(L_q' - L_q'')}{L_q' - L_l} i_d(t) \right] \tag{3.5}$$

- kq-winding on q-axis

$$\frac{de_d''(t)}{dt} = \frac{1}{T_{qo}''} \frac{(L_q' - L_l)^2}{K_q(L_q' - L_q'')(L_q - L_l)} e_{d2}(t)$$

$$= -\frac{1}{T_{qo}'' K_q} \left[K_q e_d''(t) - e_d'(t) + \omega(t)(L_q' - L_l) i_d(t) \right] \tag{3.6}$$

where K_d and K_q are defined in (2.130).

3.1.2 Characteristics of the Fundamental Equations

3.1.2.1 Voltage Equations only with the DC Component of Flux Linkage

Assuming

$$\frac{d\theta(t)}{dt} = \omega(t) = \omega_0 = \text{constant} \tag{3.7}$$

$v_d(t)$, the d-axis component of the generator terminal voltage, can be written from (2.128) and (2.131) as follows:

$$v_d(t) = \frac{d\psi_d(t)}{dt} - \omega_0\,\psi_q(t) - R_a\,i_d(t)$$

$$= \frac{d\psi_d(t)}{dt} + \omega_0\,L_q\,i_q(t) + e_{d1}(t) + e_{d2}(t) - R_a\,i_d(t) \tag{3.8}$$

Substituting $e_{d1}(t)$ and $e_{d2}(t)$, for $e_d'(t)$ and $e_d''(t)$ using (2.129), (3.8) can be written as

$$v_d(t) = \frac{d\psi_d(t)}{dt} + \left\{ K_q \frac{L_q' - L_q''}{L_q' - L_l}\,e_d''(t) + \frac{L_q'' - L_l}{L_q' - L_l}\,e_d'(t) \right\} + \omega_0\,L_q''\,i_q(t) - R_a\,i_d(t) \tag{3.9}$$

Similarly, $v_q(t)$, the q-axis component of the generator terminal voltage, can be obtained as

$$v_q(t) = \frac{d\psi_q(t)}{dt} + \left\{ K_d \frac{L_d' - L_d''}{L_d' - L_l}\,e_q''(t) + \frac{L_d'' - L_l}{L_d' - L_l}\,e_q'(t) \right\} + \omega_0\,L_d''\,i_d(t) - R_a\,i_q(t) \tag{3.10}$$

Since $e_d'(t)$, $e_d''(t)$, $e_q'(t)$, and $e_q''(t)$ are the variables proportional to the flux linkages $\psi_{gq}(t)$, $\psi_{kq}(t)$, $\psi_{fd}(t)$, and $\psi_{kd}(t)$ respectively, they always change continuously. $\psi_d(t)$ and $\psi_q(t)$, the stator flux linkages on the d- and q-axis respectively, also vary continuously while the stator currents $i_d(t)$ and $i_q(t)$ can change instantaneously with system change.

For a balanced system fault, $v_d(t)$ and $v_q(t)$ have almost dc component before or after the contingency. With this assumption, it can be shown that $\psi_d(t)$ and $\psi_q(t)$ have dc component and ac component with angular speed ω_0. Considering only the dc components of $\psi_d(t)$ and $\psi_q(t)$,

$$\frac{d}{dt}\,\psi_d(t) = 0 \qquad \text{and} \qquad \frac{d}{dt}\,\psi_q(t) = 0 \tag{3.11}$$

(3.9) and (3.10) become

$$v_d(t) = e_{gd}(t) + \omega_0\,L_q''\,i_q(t) - R_a\,i_d(t) \tag{3.12}$$

$$v_q(t) = e_{gq}(t) - \omega_0\,L_d''\,i_d(t) - R_a\,i_q(t) \tag{3.13}$$

where

$$e_{gd}(t) = K_q \frac{L_q' - L_q''}{L_q' - L_l}\,e_d''(t) + \frac{L_q'' - L_l}{L_q' - L_l}\,e_d'(t) \tag{3.14}$$

$$e_{gq}(t) = K_d \frac{L_d' - L_d''}{L_d' - L_l}\,e_q''(t) + \frac{L_d'' - L_l}{L_d' - L_l}\,e_q'(t) \tag{3.15}$$

3.1.2.2 Voltage Behind the Subtransient Reactance

In (3.12) and (3.13) where $\psi_d(t)$ and $\psi_q(t)$ are assumed to be dc components, $e_d(t)$ and $e_q(t)$ are also dc voltage since $v_d(t)$, $v_q(t)$, $i_d(t)$, and $i_q(t)$ can always be considered as dc quantities. Thus, those dc voltages and currents are written as

$$v_d(t) = V_d \qquad e_{gd}(t) = E_d \qquad i_d(t) = I_d$$
$$v_q(t) = V_q \qquad e_{gq}(t) = E_q \qquad i_q(t) = I_q \qquad (3.16)$$

Then, (3.12) and (3.13) become

$$V_d = E_{gd} + \omega_0 L_q" I_q - R_a I_d$$
$$V_q = E_{gq} - \omega_0 L_d" I_d - R_a I_q \qquad (3.17)$$

or, in vector form,

$$\begin{bmatrix} V_d \\ V_q \end{bmatrix} = \begin{bmatrix} E_{gd} \\ E_{gq} \end{bmatrix} - \begin{bmatrix} R_a & -\omega_0 L_q" \\ \omega_0 L_d" & R_a \end{bmatrix} \begin{bmatrix} I_d \\ I_q \end{bmatrix} \qquad (3.18)$$

Now, the generator voltage and current are described by phasors in steady state conditions:

- generator terminal voltage: $\widetilde{V}_t = V_d + j V_q$ $\qquad (3.19)$
- generator terminal current: $\widetilde{I}_t = I_d + j I_q$ $\qquad (3.20)$
- generator internal voltage: $\widetilde{E}_g = E_{gd} + j E_{gq}$ $\qquad (3.21)$

Generally, $L_q" > L_d"$, but for many generators, it is quite plausible assumption to consider $L_q" = L_d"$, and write the generator terminal voltage as

$$\widetilde{V}_t = \widetilde{E}_g - R_a \widetilde{I}_t - j \omega_0 L_d" \widetilde{I}_t \qquad (3.22)$$

whose phasor diagram is shown in Figure 3.1.

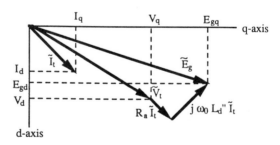

Figure 3.1 Phasor Diagram of the Generator Voltage and Current when $L_d"=L_q"$

The equivalent circuit of (3.22) is shown in Figure 3.2, where $\widetilde{E}_g$ is called the internal voltage behind the subtransient reactance $x_d"$ which is equal to $\omega_0 L_d"$ or $\omega_0 L_q"$.

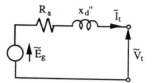

Figure 3.2 The Equivalent Circuit of a Generator

3.1.3 System Network Equations

The generators in a system are connected to the loads through the grid of transmission lines. The generator voltages and currents are balanced three phase sinusoidal waves except the moments of system changes, and the generator voltages, currents, and fluxes referred to the d-q axes are all dc quantities. Therefore, it is possible to consider the d-q axis plane as a complex plane and treat the variables in complex plane. This complex plane has totally different meaning from the complex notation for the sinusoidal ac values in the ac circuit analysis, it is convenient to consider them to be identical as far as computation concerns.

It is desirable to use $e_{gd}(t)$ and $e_{gq}(t)$ as the generator internal voltage because they are associated with the rotor flux linkages and vary only continuously. Concerned with the internal buses which have $\widetilde{E}_g$'s for their internal voltages, the next network equation of voltage and current can be obtained:

$$
\begin{bmatrix}
\widetilde{E}_{G1} \\
\widetilde{E}_{G2} \\
\vdots \\
\widetilde{E}_{GN} \\
\hline
\widetilde{E}_{L1} \\
\widetilde{E}_{L2} \\
\vdots \\
\widetilde{E}_{LM}
\end{bmatrix}
=
\begin{bmatrix}
[\widetilde{Z}_{GG}] & \vdots & [\widetilde{Z}_{GL}] \\
\cdots & \vdots & \cdots \\
[\widetilde{Z}_{LG}] & \vdots & [\widetilde{Z}_{LL}]
\end{bmatrix}
\begin{bmatrix}
\widetilde{I}_{G1} \\
\widetilde{I}_{G2} \\
\vdots \\
\widetilde{I}_{GN} \\
\hline
\widetilde{I}_{L1} \\
\widetilde{I}_{L2} \\
\vdots \\
\widetilde{I}_{LM}
\end{bmatrix}
\tag{3.23}
$$

where $\widetilde{E}_{Gi}$: Internal voltage of the i-th generator behind subtransient reactance in D-Q axis
base explained later.

$\widetilde{E}_{Lj}$: Bus voltage at the j-th load.

$\widetilde{I}_{Gi}$: Injected current from the i-th generator to the system.

$\widetilde{I}_{Lj}$: Injected current from j-th load to the system.

3.1.3.1   System Changes

When the system has changes such as short-circuits and line-opens, the impedance matrices $\widetilde{Z}_{GG}$, $\widetilde{Z}_{GL}$, $\widetilde{Z}_{LG}$, and $\widetilde{Z}_{LL}$ in (3.23) also change, and the modification can be simply achieved by using the admittance matrix as in (3.24) and the technique to build a new one in Appendix A:

$$
\begin{bmatrix}
\widetilde{I}_{G1} \\
\widetilde{I}_{G2} \\
\vdots \\
\widetilde{I}_{GN} \\
\hline
\widetilde{I}_{L1} \\
\widetilde{I}_{L2} \\
\vdots \\
\widetilde{I}_{LM}
\end{bmatrix}
=
\left[
\begin{array}{c|c}
[\widetilde{Y}_{GG}] & [\widetilde{Y}_{GL}] \\
\hline
[\widetilde{Y}_{LG}] & [\widetilde{Y}_{LL}]
\end{array}
\right]
\begin{bmatrix}
\widetilde{E}_{G1} \\
\widetilde{E}_{G2} \\
\vdots \\
\widetilde{E}_{GN} \\
\hline
\widetilde{E}_{L1} \\
\widetilde{E}_{L2} \\
\vdots \\
\widetilde{E}_{LM}
\end{bmatrix}
\tag{3.24}
$$

where $\widetilde{Y}_{GG}$, $\widetilde{Y}_{GL}$, $\widetilde{Y}_{LG}$, and $\widetilde{Y}_{LL}$ are sub-admittance matrices.

3.1.3.2   Constant Load Impedance

When the load impedances are constant, it is not necessary to use the load voltage $\widetilde{E}_{Lj}$ and current $\widetilde{I}_{Lj}$, and it is desirable to prepare the load impedances while building admittance or impedance matrices. Then, the following equation (3.25) can be derived instead of (3.24), where the generator currents can be calculated with the known $\widetilde{E}_{Gi}'s$.

$$
\begin{bmatrix}
\widetilde{E}_{G1} \\
\widetilde{E}_{G2} \\
\vdots \\
\widetilde{E}_{GN}
\end{bmatrix}
= [\widetilde{Y}_g]
\begin{bmatrix}
\widetilde{I}_{G1} \\
\widetilde{I}_{G2} \\
\vdots \\
\widetilde{I}_{GN}
\end{bmatrix}
\tag{3.25}
$$

3.1.3.3 Axis Transformation

The network equations in (3.23)-(3.25) are based on the referenced stator bus axis which is rotating at a synchronous speed ω_0. The direct axis and quadrature axis of the stator reference frame will be called as D-axis and Q-axis respectively. In case the q-axis of a generator precedes the reference axis by δ, then the phasor diagram can be illustrated as shown in Figure 3.3, where the direct and quadrature axes are indicated by the subscripts D and Q for the stator reference respectively while those for the geneator rotor by the subscripts d and q. For a arbitrary voltage $\tilde{V}$ in Figure 3.4, the following relations are obtained between the D-Q axis and d-q axis .

$$V_D + j \, V_Q = (V_d + j \, V_q) \, e^{j(\delta - \pi/2)} \tag{3.26}$$

or

$$\begin{bmatrix} V_D \\ V_Q \end{bmatrix} = \begin{bmatrix} \sin \delta & \cos \delta \\ -\cos \delta & \sin \delta \end{bmatrix} \begin{bmatrix} V_d \\ V_d \end{bmatrix} \tag{3.27}$$

Inversely,

$$V_d + j \, V_q = (V_D + j \, V_Q) \, e^{j(\pi/2 - \delta)} \tag{3.28}$$

or

$$\begin{bmatrix} V_d \\ V_q \end{bmatrix} = \begin{bmatrix} \sin \delta & -\cos \delta \\ \cos \delta & \sin \delta \end{bmatrix} \begin{bmatrix} V_D \\ V_Q \end{bmatrix} \tag{3.29}$$

For current or other quantities, the same relations as in (3.26) - (3.29) can be obtained.

3.1.3.4 Generator Internal Voltage and Current

The equation (3.25) represents the relation between the generator internal voltages and currents. The admittance matrix $[\tilde{Y}_g]$ includes generator internal impedances, transmission lines, and load impedances. The relation between the generator terminal voltages and currents based on D-Q axis is described by the admittance matrix including transmission lines and load impedances as follows:

$$\begin{bmatrix} I_{D1} + j \, I_{Q1} \\ I_{D2} + j \, I_{Q2} \\ \vdots \\ I_{DN} + j \, I_{QN} \end{bmatrix} = \begin{bmatrix} G_{11} + j \, B_{11} & G_{12} + j \, B_{12} & \cdots & G_{1N} + j \, B_{1N} \\ G_{21} + j \, B_{21} & G_{22} + j \, B_{22} & & \vdots \\ \vdots & \vdots & \ddots & \\ G_{N1} + j \, B_{N1} & \cdots & \cdots & G_{NN} + j \, B_{NN} \end{bmatrix} \begin{bmatrix} V_{D1} + j \, V_{Q1} \\ V_{D2} + j \, V_{Q2} \\ \vdots \\ V_{DN} + j \, V_{QN} \end{bmatrix} \tag{3.30}$$

or

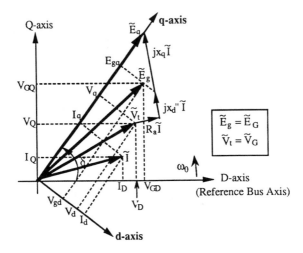

Figure 3.3 Phasor Diagram of Generator Voltage and Current on Reference Axes
(Steady State)

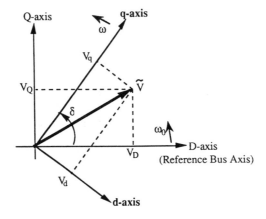

Figure 3.4 Axis Transformation between D-Q axis d-q axis

$$
\begin{bmatrix}
I_{D1} \\
I_{Q1} \\
\hline
I_{D2} \\
I_{Q2} \\
\vdots \\
\hline
I_{DN} \\
I_{QN}
\end{bmatrix}
=
\left[
\begin{array}{cc|cc|c|cc}
G_{11} & -B_{11} & G_{12} & -B_{12} & & G_{1N} & -B_{1N} \\
B_{11} & G_{11} & B_{12} & G_{12} & \cdots & B_{1N} & G_{1N} \\
\hline
G_{21} & -B_{21} & G_{22} & -B_{22} & & & \vdots \\
B_{21} & G_{21} & B_{22} & G_{22} & \ddots & & \\
\hline
& \vdots & & \vdots & \ddots & & \vdots \\
\hline
G_{N1} & -B_{N1} & & & & G_{NN} & -B_{NN} \\
B_{N1} & G_{N1} & \cdots & & \cdots & B_{NN} & G_{NN}
\end{array}
\right]
\begin{bmatrix}
V_{D1} \\
V_{Q1} \\
\hline
V_{D2} \\
V_{Q2} \\
\vdots \\
\hline
V_{DN} \\
V_{QN}
\end{bmatrix}
\tag{3.31}
$$

or simply, $\mathbf{I} = [Y]\, \mathbf{V}$ (3.32)

where

$$
\mathbf{I} =
\begin{bmatrix}
I_{D1} \\
I_{Q1} \\
\hline
I_{D2} \\
I_{Q2} \\
\vdots \\
\hline
I_{DN} \\
I_{QN}
\end{bmatrix}
\qquad
\mathbf{V} =
\begin{bmatrix}
V_{D1} \\
V_{Q1} \\
\hline
V_{D2} \\
V_{Q2} \\
\vdots \\
\hline
V_{DN} \\
V_{QN}
\end{bmatrix}
\tag{3.33}
$$

When $L_d'' \neq L_q''$, the equation for the generator voltage and current based on d-q axis is

$$
\begin{bmatrix} V_d \\ V_q \end{bmatrix}
=
\begin{bmatrix} E_{gd} \\ E_{gq} \end{bmatrix}
-
\begin{bmatrix} R_a & -\omega_0 L_q'' \\ \omega_0 L_d'' & R_a \end{bmatrix}
\begin{bmatrix} I_d \\ I_q \end{bmatrix}
\tag{3.34}
$$

Multiplying $\begin{bmatrix} \sin\delta & \cos\delta \\ -\cos\delta & \sin\delta \end{bmatrix}$ on both sides of (3.34) and using the following axis transformation

$$
\begin{bmatrix} V_D \\ V_Q \end{bmatrix}
=
\begin{bmatrix} \sin\delta & \cos\delta \\ -\cos\delta & \sin\delta \end{bmatrix}
\begin{bmatrix} V_d \\ V_d \end{bmatrix}
\tag{3.35}
$$

$$
\begin{bmatrix} E_{GD} \\ E_{GQ} \end{bmatrix}
=
\begin{bmatrix} \sin\delta & \cos\delta \\ -\cos\delta & \sin\delta \end{bmatrix}
\begin{bmatrix} E_{gd} \\ E_{gq} \end{bmatrix}
\tag{3.36}
$$

$$
\begin{bmatrix} I_d \\ I_d \end{bmatrix}
=
\begin{bmatrix} \sin\delta & -\cos\delta \\ \cos\delta & \sin\delta \end{bmatrix}
\begin{bmatrix} I_D \\ I_Q \end{bmatrix}
\tag{3.37}
$$

(3.34) becomes

$$\begin{bmatrix} V_D \\ V_Q \end{bmatrix} = \begin{bmatrix} E_{GD} \\ E_{GQ} \end{bmatrix} - \begin{bmatrix} \sin\delta & \cos\delta \\ -\cos\delta & \sin\delta \end{bmatrix} \begin{bmatrix} R_a & -\omega_0 L_q'' \\ \omega_0 L_d'' & R_a \end{bmatrix} \begin{bmatrix} I_d \\ I_q \end{bmatrix}$$

$$= \begin{bmatrix} E_{GD} \\ E_{GQ} \end{bmatrix} - \begin{bmatrix} \sin\delta & \cos\delta \\ -\cos\delta & \sin\delta \end{bmatrix} \begin{bmatrix} R_a & -\omega_0 L_q'' \\ \omega_0 L_d'' & R_a \end{bmatrix} \begin{bmatrix} \sin\delta & -\cos\delta \\ \cos\delta & \sin\delta \end{bmatrix} \begin{bmatrix} I_D \\ I_Q \end{bmatrix}$$

$$= \begin{bmatrix} E_{GD} \\ E_{GQ} \end{bmatrix} - \begin{bmatrix} Z_{11} & Z_{12} \\ Z_{21} & Z_{22} \end{bmatrix} \begin{bmatrix} I_D \\ I_Q \end{bmatrix} \tag{3.38}$$

where

$$\begin{bmatrix} Z_{11} & Z_{12} \\ Z_{21} & Z_{22} \end{bmatrix} \equiv \begin{bmatrix} \sin\delta & \cos\delta \\ -\cos\delta & \sin\delta \end{bmatrix} \begin{bmatrix} R_a & -\omega_0 L_q'' \\ \omega_0 L_d'' & R_a \end{bmatrix} \begin{bmatrix} \sin\delta & -\cos\delta \\ \cos\delta & \sin\delta \end{bmatrix} \tag{3.39}$$

Extending (3.38) for N generators to get **V** in (3.33), and substituting it for (3.31),

$$\mathbf{I} = [Y] \{ \mathbf{E}_G - [Z] \mathbf{I} \} \tag{3.40}$$

where

$$\mathbf{E}_G = \begin{bmatrix} E_{GD1} \\ E_{GQ1} \\ \hline E_{GD2} \\ E_{GQ2} \\ \vdots \\ \hline E_{GDN} \\ E_{GQN} \end{bmatrix} \tag{3.41}$$

$$[Z] = \begin{bmatrix} \begin{matrix} Z_{11}^{(1)} & Z_{12}^{(1)} \\ Z_{21}^{(1)} & Z_{22}^{(1)} \end{matrix} & & \mathbf{0} \\ & \ddots & \\ \mathbf{0} & & \begin{matrix} Z_{11}^{(N)} & Z_{12}^{(N)} \\ Z_{21}^{(N)} & Z_{22}^{(N)} \end{matrix} \end{bmatrix} \tag{3.42}$$

From (3.40),

$$\{[I] + [Y][Z]\} \mathbf{I} = [Y] \mathbf{E}_G \tag{3.43}$$

where [I] is a (2N×2N) identity matrix.

Solving for **I** in (3.41),

$$\mathbf{I} = [Y_G] \, \mathbf{E}_G \tag{3.44}$$

where $\qquad [Y_G] \equiv \{[I] + [Y][Z]\}^{-1} [Y]$ (3.45)

$[Y_G]$ is the admittance matrix seen from the generator internal voltage bus. Therefore, it represents all the system including generators, and changes with rotor angle δ. However, if

$$L_d{}'' = L_q{}'' \tag{3.46}$$

for all the generators, then $[Z]$ in (3.42) and $[Y_G]$ are not influenced by rotor angle δ.

3.1.4 Calculation of Generator Internal States

3.1.4.1 Voltages and Currents Determined from Network Solution

When each angular position δ_i is obtained from the network solution, each generator internal voltage $\widetilde{E}_{Gi}$ based on D-Q axis can be calculated from (3.44). Using the coordinate transformation in (3.35)-(3.37), the next three voltages and three currents in rotor side can be determined:

- Generator Current $\qquad\qquad$: $i_d(t), i_q(t)$
- Voltage behind subtransient impedance : $e_{gd}(t), e_{gq}(t)$
- Generator terminal voltage $\qquad$: $v_d(t), v_q(t)$

(1) Generator current; $i_d(t), i_q(t)$ [p.u.]

The generator current $\widetilde{I}_G$, based on D-Q axis, is calculated from the system network solution, and its components on d-q axis can be obtained from (3.37):

$$i_d(t) = I_d = \mathrm{Re}\left\{\widetilde{I}_G \, e^{j(\pi/2 - \delta)}\right\} \tag{3.47}$$

$$i_q(t) = I_q = \mathrm{Im}\left\{\widetilde{I}_G \, e^{j(\pi/2 - \delta)}\right\} \tag{3.48}$$

where $\qquad \widetilde{I}_G = I_D + j \, I_Q$ (3.49)

(2) Generator internal voltage; $e_{gd}(t), e_{gq}(t)$ [p.u.]

From (3.36),

$$e_{gd}(t) = E_{gd} = \mathrm{Re}\left\{\widetilde{E}_G \, e^{j(\pi/2 - \delta)}\right\} \tag{3.50}$$

$$e_{gq}(t) = E_{gq} = \mathrm{Im}\left\{\widetilde{E}_G \, e^{j(\pi/2 - \delta)}\right\} \tag{3.51}$$

where $\qquad \widetilde{E}_G = E_{GD} + j \, E_{GQ}$ (3.52)

(3) Generator terminal voltage; $v_d(t)$, $v_q(t)$ [p.u.]

From (3.35),

$$v_d(t) = V_d = \text{Re}\left\{\widetilde{V}_G \, e^{j(\pi/2-\delta)}\right\} \tag{3.53}$$

$$v_q(t) = V_q = \text{Im}\left\{\widetilde{V}_G \, e^{j(\pi/2-\delta)}\right\} \tag{3.54}$$

where $\widetilde{V}_G = V_D + j\,V_Q$ \hfill (3.55)

3.1.4.2 Electrical Power Output, Flux Linkages, and Rotor Winding Currents

(1) Generator electrical output; $P_e(t)$ [p.u.]

$$
\begin{aligned}
P_e(t) &\equiv T_e(t)\,\omega(t)/\omega_0 \\
&= \{\psi_d(t)\,i_q(t) - \psi_q(t)\,i_d(t)\}\,\omega(t)/\omega_0 \\
&= \{e_{gd}(t)\,i_d(t) + e_{gq}(t)\,i_q(t) - \omega_0\,(L_d'' - L_q'')\,i_d(t)\,i_q(t)\}\,\omega(t)/\omega_0 \tag{3.56}
\end{aligned}
$$

(2) Exciter voltage; $e_{fd}(t)$ [p.u.]

The exciter voltage $e_{fd}(t)$ of the rotor field winding in rotor is determined by the AVR characteristics, and usually it is described in the following form:

$$e_{fd}(t) = F_{AVR}\,(\,v_d(t),\,v_q(t),\,\omega(t),\,P_e(t)) \tag{3.57}$$

(3) Exciter field winding current ($i_{fd}(t)$); $e_{q1}(t)$ [p.u.]

$$
\begin{aligned}
e_{q1}(t) = &- \frac{(L_d - L_d')\,(L_d' - L_d'')}{(L_d' - L_l)^2}\,K_d\,e_q''(t) + \left\{1 + \frac{(L_d - L_d')\,(L_d' - L_d'')}{(L_d' - L_l)^2}\right\}\,e_q'(t) \\
&+ \omega(t)\,\frac{(L_d - L_d')\,(L_d' - L_d'')}{(L_d' - L_l)}\,i_d(t) \tag{3.58}
\end{aligned}
$$

(4) kd-damper winding current ($i_{kd}(t)$); $e_{q2}(t)$ [p.u.]

$$e_{q2}(t) = -\frac{(L_d' - L_d'')\,(L_d - L_l)}{(L_d' - L_l)^2}\,\{K_d\,e_q''(t) - e_q'(t) + \omega(t)\,(L_d' - L_l)\,i_d(t)\} \tag{3.59}$$

(5) gq-damper winding current ($i_{gq}(t)$); $e_{d1}(t)$ [p.u.]

$$e_{d1}(t) = - \frac{(L_q - L_q')(L_q' - L_q'')}{(L_q' - L_l)^2} K_q \, e_d''(t) + \left\{ 1 + \frac{(L_q - L_q')(L_q' - L_q'')}{(L_q' - L_l)^2} \right\} e_d'(t)$$

$$- \omega(t) \frac{(L_q - L_q')(L_q' - L_q'')}{(L_q' - L_l)} i_q(t) \tag{3.60}$$

(6) kq-damper winding current ($i_{kq}(t)$); $e_{d2}(t)$ [p.u.]

$$e_{d2}(t) = - \frac{(L_q' - L_q'')(L_q - L_l)}{(L_q' - L_l)^2} \{ K_q \, e_d''(t) - e_d'(t) - \omega(t)(L_q' - L_l) i_q(t) \} \tag{3.61}$$

3.2 Transient Stability Analysis

3.2.1 Computational Sequence

(1) Obtain system initial states

In order to be able to solve the fundamental equations, it is necessary to determine the initial generator conditions in prefault state. From the load flow solution, the following initial values are obtained:

- Generator terminal voltage; $\quad \tilde{V}_G = V_D + j \, V_Q \tag{3.62}$

- Generator current; $\quad \tilde{I}_G = I_D + j \, I_Q \tag{3.63}$

- Generator q-axis voltage; $\quad \tilde{E}_q = V_G + (R + j \, \omega_0 L_q) \tilde{I}_G \tag{3.64}$

- Initial angular position relative to the reference; $\delta_0 = \angle \, \tilde{E}_q \tag{3.65}$

Thus, the rotor angle of each generator will be determined by the phase of the voltage $\tilde{E}_q$ which is on the q-axis as shown in Figure 3.3. The phase angle of the internal voltage is not used because it is not related with the rotor position.

(2) Calculate generator initial states

Based on the d-q axes in the rotor side, the generator current and voltage are calculated:

- Generator initial current; $i_{d0} = \text{Re}\left\{\tilde{I}_G\, e^{j(\pi/2-\delta_0)}\right\}$ (3.66)

$$i_{q0} = \text{Im}\left\{\tilde{I}_G\, e^{j(\pi/2-\delta_0)}\right\}$$ (3.67)

- Generator initial voltage; $v_{d0} = \text{Re}\left\{\tilde{V}_G\, e^{j(\pi/2-\delta)}\right\}$ (3.68)

$$v_{q0} = \text{Im}\left\{\tilde{V}_G\, e^{j(\pi/2-\delta)}\right\}$$ (3.69)

Since no damping current is induced in kd-, gq-, and kq-circuit during steady state,

- Damper winding currents; $e_{d10} = e_{d20} = e_{q20} = 0$ (3.70)

The left side of (3.3) is 0 before faults. Using (2.128), (2.131), and $e_{q20} = 0$,

- Field winding current; $e_{fd0} = e_{q10} = v_{q0} + \omega_0\, L_d\, i_{d0} + R\, i_{q0}$ (3.71)
- Electrical output; $P_{e0} = v_{d0}\, i_{d0} + v_{q0}\, i_{q0} + R\left(i_{d0}^2 + i_{q0}^2\right)$ (3.72)

Using (2.129), (3.70), and (3.72), the initial voltages associated with rotor flux linkages are

- fd-winding flux linkage; $e_{q0}' = v_{q0} + \omega_0\, L_d'\, i_{d0} + R\, i_{q0}$ (3.73)

- kd-winding flux linkage; $e_{q0}'' = \dfrac{1}{K_d}\left\{e_{q0}' - \omega_0\left(L_d' - L_l\right) i_{d0}\right\}$ (3.74)

- gq-winding flux linkage; $e_{d0}' = \omega_0\left(L_q - L_q'\right) i_{q0}$ (3.75)

- kq-winding flux linkage; $e_{d0}'' = \dfrac{1}{K_q}\omega_0\left(L_q - L_l\right) i_{q0}$ (3.76)

The initial conditions to solve the differential equations in (3.1)-(3.6) are obtained from (3.70)-(3.76)

The internal voltage for the network equation is obtained from (3.14) and (3.15):

- Generator internal voltage; $e_{gd0} = K_q\, \dfrac{L_q' - L_q''}{L_q' - L_l}\, e_{d0}'' + \dfrac{L_q'' - L_l}{L_q' - L_l}\, e_{d0}'$ (3.77)

$$e_{gq0} = K_d\, \dfrac{L_d' - L_d''}{L_d' - L_l}\, e_{q0}'' + \dfrac{L_d'' - L_l}{L_d' - L_l}\, e_{q0}'$$ (3.78)

(3) Modify admittance matrix for network changes

With changes in the network, the impedance matrix or admittance matrix in (3.23)-(3.25) have to be modified. By the way, the generator internal inpedance $(Ra+jw[Ld", Lq"])$ is included in the impedance matrix and the admittance matrix.

(4) Solve the network equations

In order to solve the network equations in (3.23)-(3.25) and to get all the system voltages and currents, the generator internal voltage $\widetilde{E}_G$, which is $E_{GD} + j\,E_{GQ}$ is necessary. From (3.36),

$$E_{GD} = e_{gd} \sin\delta + e_{gq} \cos\delta \tag{3.79}$$

$$E_{GQ} = -e_{gd} \cos\delta + e_{gq} \sin\delta \tag{3.80}$$

Therefore, in the k-th step of computation when $t = k\Delta t$, it is possible to calculate all the system voltages and currents including the generator terminal voltage $\widetilde{V}_G^{(k)}$ and its current $\widetilde{I}_G^{(k)}$.

(5) Solve the differential equations for generators

Substituting $\widetilde{E}_G^{(k)}$, $\widetilde{V}_G^{(k)}$, $\widetilde{I}_G^{(k)}$, and $\delta^{(k)}$ obtained in the step (4), the following values which are required to solve the k-th step differential equations can be calculated from the equations in the section 3.1.4:

$$i_d^{(k)}, i_q^{(k)}, e_{gd}^{(k)}, e_{gq}^{(k)}, v_d^{(k)}, v_q^{(k)}, P_e^{(k)}, e_{fd}^{(k)}, e_{q1}^{(k)}, e_{q2}^{(k)}, e_{d1}^{(k)}, e_{d2}^{(k)} \tag{3.81}$$

By solving (3.1) - (3.6), the state variables at $t = (k+1)\Delta t$ are thus updated:

$$\delta^{(k+1)}, \omega^{(k+1)}, e_q^{'(k+1)}, e_q^{''(k+1)}, e_d^{'(k+1)}, e_d^{''(k+1)} \tag{3.82}$$

(6) Update the generator internal voltage $\widetilde{E}_G^{(k+1)}$

Substituting $e_q^{'(k+1)}, e_q^{''(k+1)}, e_d^{'(k+1)}, e_d^{''(k+1)}$ obtained in (3.82) into (3.14) and (3.15), $e_{gd}^{(k+1)}, e_{gq}^{(k+1)}$ can be obtained.

From (3.79) and (3.80),

$$E_{GD}^{(k+1)} = e_{gd}^{(k+1)} \sin\delta^{(k+1)} + e_{gq}^{(k+1)} \cos\delta^{(k+1)} \tag{3.83}$$

$$E_{GQ}^{(k+1)} = -e_{gd}^{(k+1)} \cos\delta^{(k+1)} + e_{gq}^{(k+1)} \sin\delta^{(k+1)} \tag{3.84}$$

and $\widetilde{E}_G^{(k+1)}$ can be calculated. Until the network has another change, return to the step (4) and repeat the routines. In case system has changed, the routine have to include the step (2)

3.2.2 Exciter and Governor-Turbine System

In the case of that the control systems of exciter and governor system are included in the generator modeling, their dynamic characteristics are described by differential equations. For the models of exciter and governor, the simple transfer functions in Figure 3.5 are used, where each model has first order transfer function.

(a) Transfer Function of an AVR

(b) Transfer Function of a Governor

Figure 3.5 Simple Exciter and Governor Models

From the transfer functions in Figure 3.8, the equations regarding the outputs of the control systems are

• Exciter System $\quad \Delta e_{fd} = \dfrac{K_A}{T_A s + 1}(V_{ref} - V_t)$ (3.85)

• Governor System $\quad \Delta P_m = \dfrac{K_G}{T_G s + 1}\dfrac{(\omega_0 - \omega)}{\omega_0}$ (3.86)

where V_{ref} is the reference voltage to be maintained, and ω_0 is the synchronous angular speed. Since the terminal voltage V_t is calculated as

$$V_t = \sqrt{v_d^2(t) + v_q^2(t)}$$ (3.87)

The differential equations for the transfer functions in (3.85) and (3.86) become

$$\frac{de_{fd}(t)}{dt} = -\frac{1}{T_A}\left\{ e_{fd}(t) + K_A\sqrt{v_d^2(t) + v_q^2(t)} - (e_{fd0}(t) + K_A V_{t0}) \right\} \tag{3.88}$$

$$\frac{dP_m(t)}{dt} = -\frac{1}{T_G}\left\{ P_m(t) + \frac{K_A}{\omega_0}\,\omega(t) - (P_{m0}(t) + K_G) \right\} \tag{3.89}$$

where V_{t0} and P_{m0} are the generator terminal voltage and the mechanical input in prefault state respectively. The differential equations in (3.1) - (3.6) and (3.88) - (3.89) will be solved simultaneously in the step (4) of Section 3.2.1 to solve for generators. Even when complex transfer functions are adapted for more accurate control systems, the basic algorithms are the same as in this simple model simulation. In addition, when the dynamic load model are considered, the differential equations for that model will be combined to those for the generators and control systems and will solved simultaneously at each tine step.

3.2.3   Load Model

Load model is a crucial part in transient stability simulation because it may change system response dramatically depending on the systems. Conventionally, load has been represented as constant power, constant current, or constant impedance. In order to get results closer to real system response, however, load characteristics depending on system states such as voltage and frequency should be included in the simulation. Efforts have been made to represent loads as a polynomial function of voltage and frequency, of which the coefficients can be obtained from experimental data. On the other hand, the dynamic load model representing the time-variant characteristic of a large induction motor is also used. In this section, the composite load model which is static and the dynamic load model will be described.

3.2.3.1   Static Composite Load

In order to make the static composition load model based on the load component method, the following three sets of data are required:
(1) Load Characteristic Data

A load component at a voltage and a frequency can be calculated from testing. For example, the real and reactive load of each component can be represented as a function of voltage and frequency deviation as follows:

$$P = 1.0 + k_1 \cdot \Delta V + k_2 \cdot \Delta V^2 + k_3 \cdot \Delta V^3 + k_4 \cdot \Delta V^4 + k_5 \cdot \Delta F + k_6 \cdot \Delta V \Delta F \qquad (3.90)$$

$$Q = k_7 + k_8 \cdot \Delta V + k_9 \cdot \Delta V^2 + k_{10} \cdot \Delta V^3 + k_{11} \cdot \Delta V^4 + k_{12} \cdot \Delta F + k_{13} \cdot \Delta V \Delta F \qquad (3.91)$$

where k_i's are the coefficients. This load component model makes equipment tests to collect enough data and use curve fitting techniques to obtain polynomial functions to represent its characteristics. The coefficients of the polynomial functions for principal loads are shown in Table 3.1.

(2) Load Class Data

In a system operating condition, the loads at each bus can be sorted by the load class data which are residential load, commercial load, industrial load, and other load with their ratios in that bus.

(3) Load Composition Data

The load composition data for each load class data in a bus can be prepared with the percentage of the components in Table 3.1. Then, the we can make an aggregation of loads without getting detailed data on each bus assuming all load components are connected at that bus in parallel.

The static load composition method is based on the real power consumption of the load components at rated voltage and frequency. Most of the load components, however, actually have different reactive power consumption although they have same brand and rating. Since the reactive power characteristic at each bus is determined from the ratios of the included components, the reactive power consumption may not match with that from the load flow data. To adjust this mismatch between the calculated value and the initial actual load, several methods are used to compensate it:

- The mismatch between the calculated value and the actual load when the voltage is 1.0 p.u. is treated as a capacitor and its variation will be proportional to the square of the voltage[].

• The mismatch between the calculated value and the actual load will be treated as a constant multiplier for the reactive power in each calculation step.

Table 3.1 Polynomial Coefficients for Principal Load Elements

Load Type		Const. 1 / k_7	ΔV k_1 / k_8	ΔV^2 k_2 / k_9	ΔV^3 k_3 / k_{10}	ΔV^4 k_4 / k_{11}	ΔF k_5 / k_{12}	$\Delta V \cdot \Delta F$ k_6 / k_{13}
3φ Central Air Conditioner	P	1.0000	0.0883	0.3510	0.0000	0.0000	0.9800	-2.3560
	Q	0.5330	1.3360	5.2100	2.3380	0.0000	-0.7030	-10.2820
1φ Central Air Conditioner	P	1.0000	0.2016	1.6598	-9.1079	0.0000	0.9015	-2.1680
	Q	0.2427	0.5581	7.0228	-6.5456	0.0000	-0.6473	-9.4606
Air Conditioner (Window Type)	P	1.0000	0.4675	1.9601	0.0000	0.0000	0.5628	-2.8865
	Q	0.6896	1.6993	3.8889	7.6570	53.7198	-1.9372	-14.1787
Duct Heater (Incl. Blowers)	P	1.0000	1.5655	0.8548	0.0000	0.0000	0.5121	-0.7530
	Q	0.1472	0.3518	1.1825	0.0000	0.0000	-0.1715	-3.4677
Water Heater	P	1.0000	2.0000	1.0000	0.0000	0.0000	0.0000	0.0000
	Q	0.0000	0.0000	0.0000	0.0000	0.0000	0.0000	0.0000
Clothes Dryer	P	1.0000	2.0400	0.9950	-0.5930	0.0000	0.0000	0.0000
	Q	0.1307	0.4271	0.6274	0.4690	0.0000	-0.3437	-0.6734
Refrigerator	P	1.0000	0.7594	1.4361	0.0000	0.0000	0.5238	-3.3710
	Q	0.7820	1.9298	4.2231	0.0000	0.0000	-1.1266	-9.2356
Incandescent Light	P	1.0000	1.5520	0.4590	0.0000	0.0000	0.0000	0.0000
	Q	0.0000	0.0000	0.0000	0.0000	0.0000	0.0000	0.0000
Fluorescent Light	P	1.0000	0.6534	-1.6500	0.0000	0.0000	0.0000	0.0000
	Q	-0.1535	-0.0403	2.7340	0.0000	0.0000	0.0000	0.0000

3.2.3.1 Dynamic Load

In the dynamic Load model, the effects of inertia and mechanical load characteristics should be taken into account while hysterisis effects and magnetic saturation are neglected. In addition, the resistance and reactance parameters are assumed to be constant in the dynamic load model.

(1) Induction Motor

Figure 3.6 shows the equivalent circuit of an ordinary induction motor (first order model). Although the motor model is simple, it is believed that it represents the dynamic characteristic close to the real performance.

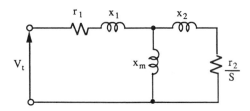

Figure 3.6 Equivalent Circuit for an Induction Motor

The equations for the real and reactive power delivered to the induction motor are derived as follows:

$$P = \frac{\left\{ r_1 \left(x_2 + x_m \right)^2 + \frac{r_2}{S} \left(\frac{r_1 r_2 + x_m^2}{S} \right) \right\} |V_t|^2}{A} \tag{3.92}$$

$$Q = \frac{\left\{ \left(x_2 + x_m \right) \left(x_2^2 + \frac{r_2^2}{S} \right) + \left(x_1 + x_2 \right) x_m^2 + 2 \, x_1 \, x_2 \, x_m \right\} |V_t|^2}{A} \tag{3.93}$$

where V_t : Terminal voltage

x_1, x_2 : Leakage reactances of stator and rotor respectively

r_1, r_2 : Resistances of stator and rotor respectively

x_m : Magnetizing reactance

S : Slip; $S \equiv (\omega_0 - \omega)/\omega_0$

ω_0 : Synchronous rotor velocity

$$A = \left(\frac{r_1 r_2}{S} - x_2 x_m - x_1 x_2 - x_1 x_m \right)^2 + \left(r_1 x_m + \frac{r_2 x_m}{S} + r_1 x_2 + \frac{r_2 x_1}{S} \right)^2 \tag{3.94}$$

From the equivalent circuit in Figure 3.6, the open circuit transient time constant T_{do}' is obtained as

$$T_{do}' = \frac{x_2 + x_m}{\omega\,(r_2/S)} \tag{3.95}$$

and the electrical torque in induction motor T_e is also calculated as a function of the motor slip

S:

$$T_e = \frac{x_m^2\, r_2\, |V_t|^2}{A \cdot S \cdot \omega_0} \tag{3.96}$$

On the other hand, the mechanical torque T_m in each induction motor is described as a function the rotor speed on the assumption that load characteristics are of exponential form as

$$T_m = K\,\omega^n \tag{3.97}$$

where K is a coefficient and the exponent n is a constant depending on load conditions. The difference between the electrical and mechanical torque is used to calculate the change in the motor speed at each time step as follows:

$$\frac{\Delta\omega}{\Delta t} = \frac{T_e - T_m}{2\,H} \tag{3.98}$$

(2) Aggregation of Induction Motors

In the real power system, a number of motors with different sizes and ratings are existing in the loads. They are distributed geographically and operated to drive different types of mechanical apparatus. Obviously, it is impossible to simulate every motor as a dynamic load model. Therefore, it it necessary to aggregate a group of induction motors and represent them as a single unit equivalent motor. The single unit model for a group of motors is assumed to have an electrical circuit structure identical to the conventional approximate equivalent circuit of a symmetrical three phase induction motor.

Let's calculate the equivalent impedance and the equivalent inertia for a group of n-motors connecting a bus in parallel. First, we need to get the per unit impedances on the same base which is the total MVA of n-motors:

$$Z_{i\,new} = Z_i \frac{\Sigma\,MVA_i}{MVA_i} \tag{3.99}$$

where Z_i is the per unit impedance of each motor based on its MVA rating. Since n-motors are connected in parallel,

$$Z_{equiv} = \frac{1}{\sum \dfrac{1}{Z_{i\,new}}} = \frac{1}{\sum \left(\dfrac{1}{Z_i \dfrac{\sum MVA_i}{MVA_i}} \right)} \qquad (3.100)$$

Assuming $Z_1 = Z_2 = \cdots = Z_n$, then $Z_{equiv} = Z_1 = Z_2 = \cdots = Z_n$ from (3.100). This means that when all the per unit impedances based on their machine MVA ratings are same, the per unit equivalent impedance of n-motors based on the total MVA is the same as them.

Regarding the moment of inertia of the equivalent motor, the motor at its synchronous speed is assumed to have the equal amount of kinetic energy to the total kinetic energies of individual motor at synchronous speed. As defined in (2.50), the inertia constant H of a machine is the stored kinetic energy at synchronous speed divided by machine rating. Thus,

$$H_{equiv} \cdot \Sigma(MVA_i) = H_1 \cdot MVA_1 + H_2 \cdot MVA_2 + \cdots + H_n \cdot MVA_n \qquad (3.101)$$

Therefore, the equivalent inertia constant for n-motors on the same bus can be obtained as

$$H_{equiv} = \frac{\Sigma\,(H_i \cdot MVA_i)}{\Sigma\,MVA_i} \qquad (3.102)$$

If the inertia constants $H_1 = H_2 = \cdots = H_n$ in per unit based on each machine MVA rating, then H_{equiv} will have the same value based on the total MVA.

3.3 State Equations for Dynamic Stability Analysis

3.3.1 Network Equations

3.3.1.1 Network Equation with D-Q Axis

In a multi-machine system, the terminal voltage and current of each generator can be expressed using the reference D-Q axis, which rotates at constant angular speed ω_0, as follows:

• generator terminal voltage; $V_{Gi} = V_{Di} + j\,V_{Qi}$ \qquad (3.103)

• generator current $I_{Gi} = I_{Di} + j\,I_{Qi}$ \qquad (3.104)

Assuming the driving point admittance at generator k as

$$Y_{kk} = G_{kk} + j B_{kk} \tag{3.105}$$

and the transfer admittance between generators k and m as

$$Y_{km} = G_{km} + j B_{km} \qquad (k \neq m) \tag{3.106}$$

the network equation with N-generators can be expressed as

$$
\begin{bmatrix} I_{D1} \\ I_{Q1} \\ \hline I_{D2} \\ I_{Q2} \\ \vdots \\ \hline I_{DN} \\ I_{QN} \end{bmatrix}
=
\begin{bmatrix}
G_{11} & -B_{11} & G_{12} & -B_{12} & & & G_{1N} & -B_{1N} \\
B_{11} & G_{11} & B_{12} & G_{12} & \cdots & & B_{1N} & G_{1N} \\
\hline
G_{21} & -B_{21} & G_{22} & -B_{22} & & & & \vdots \\
B_{21} & G_{21} & B_{22} & G_{22} & \cdots & & & \\
\hline
& \vdots & & \vdots & & \ddots & & \vdots \\
\hline
G_{N1} & -B_{N1} & & & & & G_{NN} & -B_{NN} \\
B_{N1} & G_{N1} & & \cdots & & & B_{NN} & G_{NN}
\end{bmatrix}
\begin{bmatrix} V_{D1} \\ V_{Q1} \\ \hline V_{D2} \\ V_{Q2} \\ \vdots \\ \hline V_{DN} \\ V_{QN} \end{bmatrix}
\tag{3.107}
$$

or simply $\mathbf{I} = [Y]\,\mathbf{V}$ $\tag{3.108}$

where

$$
\mathbf{I} = \begin{bmatrix} I_{D1} \\ I_{Q1} \\ \hline I_{D2} \\ I_{Q2} \\ \vdots \\ \hline I_{DN} \\ I_{QN} \end{bmatrix}
\qquad
\mathbf{V} = \begin{bmatrix} V_{D1} \\ V_{Q1} \\ \hline V_{D2} \\ V_{Q2} \\ \vdots \\ \hline V_{DN} \\ V_{QN} \end{bmatrix}
\tag{3.109}
$$

For small disturbance, (3.108) becomes

$$\Delta \mathbf{I} = [Y]\,\Delta \mathbf{V} \tag{3.110}$$

3.3.1.2   Axis Transformation for N-Machine System

Assuming δ_i the angle between each generator and the referenced D-Q axes, then the transformation for each generator terminal voltage expression with its own d-q axes and the referenced D-Q axes is obtained by extending the equation (3.35) for N-generator system as follows:

$$
\begin{bmatrix} V_{D1} \\ V_{Q1} \\ \hline V_{D2} \\ V_{Q2} \\ \hline \vdots \\ \hline V_{DN} \\ V_{QN} \end{bmatrix} = [\mathcal{T}] \begin{bmatrix} v_{d1} \\ v_{q1} \\ \hline v_{d2} \\ v_{q2} \\ \hline \vdots \\ \hline v_{dN} \\ v_{qN} \end{bmatrix} \qquad \text{or} \qquad V = \begin{bmatrix} \mathcal{T} \end{bmatrix} v \tag{3.111}
$$

and similarly for the generator currents

$$
\begin{bmatrix} I_{D1} \\ I_{Q1} \\ \hline I_{D2} \\ I_{Q2} \\ \hline \vdots \\ \hline I_{DN} \\ I_{QN} \end{bmatrix} = [\mathcal{T}] \begin{bmatrix} i_{d1} \\ i_{q1} \\ \hline i_{d2} \\ i_{q2} \\ \hline \vdots \\ \hline i_{dN} \\ i_{qN} \end{bmatrix} \qquad \text{or} \qquad I = \begin{bmatrix} \mathcal{T} \end{bmatrix} i \tag{3.112}
$$

where $\begin{bmatrix} \mathcal{T} \end{bmatrix}$ is called the transform matrix $(2N \times 2N)$ which is orthogonal and it is defined as

$$
[\mathcal{T}] \equiv \begin{bmatrix} \begin{matrix} \sin\delta_1 & \cos\delta_1 \\ -\cos\delta_1 & \sin\delta_1 \end{matrix} & 0 & \cdots & 0 \\ 0 & \begin{matrix} \sin\delta_2 & \cos\delta_2 \\ -\cos\delta_2 & \sin\delta_2 \end{matrix} & \cdots & 0 \\ \vdots & \vdots & \ddots & \vdots \\ 0 & 0 & \cdots & \begin{matrix} \sin\delta_N & \cos\delta_N \\ -\cos\delta_N & \sin\delta_N \end{matrix} \end{bmatrix} \tag{3.113}
$$

For a small disturbance, (3.111) becomes

$$
\Delta V = [T] \Delta v + [M_v] \Delta \delta \tag{3.114}
$$

Similarly for a small disturbance, (3.112) becomes

$$
\Delta I = [T] \Delta i + [M_i] \Delta \delta \tag{3.115}
$$

where

$$
\Delta \delta = \begin{bmatrix} \Delta \delta_1 \\ \Delta \delta_2 \\ \vdots \\ \Delta \delta_N \end{bmatrix} \tag{3.116}
$$

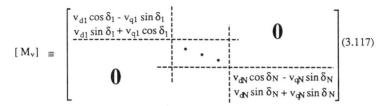

$$[M_v] \equiv \begin{bmatrix} \begin{matrix} v_{d1}\cos\delta_1 - v_{q1}\sin\delta_1 \\ v_{d1}\sin\delta_1 + v_{q1}\cos\delta_1 \end{matrix} & & & \\ & \ddots & & 0 \\ 0 & & & \begin{matrix} v_{dN}\cos\delta_N - v_{qN}\sin\delta_N \\ v_{dN}\sin\delta_N + v_{qN}\sin\delta_N \end{matrix} \end{bmatrix} \qquad (3.117)$$

and

$$[M_i] \equiv \begin{bmatrix} \begin{matrix} i_{d1}\cos\delta_1 - i_{q1}\sin\delta_1 \\ i_{d1}\sin\delta_1 + i_{q1}\cos\delta_1 \end{matrix} & & & \\ & \ddots & & 0 \\ 0 & & & \begin{matrix} i_{dN}\cos\delta_N - i_{qN}\sin\delta_N \\ i_{dN}\sin\delta_N + i_{qN}\sin\delta_N \end{matrix} \end{bmatrix} \qquad (3.118)$$

Substituting (3.114) and (3.115) for ΔV and ΔI in (3.110),

$$[\mathcal{J}]\Delta i + [M_i]\Delta\delta = [Y] ([\mathcal{J}]\Delta v + [M_v]\Delta\delta) \qquad (3.119)$$

Thus,

$$\Delta i = [\mathcal{J}]^{-1}[Y][\mathcal{J}]\Delta v + [\mathcal{J}]^{-1} ([Y][M_v] - [M_i]) \Delta\delta \qquad (3.120)$$

or $\qquad \Delta i = [Y_1]\Delta v + [K_1]\Delta\delta \qquad (3.121)$

where $\qquad [Y_1] \equiv [\mathcal{J}]^{-1}[Y][\mathcal{J}] \qquad (3.122)$

and $\qquad [K_1] \equiv [\mathcal{J}]^{-1} ([Y][M_v] - [M_i]) \qquad (3.123)$

3.3.2 State Equations for Synchronous Machines

3.3.2.1 Fundamental Equations of a Synchronous Machine

(1) Flux Linkage Equations

$$\psi_d(t) = -L_d\, i_d(t) + M_{afd}\, i_{fd}(t) + M_{akd}\, i_{kd}(t) \qquad (3.124)$$
$$\psi_q(t) = -L_q\, i_q(t) + M_{agq}\, i_{gq}(t) + M_{akq}\, i_{kq}(t) \qquad (3.125)$$
$$\psi_{fd}(t) = -M_{afd}\, i_d(t) + L_{fdfd}\, i_{fd}(t) + M_{fdkd}\, i_{kd}(t) \qquad (3.126)$$
$$\psi_{kd}(t) = -M_{akd}\, i_d(t) + M_{fdkd}\, i_{fd}(t) + L_{kdkd}\, i_{kd}(t) \qquad (3.127)$$
$$\psi_{gq}(t) = -M_{agq}\, i_q(t) + L_{gqgq}\, i_{gq}(t) + M_{gqkq}\, i_{kq}(t) \qquad (3.128)$$
$$\psi_{kq}(t) = -M_{akq}\, i_q(t) + M_{gqkq}\, i_{gq}(t) + L_{kqkq}\, i_{kq}(t) \qquad (3.129)$$

(2) Voltage-Flux Linkage-Current Equations

$$v_d(t) = \frac{d}{dt} \psi_d(t) - \psi_q(t) \frac{d}{dt} \theta(t) - R_a i_d(t)$$

(3.130)

$$v_q(t) = \frac{d}{dt} \psi_q(t) + \psi_d \frac{d}{dt} \theta(t) - R_a i_q(t)$$

(3.131)

$$v_{fd}(t) = \frac{d}{dt} \psi_{fd}(t) + R_{fd} i_{fd}(t)$$

(3.132)

$$0 = \frac{d}{dt} \psi_{kd}(t) + R_{kd} i_{kd}(t)$$

(3.133)

$$0 = \frac{d}{dt} \psi_{gq}(t) + R_{gq} i_{gq}(t)$$

(3.134)

$$0 = \frac{d}{dt} \psi_{kq}(t) + R_{kq} i_{kq}(t)$$

(3.135)

(3) Terminal Voltage

$$v^2(t) = v_d^2(t) + v_q^2(t)$$

(3.136)

(4) Electrical Torque and Swing Equation

$$T_e(t) = \psi_d(t) i_q(t) - \psi_q(t) i_d(t)$$

(3.137)

$$P_m(t) = T_m(t) \frac{d}{dt} \theta(t)$$

(3.138)

$$\overline{T_m}(t) - \overline{T_e}(t) = (2H/\omega_0) \frac{d^2}{dt^2} \theta(t) + (D/\omega_0) \frac{d}{dt} \theta(t)$$

(3.139)

3.3.2.2 Linearized Equations

Let $\theta(t) = \omega_0 t + \delta(t)$

(3.140)

then, the rotor angular speed is described as

$$\frac{d}{dt} \theta(t) = \omega_0 + \frac{d}{dt} \delta(t)$$

(3.141)

and for small changes, (3.128) becomes

$$\frac{d}{dt}\Delta\theta(t) = \frac{d}{dt}\Delta\delta(t)$$

$$(3.142)$$

In steady state, the synchronous rotor speed is

$$\frac{d}{dt}\theta(t) = \omega_0$$

$$(3.143)$$

Now, linearizing the fundamental equations in (3.111) - (3.126) at the operating points, the next equations are obtained.

(1) Flux Linkage Equations

From (3.111) - (3.116),

$$
\begin{bmatrix} \Delta\psi_d \\ \Delta\psi_q \\ \Delta\psi_{fd} \\ \Delta\psi_{kd} \\ \Delta\psi_{gq} \\ \Delta\psi_{kq} \end{bmatrix}
=
\begin{bmatrix}
-L_d & 0 & M_{afd} & M_{akd} & 0 & 0 \\
0 & -L_q & 0 & 0 & M_{bgq} & M_{bkq} \\
-(3/2)M_{afd} & 0 & L_{fdfd} & M_{fdkd} & 0 & 0 \\
-(3/2)M_{akd} & 0 & M_{kdfd} & L_{kdkd} & 0 & 0 \\
0 & -(3/2)M_{agq} & 0 & 0 & L_{gqgq} & M_{gqkq} \\
0 & -(3/2)M_{akq} & 0 & 0 & M_{kqgq} & L_{kqkq}
\end{bmatrix}
\begin{bmatrix} \Delta i_d \\ \Delta i_q \\ \Delta i_{fd} \\ \Delta i_{kd} \\ \Delta i_{gq} \\ \Delta i_{kq} \end{bmatrix}
\quad (3.144)
$$

(2) Voltage-Flux Linkage-Current Equations

From (3.130) and (3.131),

$$
\begin{bmatrix} \Delta v_d \\ \Delta v_q \end{bmatrix}
= \frac{d}{dt}\begin{bmatrix} \Delta\psi_d \\ \Delta\psi_q \end{bmatrix}
+ \begin{bmatrix} 0 & -\omega_0 \\ \omega_0 & 0 \end{bmatrix}\begin{bmatrix} \Delta\psi_d \\ \Delta\psi_q \end{bmatrix}
- R_a\begin{bmatrix} \Delta i_d \\ \Delta i_q \end{bmatrix}
+ \begin{bmatrix} -\Delta\psi_q \\ \Delta\psi_p \end{bmatrix}\frac{d}{dt}\Delta\delta
\quad (3.145)
$$

From (3.132) - (3.135)

$$
\begin{bmatrix} \Delta i_{fd} \\ \Delta i_{kd} \\ \Delta i_{gq} \\ \Delta i_{kq} \end{bmatrix}
= -\begin{bmatrix} 1/R_{fd} & & & \\ & 1/R_{kd} & & \\ & & 1/R_{gq} & \\ & & & 1/R_{kq} \end{bmatrix}
\frac{d}{dt}\begin{bmatrix} \Delta\psi_{fd} \\ \Delta\psi_{kd} \\ \Delta\psi_{gq} \\ \Delta\psi_{kq} \end{bmatrix}
+ \begin{bmatrix} 1/R_{fd} \\ 0 \\ 0 \\ 0 \end{bmatrix}\Delta v_{fd}
\quad (3.146)
$$

(3) Terminal Voltage

From (3.1136),

$$\Delta v = \begin{bmatrix} \dfrac{v_d}{v} & \dfrac{v_q}{v} \end{bmatrix} \begin{bmatrix} \Delta v_d \\ \Delta v_q \end{bmatrix} \tag{3.147}$$

(4) Electrical Torque and Swing Equation

$$\Delta T_e = \begin{bmatrix} i_q & -i_d \end{bmatrix} \begin{bmatrix} \Delta \psi_d \\ \Delta \psi_q \end{bmatrix} + \begin{bmatrix} -\psi_q & \psi_d \end{bmatrix} \begin{bmatrix} \Delta i_d \\ \Delta i_q \end{bmatrix} \tag{3.148}$$

$$\Delta P_e = \omega_0 \Delta T_e + T_e \frac{d}{dt}(\Delta \delta) \tag{3.149}$$

$$\Delta P_m = \omega_0 \Delta T_m + T_m \frac{d}{dt}(\Delta \delta) \tag{3.150}$$

$$\Delta T_m - \Delta T_e = 2H \frac{d^2}{dt^2}(\Delta \delta) + \frac{D}{\omega_0} \frac{d}{dt}(\Delta \delta) \tag{3.151}$$

3.3.2.3　State Variables for a Synchronous Machine

For the linearized fundamental equations derived in the previous section, either set of the variables in (3.152) can be chosen as a state vector to represent the generator internal states. Using (3.144) the other set can be calculated since the inverse of the transformation matrix always exists. Here, flux linkage variables will be used for building a system matirix.

$$x = \begin{bmatrix} \Delta i_d \\ \Delta i_q \\ \Delta i_{fd} \\ \Delta i_{kd} \\ \Delta i_{gq} \\ \Delta i_{kq} \\ \Delta \delta \\ \frac{d}{dt} \Delta \delta \end{bmatrix} \quad \text{or} \quad x = \begin{bmatrix} \Delta \psi_d \\ \Delta \psi_q \\ \Delta \psi_{fd} \\ \Delta \psi_{kd} \\ \Delta \psi_{gq} \\ \Delta \psi_{kq} \\ \Delta \delta \\ \frac{d}{dt} \Delta \delta \end{bmatrix} \tag{3.152}$$

3.3.2.4 State Equation for a Multi-Machine System

Extending (3.146) for the N machine system,

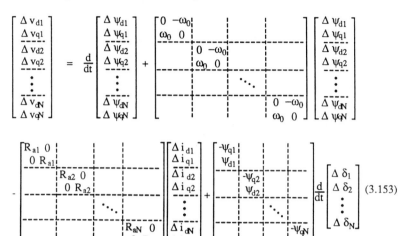

$$\Delta v = \frac{d}{dt}(\Delta \psi) + [W] \Delta \psi - [R_a] \Delta i + [\Psi] \frac{d}{dt}(\Delta \delta) \tag{3.154}$$

Combining (3.154) and (3.121),

$$
\begin{bmatrix} \Delta i_{d1} \\ \Delta i_{q1} \\ \hline \Delta i_{d2} \\ \Delta i_{q2} \\ \hline \vdots \\ \hline \Delta i_{dN} \\ \Delta i_{qN} \end{bmatrix}
=
\begin{bmatrix}
C_{11dd} & C_{11dq} & \cdots & C_{1Ndd} & C_{1Ndq} \\
C_{11qd} & C_{11qq} & \cdots & C_{1Nqd} & C_{1Nqq} \\
\hline
C_{12dd} & C_{12dq} & \cdots & C_{2Ndd} & C_{2Ndq} \\
C_{12qd} & C_{12qq} & \cdots & C_{2Nqd} & C_{2Nqq} \\
\hline
& \vdots & & \vdots & \\
\hline
C_{N1dd} & C_{N1dq} & \cdots & C_{NNdd} & C_{NNdq} \\
C_{N1qd} & C_{N1qq} & \cdots & C_{NNqd} & C_{NNqq}
\end{bmatrix}
\frac{d}{dt}
\begin{bmatrix} \Delta \psi_{d1} \\ \Delta \psi_{q1} \\ \hline \Delta \psi_{d2} \\ \Delta \psi_{q2} \\ \hline \vdots \\ \hline \Delta \psi_{dN} \\ \Delta \psi_{qN} \end{bmatrix}
$$

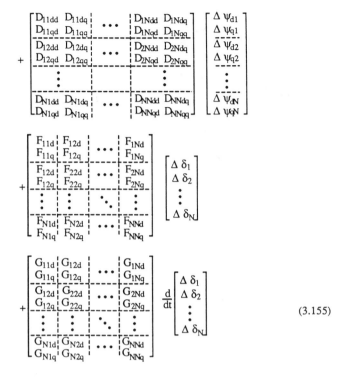

$$\Delta i = [C]\frac{d}{dt}\left(\Delta\psi\right) + [D]\,\Delta\psi + [F]\,\Delta\delta + [G]\frac{d}{dt}\left(\Delta\delta\right) \tag{3.156}$$

where

$$[C] = ([I] + [Y_1][R_a])^{-1}[Y_1] \quad : (2N \times 2N) \tag{3.157}$$

$$[D] = [C][W] \quad : (2N \times 2N) \tag{3.158}$$

$$[F] = ([I] + [Y_1][R_a])^{-1}[K_1] \quad : (2N \times N) \tag{3.159}$$

$$[G] = [C][\Psi] \quad : (2N \times N) \tag{3.160}$$

Combining (3.146) and (3.156),

$$
\begin{bmatrix} \Delta i_{d1} \\ \Delta i_{q1} \\ \Delta i_{fd1} \\ \Delta i_{kd1} \\ \Delta i_{gq1} \\ \Delta i_{kq1} \\ \hline \\ \vdots \\ \\ (N) \end{bmatrix}
=
\begin{bmatrix}
C_{11dd} & C_{11dq} & 0 & 0 & 0 & 0 \\
C_{11qd} & C_{11qq} & 0 & 0 & 0 & 0 \\
0 & 0 & -1/R_{fd1} & 0 & 0 & 0 \\
0 & 0 & 0 & -1/R_{kd1} & 0 & 0 \\
0 & 0 & 0 & 0 & -1/R_{gq1} & 0 \\
0 & 0 & 0 & 0 & 0 & -1/R_{kq1} \\
\hline
& & & \vdots \\
& & & (N)
\end{bmatrix}
\begin{matrix} \cdots & (N) \\ \\ \ddots \\ (N) \end{matrix}
\;\;
\frac{d}{dt}
\begin{bmatrix} \Delta \psi_{d1} \\ \Delta \psi_{q1} \\ \Delta \psi_{fd1} \\ \Delta \psi_{kd1} \\ \Delta \psi_{gq1} \\ \Delta \psi_{kq1} \\ \hline \\ \vdots \\ \\ (N) \end{bmatrix}
$$

$$
+
\begin{bmatrix}
D_{11dd} & D_{11dq} & 0 & 0 & 0 & 0 & F_{11d} & F_{12d} \\
D_{11qd} & D_{11qq} & 0 & 0 & 0 & 0 & F_{11q} & F_{12q} \\
0 & 0 & 0 & 0 & 0 & 0 & 0 & 0 \\
0 & 0 & 0 & 0 & 0 & 0 & 0 & 0 \\
0 & 0 & 0 & 0 & 0 & 0 & 0 & 0 \\
0 & 0 & 0 & 0 & 0 & 0 & 0 & 0 \\
\hline
& & & \vdots \\
& & & (N)
\end{bmatrix}
\begin{matrix} \cdots & (N) \\ \\ \ddots \\ (N) \end{matrix}
\;\;
\begin{bmatrix} \Delta \psi_{d1} \\ \Delta \psi_{q1} \\ \Delta \psi_{fd1} \\ \Delta \psi_{kd1} \\ \Delta \psi_{gq1} \\ \Delta \psi_{kq1} \\ \hline \Delta \delta_1 \\ \frac{d}{dt}\Delta \delta_1 \\ \hline \vdots \\ (N) \end{bmatrix}
$$

$$
+
\begin{bmatrix}
0 & & & \\
0 & & & \\
1/R_{fd1} & & \mathbf{0} & \\
0 & & & \\
0 & & & \\
0 & & & \\
\hline
& 1/R_{fd2} & & \\
& \vdots & & \\
\hline
\mathbf{0} & & \ddots & \\
& & & (N)
\end{bmatrix}
\begin{bmatrix} \Delta v_{fd1} \\ \Delta v_{fd2} \\ \vdots \\ \Delta v_{fdN} \end{bmatrix}
\qquad (3.161)
$$

or simply

$$
\Delta \hat{\mathbf{i}} = [P]\frac{d}{dt}\Delta\hat{\psi} + [Q]\,\hat{\mathbf{x}} + [S]\,\Delta\hat{\mathbf{v}}_{fd}
\qquad (3.162)
$$

where

$$\hat{x} \equiv \atop (8N\times 1) \qquad \begin{bmatrix} \Delta\ \psi_{d1} \\ \Delta\ \psi_{q1} \\ \Delta\ \psi_{fd1} \\ \Delta\ \psi_{kd1} \\ \Delta\ \psi_{gq1} \\ \Delta\ \psi_{kq1} \\ \Delta\ \delta_1 \\ \frac{d}{dt}\Delta\ \delta_1 \\ ----- \\ \vdots \\ (N) \end{bmatrix} \tag{3.163}$$

and the dimensions of the matrices are given as

$$[P] = (6N \times 6N),\ [Q] = (6N \times 8N),\ \text{and}\ [S] = (6N \times N) \tag{3.164}$$

From (3.144)

$$\Delta\hat{\psi} = [L]\,\Delta\hat{i} \tag{3.165}$$

Thus,

$$\Delta\hat{\psi} = [L][P]\frac{d}{dt}\Delta\hat{\psi} + [L][Q]\,\hat{x} + [L][S]\,\Delta\hat{v}_{fd} \tag{3.166}$$

By defining a matrix [J] which satisfies

$$\Delta\hat{\psi} = [J]\,\hat{x} \tag{3.167}$$

Therefore,

$$[L][P]\frac{d}{dt}\Delta\hat{\psi} = ([J] - [L][Q])\,\hat{x} - [L][S]\,\Delta\hat{v}_{fd} \tag{3.168}$$

or

$$\frac{d}{dt}\Delta\hat{\psi} = [A']\,\hat{x} + [B_v']\,\Delta\hat{v}_{fd} \tag{3.169}$$

where

$$[A'] = [P]^{-1}[L]^{-1}\,([J] - [L][Q]) \tag{3.170}$$

and

$$[B_v'] = -[P]^{-1}[S] \tag{3.171}$$

3.3.2.5 Swing Equation

From (3.151),

$$\frac{d}{dt}(\Delta\delta) = \frac{d}{dt}\Delta\delta \tag{3.172}$$

$$\frac{d}{dt}\left(\frac{d}{dt}\Delta\delta\right) = \frac{1}{2H}\Delta P_m - \frac{1}{2H}\left(\frac{P_m}{\omega_0} + D\right)\frac{d}{dt}\Delta\delta - \frac{\omega_0}{2H}T_e \tag{3.173}$$

As in (3.148), ΔT_e is determined with $\Delta\psi_d$, $\Delta\psi_q$, Δi_d, and Δi_q. Since Δi_d, and Δi_q are related with $\Delta\psi_d$, $\Delta\psi_q$, $\Delta\psi_{fd}$, $\Delta\psi_{kd}$, $\Delta\psi_{gq}$, and $\Delta\psi_{kq}$ from (3.144),

$$\Delta T_e \equiv [K][\Delta\psi_d \; \Delta\psi_q \; \Delta\psi_{fd} \; \Delta\psi_{kd} \; \Delta\psi_{gq} \; \Delta\psi_{kq}]^t \tag{3.174}$$

where

$$[K] = [(i_q - l_{11}\psi_q + l_{21}\psi_d), (-i_d - l_{12}\psi_q + l_{22}\psi_d), (l_{13}\psi_q + l_{23}\psi_d), (-l_{14}\psi_q + l_{24}\psi_d),$$

$$(-l_{15}\psi_q + l_{25}\psi_d), (-l_{16}\psi_q + l_{26}\psi_d)] \tag{3.175}$$

and l_{ij}'s are obtained from the inverse of the inductance matrix in (3.144)as follows;

$$\begin{bmatrix} l_{11} & l_{12} & \cdots & l_{16} \\ l_{21} & l_{22} & \cdots & l_{26} \\ \vdots & \ddots & & \vdots \\ \vdots & & \ddots & \vdots \\ l_{61} & \cdots\cdots & l_{66} \end{bmatrix} = \begin{bmatrix} -L_d & 0 & M_{afd} & M_{akd} & 0 & 0 \\ 0 & -L_q & 0 & 0 & M_{bgq} & M_{bkq} \\ -(3/2)M_{afd} & 0 & L_{fdfd} & M_{fdkd} & 0 & 0 \\ -(3/2)M_{akd} & 0 & M_{kdfd} & L_{kdkd} & 0 & 0 \\ 0 & -(3/2)M_{agq} & 0 & 0 & L_{gqgq} & M_{gqkq} \\ 0 & -(3/2)M_{akq} & 0 & 0 & M_{kqgq} & L_{kqkq} \end{bmatrix}^{-1} \tag{3.176}$$

Then, the next equation can be obtained:

$$\frac{d}{dt}\begin{bmatrix} \Delta\delta \\ \frac{d}{dt}\Delta\delta \end{bmatrix} = \begin{bmatrix} 0 \\ \frac{1}{2H} \end{bmatrix}\Delta P_m$$

$$+ \begin{bmatrix} 0 & 0 & 0 & 0 & 0 & 0 & 0 & 1 \\ \frac{-\omega_0}{2H}K_1 & \frac{-\omega_0}{2H}K_2 & \frac{-\omega_0}{2H}K_3 & \frac{-\omega_0}{2H}K_4 & \frac{-\omega_0}{2H}K_5 & \frac{-\omega_0}{2H}K_6 & 0 & \frac{-1}{2H}(D+\frac{P_m}{\omega_0}) \end{bmatrix} \begin{bmatrix} \Delta\psi_d \\ \Delta\psi_q \\ \Delta\psi_{fd} \\ \Delta\psi_{kd} \\ \Delta\psi_{gq} \\ \Delta\psi_{kq} \\ \Delta\delta \\ \frac{d}{dt}\Delta\delta \end{bmatrix} \quad (3.177)$$

Extending (3.177) for N generators in the system,

$$\frac{d}{dt} \begin{bmatrix} \Delta\psi_{d1} \\ \Delta\psi_{q1} \\ \Delta\psi_{fd1} \\ \Delta\psi_{kd1} \\ \Delta\psi_{gq1} \\ \Delta\psi_{kq1} \\ \Delta\delta_1 \\ \frac{d}{dt}\Delta\delta_1 \\ \vdots \\ (N) \end{bmatrix} = \begin{bmatrix} [A_{11}'] & \cdots & (N) \\ \begin{matrix} 0 & 0 & 0\,0\,0\,0\,0 & 1 \\ \frac{-\omega_0}{2H}K_1 & \frac{-\omega_0}{2H}K_2 & \cdots\ \ 0 & \frac{-1}{2H}(D+\frac{P_m}{\omega_0}) \end{matrix} & & \\ \vdots & \ddots & (N) \\ (N) & & \end{bmatrix} \begin{bmatrix} \Delta\psi_{d1} \\ \Delta\psi_{q1} \\ \Delta\psi_{fd1} \\ \Delta\psi_{kd1} \\ \Delta\psi_{gq1} \\ \Delta\psi_{kq1} \\ \Delta\delta_1 \\ \frac{d}{dt}\Delta\delta_1 \\ \vdots \\ (N) \end{bmatrix}$$

$$+ \begin{bmatrix} [B_{v1}'] \\ \hline 0\ \cdots\ 0 \\ 0\ \cdots\ 0 \\ \vdots \\ (N) \end{bmatrix} \begin{bmatrix} \Delta v_{fd1} \\ \Delta v_{fd2} \\ \vdots \\ \Delta v_{fdN} \end{bmatrix} + \begin{bmatrix} \begin{matrix} 0 \\ 0 \\ 0 \\ 0 \\ 0 \\ 0 \\ 0 \\ 1/2H_1 \end{matrix} & & 0 \\ & \ddots \\ & 1/2H_2 \\ 0 & & (N) \end{bmatrix} \begin{bmatrix} \Delta P_{m1} \\ \Delta P_{m2} \\ \vdots \\ \Delta P_{mN} \end{bmatrix} \quad (3.178)$$

where

$$[A'] = \begin{bmatrix} [A_{11}'] & \cdots & [A_{1N}'] \\ \vdots & \ddots & \vdots \\ [A_{N1}'] & \cdots & [A_{NN}'] \end{bmatrix} \quad \text{and} \quad [B_v'] = \begin{bmatrix} [B_{v1}'] \\ \vdots \\ [B_{vN}'] \end{bmatrix} \quad (3.179)$$

where $[A_{ij}']$ and $[B_{vi}']$ are the submatrices of $[A']$ and $[B_v']$. They have (8×6) and $(6\times N)$ dimensions, respectively.

For a simple form, (3.178) can be rewritten as

$$\frac{d}{dt}\hat{x} = [A]\,\hat{x} + [B_v]\,\Delta\hat{v}_{fd} + [B_p]\,\Delta\hat{P}_m \tag{3.180}$$

where the matrices have the following dimensions; $[A]$: $(8N \times 8N)$, $[B_v]$: $(8N \times N)$, and $[B_p]$: $(8N \times N)$. (3.167) is the state equation of the N-machine system including transmission lines with $\Delta\hat{v}_{fd}$ and $\Delta\hat{P}_m$ as control variables. The state equation consists of the following 8 variables;

- flux linkages (6): $\Delta\psi_d$, $\Delta\psi_q$, $\Delta\psi_{fd}$, $\Delta\psi_{kd}$, $\Delta\psi_{gq}$, and $\Delta\psi_{kq}$
- rotor angle   : $\Delta\delta$
- rotor angular velocity : $\frac{d}{dt}\delta$

The other variables including voltages and currents can be obtained from the associated equations as follows:

- Δi_d, Δi_q, Δi_{fd}, Δi_{kd}, Δi_{gq}, and Δi_{kq}: From (3.144)
- Δv_d, Δv_q : From the obtained Δi_d and Δi_q and (3.121)
- Terminal voltage Δv_t : From the obtained Δv_d and Δv_q and (3.147)
- ΔT_e : From (3.174)
- ΔP_e : From (3.149)

3.3.3   State Equation Including Control Systems

3.3.3.1   Excitation System

Considering the simple exciter model shown in Figure 3.7, the state equation for the excitation system in Figure 3.7 can be obtained as follows:

$$\frac{d}{dt}\begin{bmatrix} \Delta v_{fd}(t) \\ \Delta v_R(t) \\ \Delta v_S(t) \end{bmatrix} = \begin{bmatrix} -\dfrac{K_E}{T_E} & \dfrac{1}{T_E} & 0 \\ 0 & -\dfrac{1}{T_A} & -\dfrac{K_A}{T_A} \\ -\dfrac{K_E K_F}{T_E T_F} & \dfrac{K_F}{T_E T_F} & -\dfrac{1}{T_F} \end{bmatrix} \begin{bmatrix} \Delta v_{fd}(t) \\ \Delta v_R(t) \\ \Delta v_S(t) \end{bmatrix} + \begin{bmatrix} 0 \\ \dfrac{K_A}{T_A} \\ 0 \end{bmatrix} \Delta u_G(t) \tag{3.181}$$

Or simply

$$\frac{d}{dt}x_E(t) = [A_E]\,x_E(t) + [B_E]\,u_E(t) \tag{3.182}$$

where $x_E(t) = [\Delta v_{fd}(t), \Delta v_R(t), \Delta v_S(t)]^t$ \qquad (3.183)

and $u_E(t) = [u_E(t)]$ \qquad (3.184)

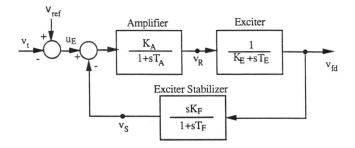

Figure 3.7 An Excitation System Similar to IEEE Type 1 Model

3.3.3.2 Governor-Turbine System

Considering a simlple governor-turbine model shown in Figure 3.8, the follwing state equation can be obtained:

$$\frac{d}{dt}\begin{bmatrix} \Delta P_m(t) \\ \Delta y(t) \\ \Delta Y(t) \end{bmatrix} = \begin{bmatrix} -\frac{1}{T_3} & 0 & \frac{K_3}{T_3} \\ 0 & -\left(\frac{1}{T_1}+\frac{1}{T_2}\right) & -\frac{1}{T_1 T_2} \\ 0 & 1 & 0 \end{bmatrix} \begin{bmatrix} \Delta P_m(t) \\ \Delta y(t) \\ \Delta Y(t) \end{bmatrix} + \begin{bmatrix} 0 \\ \frac{K_2}{T_1 T_2} \\ 0 \end{bmatrix} \Delta u_G(t)$$

(3.185)

or

$$\frac{d}{dt} x_G(t) = [A_G]\, x_G(t) + [B_G]\, u_G(t)$$ (3.186)

where $x_G(t) = [\Delta P_m(t), \Delta y(t), \Delta Y(t)]^t$ (3.187)

and $u_G(t) = [u_G(t)]$ (3.188)

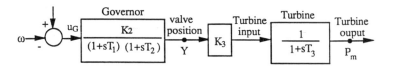

Figure 3.8 A Simplified Governor-Turbine System Model

3.3.3.3 State Equation of Multi-Machine System with Control Systems

Rewriting (3.180),

$$\frac{d}{dt} \widehat{x} = [A] \widehat{x} + [B_v] \Delta \widehat{x}_{fd} + [B_p] \Delta \widehat{P}_m \tag{3.189}$$

whree

$$[A] = \begin{bmatrix} [A_{11}] & [A_{12}] & \cdots & [A_{1N}] \\ [A_{21}] & [A_{22}] & \cdots & [A_{2N}] \\ \vdots & \vdots & \ddots & \vdots \\ [A_{N1}] & [A_{N2}] & \cdots & [A_{NN}] \end{bmatrix} \qquad \text{where } [A_{ij}] : (8 \times 8) \tag{3.190}$$

$$[B_v] = \begin{bmatrix} [B_{v1}] & & \cdots & 0 \\ 0 & [B_{v2}] & \cdots & 0 \\ \vdots & \vdots & \ddots & \vdots \\ 0 & 0 & \cdots & [B_{vN}] \end{bmatrix} \qquad \text{where } [B_{vi}] : (8 \times 8) \tag{3.191}$$

and

$$[B_p] = \begin{bmatrix} [B_{p1}] & & \cdots & 0 \\ 0 & [B_{p2}] & \cdots & 0 \\ \vdots & \vdots & \ddots & \vdots \\ 0 & 0 & \cdots & [B_{pN}] \end{bmatrix} \qquad \text{where } [B_{pi}] : (8 \times 8) \tag{3.192}$$

In additon, the state variables can be expressed as follows:

$$\widehat{x} = \begin{bmatrix} \widehat{x}_1 \\ \widehat{x}_2 \\ \vdots \\ \widehat{x}_N \end{bmatrix} \tag{3.193}$$

where $\widehat{x}$ is defined in (3.163).

$$\Delta \widehat{v}_{fd} = \begin{bmatrix} \Delta \widehat{v}_{fd1} \\ \Delta \widehat{v}_{fd2} \\ \vdots \\ \Delta \widehat{v}_{fdN} \end{bmatrix} = \begin{bmatrix} [M_E] \, x_{E1} \\ [M_E] \, x_{E2} \\ \vdots \\ [M_E] \, x_{EN} \end{bmatrix} \tag{3.194}$$

$$\Delta \hat{P}_m = \begin{bmatrix} \Delta \hat{P}_{m1} \\ \Delta \hat{P}_{m2} \\ \vdots \\ \Delta \hat{P}_{mN} \end{bmatrix} = \begin{bmatrix} [M_G]\, x_{G1} \\ [M_G]\, x_{G2} \\ \vdots \\ [M_G]\, x_{GN} \end{bmatrix} \tag{3.195}$$

where $[M_E] \equiv [1\ 0\ 0]$ and $[M_G] \equiv [1\ 0\ 0]$.

$$\frac{d}{dt} \begin{bmatrix} \hat{x}_1 \\ \hat{x}_2 \\ \vdots \\ \hat{x}_N \end{bmatrix} = \begin{bmatrix} [A_{11}] & [A_{12}] & \cdots & [A_{1N}] \\ [A_{21}] & [A_{22}] & \cdots & [A_{2N}] \\ \vdots & \vdots & \ddots & \vdots \\ [A_{N1}] & [A_{N2}] & \cdots & [A_{NN}] \end{bmatrix} \begin{bmatrix} \hat{x}_1 \\ \hat{x}_2 \\ \vdots \\ \hat{x}_N \end{bmatrix}$$

$$+ \begin{bmatrix} [B_{v1}][M_E] & 0 & \cdots & 0 \\ 0 & [B_{v2}][M_E] & \cdots & 0 \\ \vdots & \vdots & \ddots & \vdots \\ 0 & 0 & \cdots & [B_{vN}][M_E] \end{bmatrix} \begin{bmatrix} x_{E1} \\ x_{E2} \\ \vdots \\ x_{EN} \end{bmatrix}$$

$$+ \begin{bmatrix} [B_{P1}][M_G] & 0 & \cdots & 0 \\ 0 & [B_{P2}][M_G] & \cdots & 0 \\ \vdots & \vdots & \ddots & \vdots \\ 0 & 0 & \cdots & [B_{PN}][M_G] \end{bmatrix} \begin{bmatrix} x_{G1} \\ x_{G2} \\ \vdots \\ x_{GN} \end{bmatrix} \tag{3.196}$$

(3.182) and (3.186) can be modified for N generators as follows;

• excitation system

$$\frac{d}{dt} x_{E1} = [A_{E1}]\, x_{E1} + [B_{E1}]\, u_{E1}$$
$$\frac{d}{dt} x_{E2} = [A_{E2}]\, x_{E2} + [B_{E2}]\, u_{E2}$$
$$\vdots$$
$$\frac{d}{dt} x_{EN} = [A_{EN}]\, x_{EN} + [B_{EN}]\, u_{EN} \tag{3.197}$$

• governor-turbine system

$$\frac{d}{dt} x_{G1} = [A_{G1}]\, x_{G1} + [B_{G1}]\, u_{G1}$$
$$\frac{d}{dt} x_{G2} = [A_{G2}]\, x_{G2} + [B_{G2}]\, u_{G2}$$
$$\vdots$$
$$\frac{d}{dt} x_{GN} = [A_{GN}]\, x_{GN} + [B_{GN}]\, u_{GN} \tag{3.198}$$

$$
\frac{d}{dt}
\begin{bmatrix}
\hat{x}_1 \\ x_{E1} \\ x_{G1} \\ \hline \hat{x}_2 \\ x_{E2} \\ x_{G2} \\ \hline \vdots \\ (N)
\end{bmatrix}
=
\begin{bmatrix}
[A_{11}] & [B_{v1}][M_E] & [B_{P1}][M_G] & | & [A_{12}] & 0 & 0 & | & [A_{13}] \\
0 & [A_{E1}] & 0 & | & 0 & 0 & 0 & | & 0 & \cdots \\
0 & 0 & [A_{G1}] & | & 0 & 0 & 0 & | & 0 \\
\hline
[A_{21}] & 0 & 0 & | & [A_{22}] & [B_{v2}][M_E] & [B_{P2}][M_G] & | & [A_{23}] \\
0 & 0 & 0 & | & 0 & [A_{E2}] & 0 & | & 0 & \cdots \\
0 & 0 & 0 & | & 0 & 0 & [A_{G2}] & | & 0 \\
\hline
[A_{31}] & 0 & 0 & | & & & & & \\
& \vdots & & & & & (N) & &
\end{bmatrix}
\begin{bmatrix}
\hat{x}_1 \\ x_{E1} \\ x_{G1} \\ \hline \hat{x}_2 \\ x_{E2} \\ x_{G2} \\ \hline \vdots \\ (N)
\end{bmatrix}
$$

$$
+
\begin{bmatrix}
0 & & \\ [B_{v1}] & & \\ 0 & & \Huge 0 \\ \hline & 0 & \\ & [B_{v2}] & \\ & 0 & \\ \hline \Huge 0 & & \ddots \\ & & (N)
\end{bmatrix}
\begin{bmatrix}
u_{E1} \\ u_{E2} \\ \vdots \\ u_{EN}
\end{bmatrix}
+
\begin{bmatrix}
0 & & \\ 0 & & \\ [B_{P1}] & & \Huge 0 \\ \hline & 0 & \\ & 0 & \\ & [B_{P2}] & \\ \hline \Huge 0 & & \ddots \\ & & (N)
\end{bmatrix}
\begin{bmatrix}
u_{G1} \\ u_{G2} \\ \vdots \\ u_{GN}
\end{bmatrix}
\tag{3.199}
$$

In a simple form,

$$
\frac{d}{dt}\boxtimes = [\mathbb{A}]\boxtimes + [\mathbb{B}_E]\,\mathbf{u}_E + [\mathbb{B}_G]\,\mathbf{u}_G
\tag{3.200}
$$

where $\mathbf{u}_E$ and $\mathbf{u}_G$ are the control signals of exciter and governor respectively. In each generator, usually the control signal of exciter is proportional to the change of the generator terminal voltage Δv_i while that of governor proportional to the change of rotor speed $d(\Delta\delta_i)/dt$. Thus, $\mathbf{u}_E$ and $\mathbf{u}_G$ can be expressed in the linear combination of the state variable, $\boxtimes$ as follows:

$$
\mathbf{u}_E = [\mathbb{E}]\boxtimes \quad \text{and} \quad \mathbf{u}_G = [\mathbb{G}]\boxtimes
\tag{3.201}
$$

Then, (3.200) becomes

$$
\begin{aligned}
\frac{d}{dt}\boxtimes &= [\mathbb{A}]\boxtimes + [\mathbb{B}_E][\mathbb{E}]\boxtimes + [\mathbb{B}_G][\mathbb{G}]\boxtimes \\
&= \{[\mathbb{A}] + [\mathbb{B}_E][\mathbb{E}] + [\mathbb{B}_G][\mathbb{G}]\}\,\boxtimes
\end{aligned}
\tag{3.202}
$$

Finally,

$$\frac{d}{dt} \mathbf{x} = [\mathbf{M}] \, \mathbf{x}$$

(3.203)

where

$$[\mathbf{M}] = [\mathbf{A}] + [\mathbf{B}_Z] [\mathbf{E}] + [\mathbf{B}_G] [\mathbf{G}]$$

(3.204)

To obtain the response of the system, the eigenvalues of the characteristic matrix $[\mathbf{M}]$ will be examined for dynamic stability. This is obtained from the characteristic equation derived from equating the determinant of the matrix to zero, as follows:

$$\det (\, [\mathbf{M}] - \lambda \, [\mathbf{I}] \,) = 0$$

(3.205)

where λ is the eigenvalue and $[\mathbf{I}]$ is an identity matrix. Usually, each eigenvalue has a complex number of which the real part represents the damping ratio and the imaginary part gives the frequency of oscillation which will result from small disturbances.

3.3.3.4 Power System Stabilizer

A power system may become unstable if there is insufficient damping. It has been found that system damping can be improved by adding special supplementary stabilizing control signals to the voltage regulator at selected plants. There are many signals which can be picked up and used as stabilizer input signals. Though there are some disadvantages, it is convenient for us to select the shaft speed or output power deviation as a supplementary input signal of a voltage regulator.

In this section, only the classical stabilizer is discussed. It uses a lead-lag compensation network to compensate for the phase shift caused by the low frequency oscillation of the system during some small perturbations. The theory and design criteria of the stabilizer will be discussed in the following.

As shown in figure 3.9, the lead-lag compensation network consists of two operation amplifiers with some resistors and capacitors. The derivation of transfer function of the network is shown below:

$$E_i * (1/R + K_1/(R_1 + 1/SC_1)) = E_o * (1/R + SC_F + K_2SC_2)$$

(3.206)

$$E_i * (((1 + SR_1C_1) + SK_1RC_1)/(R * (1 + SR_1C_1))$$
$$= E_o * ((1 + S * (K_2RC_2 + RC_F))/R)$$

(3.207)

$$\frac{E_o}{E_i} = \frac{1 + S * (K_1RC_1 + R_1C_1)}{(1 + SR_1C_1) * (1 + S * (K_2RC_2 + RC_F))}$$

(3.208)

$$\frac{E_o}{E_i} = \frac{1 + S * (T_A + T_B)}{(1 + ST_B) * (1 + S * (T_C + T_D))}$$

(3.209)

where

$K_1 = R_B/(R_A + R_B)$

$K_2 = R_D/(R_C + R_D)$

$T_A = K_1RC_1$ ---------   Lead time constant

$T_B = R_1C_1$ -----------   Noise filter time constant

$T_C = K_2RC_2$ ---------   Lag time constant

$T_D = RC_F$ ------------   Amplifier stabilizing circuit time constant

Usually, $T_A \gg T_B$ and $T_C \gg T_D$, so that we can change Eq. 3.209 into Eq. 3.210 without creating too large an error.

$$\frac{E_o}{E_i} = \frac{1 + ST_A}{1 + ST_B} = \frac{1 + SaT}{1 + ST}$$

(3.210)

where $a = K_1C_1/K_2C_2 > 1$

For any single stage lead-lag network, if $a > 1$, its Bode plot diagram is shown in figure 3.10. The maximum phase lead ϕ_m occurs at the medium frequency ω_m, which is the geometric mean of the corner frequencies; i.e.,

$$\log \omega_m = \frac{1}{2} * (\log \frac{1}{aT} + \log \frac{1}{T}) = \frac{1}{2} * \log \frac{1}{aT^2} = \log \frac{1}{\sqrt{a}T}$$

(3.211)

then

$$\omega_m = \frac{1}{\sqrt{a}T}$$

(3.212)

The magnitude of the maximum phase lead ϕ_m is computed as follows:

$$\phi_m = \arg (\frac{1 + j\omega_maT}{1 + j\omega_mT}) = \tan^{-1}(\omega_maT) - \tan^{-1}(\omega_mT) = x - y$$

(3.213)

then

$$\tan \phi_m = \frac{\omega_maT - \omega_mT}{1 + \omega_m^2aT^2} = \frac{a-1}{2\sqrt{a}}$$

(3.214)

Now, visualizing a right triangle with base $2\sqrt{a}$, height $(a - 1)$ and hypotenuse b, we can compute $b^2 = (a + 1)^2$ or $\sin \phi_m = (a - 1)/(a + 1)$. This expression can be used to solve the value of a; i.e., $a = (1 + \sin \phi_m)/(1 - \sin \phi_m)$. This gives the desired relation between maximum phase lead ϕ_m and the parameter a. We can fix the parameter a by determining the desired phase lead ϕ_m. Knowing both a and ω_m, we can determine T from Eq. 3.212.

In many practical cases, the required phase shift is larger than that obtainable from a single stage network. Then two or more cascadeed stages must be used. But between these stages, some buffer circuits have to be added to eliminate the load effect and get the performance desired. We often write the transfer function of the overall lead-lag compensation network as:

$$G_s(s) = \frac{K_oT_oS}{1 + T_oS} * (\frac{1 + aTS}{1 + TS})^n$$

$$(3.215)$$

where n is the number of the cascaded stages, T_o is the washout time constant, and $K_oT_oS/(1 + ST_o)$ is the transfer function of signal reset circuit .

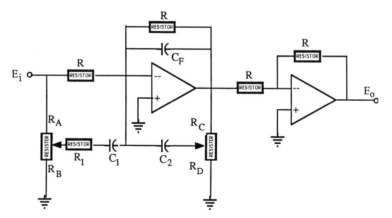

Figure 3.9 Circuit Diagram of the Lead-Lag Compensation Network

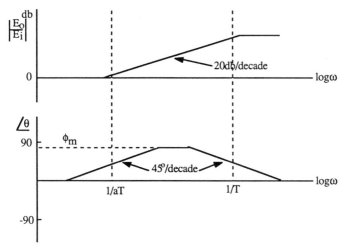

Figure 3.10 Bode Plot Diagram of the Lead-Lag Network

3.3.3.5 Static Var Compensator

Shunt capacitors, shunt reactors, series capacitors, and synchronous condensors have been applied for reactive power compensation for many years. Each of them has its unique functions and covers a different range of applications. Shunt capacitors and shunt reactors can control steady state voltage and reactive power flow. Series capacitors can improve power transfer capability. Synchronous condensors not only can supply reactive power but also control voltage variation. However, the Static Var Compensators (SVC) can perform nearly all the functions of other compensating methods with very high speed.

The first static var compensator in the form of saturated reactors in the power industry was introduced by E. Fredlander in the early 60s. He tried to find a device which could keep the voltage within a certain range at a strategic location on the transmission line, regardless of the power flow. Thyristor, or Gate-Turn-Of Thyristor (GTO), controlled reactors and capacitors were developed next to overcome some of the technical limitations of the saturated reactor, and to assimilate, in principle, its basic voltage control function. Using standard or modular components in such devices provided the basic ground for their economic competitiveness with external sophisticated controls. Applying thyristors in the SVC control schemes show great flexibility and reliability in operations in the ever changing power systems. With the employment of microprocessor based control, it is now possible to introduce new control algorithms to enable SVC to enhance its potential in the electric power system. With appropriate control scheme, Static Var Compensators could be the best chosen tools to compensate the rapidly changing demands of reactive power and improve the dynamic stability of the power system.

Fix-Capacitor, Thyristor-Controlled-Reactor (FC-TCR) type SVC is one of the most popular SVC configurations. Its steady state characteristic is shown in figure 3.11. The operation of FC-TCR type SVC is shown in figure 3.12. If the system voltage is operated within the regulation band, the voltage at point a (without compensation) will be raised to point b. Similarly, in the load line one, the voltage at point c (without compensation) will be moved to point d. The voltage variation with compensation is determined by regulation slope X_K as indicated in figure 3.12. For voltage beyond regulation band, the SVC will act either as a capacitive (thyristor turn-off) or as an inductive susceptance (thyristor fully turn-on).

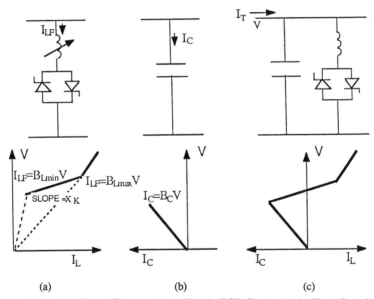

(a)   (b)   (c)

Figure 3.11 The v-i Characteristics of (a) the TCR Circuit, (b) the Fixed Capacitor, and (c) the Combined Circuit.

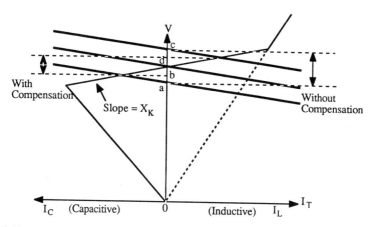

Figure 3.12 v-i Characteristic of FC-TCR Static Var Compensator with Load Lines

CHAPTER 4

BLOCK DIAGRAM CONVERSION AND RESULTS INTERPRETATION

4.1 Block Diagram Conversion

The phenomena of stability and damping of a synchronous machine for the mode of a small disturbance can be examined with the aid of the block diagrams. We can use the Continuous System Modelling Program (CSMP) or Easy 5[TM],a package newly developed by Boeing Computer, to check the response. These programs transfer block diagrams into state variable representations and use numerical methods to solve these equations directly. However, for a large system, a direct numerical analysis method is a more feasible approach. Frequency domain eigenvalue analysis and time domain numerical analysis are two most useful tools. No matter what kind of method is selected, the first step is to develop a set of first order differential equations from the block diagram which can represent the fundamental system characteristics. The control block diagram can be categorized into four types as follows:

(a) **Low Pass Filter:** This circuit is used to reduce the noise of the system and prevent control equipment from chattering due to sudden and temporary variations of the input signal. The general block diagram of the low pass filter is shown in figure 4.1. This diagram can be decomposed into figure 4.2 and the state variable is chosen as X(i).

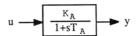

Figure 4.1 The Block Diagram of Low Pass Filter

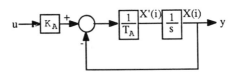

Figure 4.2 The Decomposed Diagram of Low Pass Filter

Where K_A is the gain and T_A is the time constant of the filter. The state function of the block diagram is

$$X'(i) = \frac{1}{T_A} * (K_A * u - X(i))$$

(4.1)

and

$$y = X(i)$$

(4.2)

(b) Signal Reset Circuit: As shown in figure 4.3, this block can reset the input signal at a certain time. When there is a sudden change on the system status, the disturbance signal will appear at the output of the diagram for a period of time and then die out. This circuit can block all the steady state signals. Its decomposed diagram is shown in figure 4.4 and the state variable is chosen as $X(i)$.

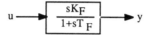

Figure 4.3 The Block Diagram of Signal Reset Circuit

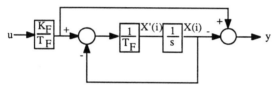

Figure 4.4 The Decomposed Diagram of Signal Reset Circuit

Where K_F is the gain and T_F is the time constant and the overall transfer function is:

$$X'(i) = \frac{1}{T_F} * (\frac{K_F}{T_F} * u - X(i))$$

(4.3)

and

$$y = \frac{K_F}{T_F} * u - X(i)$$

(4.4)

(c) Lead Lag Compensation Network: The purpose of this block diagram is to compensate the phase shift of the signal during transmission. It is important in the signal feedback and/or stabilization circuit. The block diagram of the circuit is shown in figure 4.5. We can divide this block diagram into two part which are equivalent to the summation of the first two circuits. However, this approach needs two differential equations to represent this diagram. Another approach is to rearrange the equation and let $T_1/T_2 = K$. The rearranged block diagram is shown in figure 4.6 and its transfer function of the block diagram is

$$X'(i) = \frac{1}{T_2} * (\frac{T_2}{T_1} * (1 - K) * u - X(i))$$

(4.4)

and

$$y = \frac{T_2}{T_1} * u - X(i)$$

(4.5)

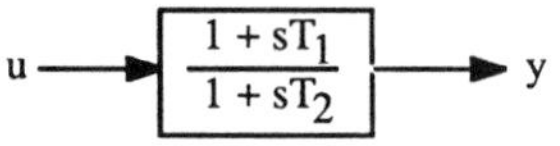

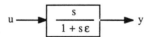

Figure 4.5 The Block Diagram of Lead-Lag Compensation Network

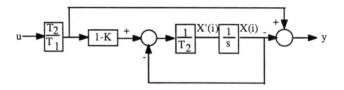

Figure 4.6 The Decomposed Diagram of Lead-Lag CompensationNetwork

(d) Differentiator: Not too many circuit consists differentiator because it will pick up and amplify the noise and create unstable control condition. However, in some control loop, i.e., PID controller, rate of change is an important information for the system control. Theoretically, the output of the differentiator is an impulse if a step function is applied at the input terminal. In the real world this output signal is limited by the power supply or other limiter circuits. An approximate circuit for the differentiator is shown in figure 4.7. It is similar to washout circuit and the value of ε will determine the peak of the impulse signal.

Figure 4.7 The Approximate Circuit of the Differentiator

4.2 Results Interpretation

The dynamic stability problem becomes a critical issue for the secure operation of large power systems. Presence of long transmission lines in the system network, used to transfer gigantic power from generation plants to the load center, is one of the main factors for the occurrence of dynamic oscillations. The oscillating phenomena of system generators are complicated and have been recognized as dynamic stability problems. A comprehensive study of these dynamic problems allow engineers to detect and solve them before they may happen. Utilization of the study results depend upon the the understanding of the engineers.

(a) **Eigenvalue Method**: The Eigenvalue method has been widely used for analyzing the dynamic behavior of power systems. A sample system with a simple machine model, which only the swing equation to build the system matrix is used in this example. The sample system is shown in figure 4.8 and the step of calculation are explained in figure 4.9. Table 4.1 shows the system eigenvalues and the associated eigenvectors. It means that each generator was represented by one state variable. Figure 4.10 is the graphical representation of Table 4.1. Figure 4.10(a) shows that every eigenvector moves toward the same direction with the same magnitude, i.e., no oscillation at this frequency. In figure 4.10(b) and (c), there are different directions and values on the eigenvectors, especially between generator #1 and generator #3 in (c). We have to pay more attention to this oscillation frequency (2.33 in this example). The most effective control points to this oscillation frequency are both generator #1 and generator #3. The eigenvalue methods make it possible to explain physically the very complex behavior of a multi-machine system and the relation between the system behavior and its control parameters. This information is useful in clarifying the main causes of instability of a power system and provide tools to study the most effective countermeasures.

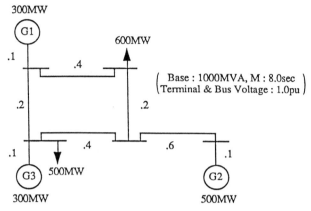

Figure 4.8 Sample Power System

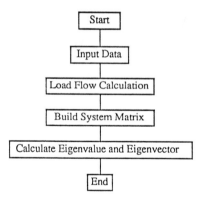

Figure 4.9 Flow Chart of Calculation

Table 4.1 Oscillation Frequency and Eigenvectors

Osc. Freq. Name of Gen.	0.0000	1.2087	2.3300
Gen. #1	1.0000	-0.5169	0.9778
Gen. #2	1.0000	-0.4831	-1.0000
Gen. #3	1.0000	1.0000	0.0222

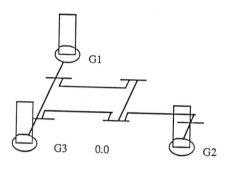

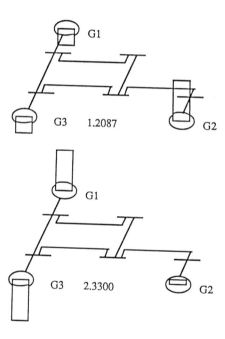

Figure 4.10 Oscillation Phenomena of the Sample Power System

(b) Time Domain Analysis: Time domain analysis can illustrate the system response with respect to time. Power engineers can evaluate the system condition through these out waveforms. For a normal power system, small perturbations should die out in a few seconds. However, in certain situations, even the system will reach steady state eventually, but the system response curves are somewhat unusual. Figure 4.11 shows that if two eigenvalues in the system are too close and their magnitudes are similar, the oscillation will decay during the first six (6) seconds and sustain till nine (9) seconds. After nine seconds, the system gradually reaches its steady state condition. If the magnitude of the oscillation signals are different, as shown in figure 4.12, it decays very fast in the first four (4) seconds and then with an unusual transition the oscillation comes back and sustains for a period of time before it disappears.

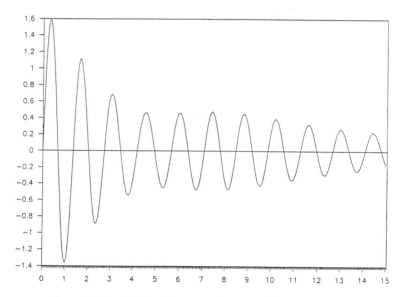

Figure 4.11 The Combined Frequency Response of Two Eigenvalues:
$0.7e^{-0.3t}\sin(2\pi*0.81t)$ and $1.0e^{-0.1t}\sin(2\pi*0.71t)$

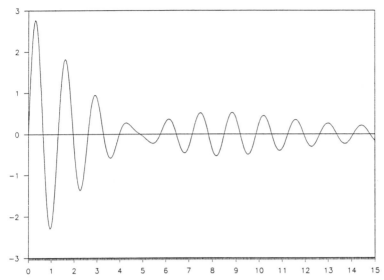

Figure 4.12 The Combined Frequency Response of Two Eigenvalues:
$2.0e^{-0.3t}\sin(2\pi*0.81t)$ and $1.0e^{-0.1t}\sin(2\pi*0.71t)$

CHAPTER 5

CONCLUSIONS

The dynamic stability study has become a critical issue for the secure operation of large electric power systems. Presence of long transmission line in the systems network, used to transfer gigantic power from one location to another, is one of the main factors for the occurrence of dynamic power flow or frequency oscillations. A comprehensive study of these dynamic problems allows the parameters to be detected and countermeasures to be designed to solve the problems before they may happen.

Frequency domain (eigenvalue) analysis and time domain analysis are two of the most popular methods for the stability analysis. However, each of them has its advantages and drawbacks. It would be better to use both methods to compensate the other's drawback and give system a thorough analysis. This book the general concepts about the dynamic stability phenomena of the power system , the derivation   of the linearized models, and the interpretation of the study results.

Analytical Results on Direct Methods for Power System Transient Stability Analysis

Hsiao-Dong Chiang
School of Electrical Engineering
Cornell University Ithaca, NY 14850

I. Introduction

A major activity in utility system planning and operations is to test system stability relative to disturbances. The stability simulation results are used to choose needed equipment and set operating limits. The disturbances that continually occur in power systems may be classified into event disturbances and load disturbances. The event disturbances contain the generator outages, short-circuit caused by lightning, sudden large load changes, or a combination of such events. An event disturbance usually leads to the change in the configuration of the power system. Load disturbances are the small random fluctuations of load demands. The system configuration usually remains unchanged after load disturbances.

The present-day power system operating environment has contributed to the increasing importance of the problems associated with the dynamic security assessment of power systems. To a large extent, this is due to the fact that most of the major power system breakdowns are caused by problems relating to the system dynamic responses [1] which mainly come from disturbances on the power system. Power system dynamic security assessment is concerned with the system's ability to reach an acceptable steady-state (operating condition) following a disturbance. Because of the inherent difference between load disturbance and event disturbance, the task of dynamic security assessment is divided into transient stability analysis and small-disturbance stability analysis [2].

Transient stability analysis is tailored to the system's ability to reach an acceptable steady-state (operating condition) following an event disturbance. The power system under this circumstance can be considered going through changes in configuration in three stages: from pre-fault, fault-on to post-fault systems. The pre-fault system is in a stable steady state. The fault occurs (e.g., a short circuit), and the system is then in the fault-on condition before it is cleared by the protective system operation. The transient stability analysis is the study of whether the post-fault trajectory will converge (tend) to an acceptable steady-state as time increases.

CONTROL AND DYNAMIC SYSTEMS, VOL. 43

Mathematically the problem can be phrased as follows. In the pre-fault regime the system is at a known stable equilibrium point, say x_i. At some time t_f, the system undergoes a fault, which results in a structural change in the system. Suppose the fault duration is confined to the time interval $[t_f, t_p]$. During this interval, the system is governed by a fault-on dynamics described by

$$\dot{x}(t) = f_F(x(t)), \qquad t_f \leq t < t_p \qquad (1)$$

where x(t) is the vector of state variables of the system at time t. The fault is cleared at time t_p and the system is henceforth governed by a post-fault dynamics described by

$$\dot{x}(t) = f(x(t)), \qquad t_p \leq t < \infty \qquad (2)$$

Next, assume that the post-fault system (2) has a (asymptotically) stable equilibrium point x_s which is acceptable relative to operational constraints. The fundamental problem of transient stability is the following. Starting from the post-fault initial state x(t_p), will the post-fault settle down to the steady state condition x_s ? In other words, the transient stability analysis is to determine whether the initial point of post-fault trajectory (i.e. the final point of fault-on trajectory) is located inside the stability region (domain of attraction) of an acceptable stable equilibrium point (acceptable steady state).

The study of transient stability has been playing an important role in the planning and operation of electric power systems. Until recently, transient stability analysis has been performed in utilities exclusively by means of numerical integration of the nonlinear differential equations describing the fault-on system and post-fault system. The computation associated with this conventional approach is known to be very involved. Furthermore, the cost of transient stability analysis using this conventional approach has been increasing significantly with the larger dimension of power systems.

An alternative approach to transient stability analysis employing Lyapunov function theory, called the *direct methods*, was proposed in 1966 by Gless [3], and El-abiad and Nagappan [4]. The growing need for tools suitable for dynamic security assessment, as well as a fast screening tool for planning stability studies, has generated renewed interests in such direct methods [5]. Recently, significant progress has been made in the direct methods for transient stability analysis. The details may be found, for example, in [6] - [10]. Among the direct methods, the classical method using the concept of closest u.e.p. (unstable equilibrium point) gives very conservative estimates of stability because this method is independent of the fault location. The potential energy boundary surface (PEBS) method proposed by Kakimoto et al. [11] gives fairly fast and accurate stability assessments but may give inaccurate (both over-estimate and under-estimate) results [12]. A fault-dependent method using the concept of controlling u.e.p. makes the direct methods more applicable in practical systems [13]. It is believed that the controlling u.e.p. method will continue to be a viable method, in terms of its accuracy and reliability, among the direct methods for transient stability analysis. However, a great majority of work on the controlling u.e.p. method is based on physical reasonings, heuristics and simulations without much theoretical support. For instance, a precise definition of the controlling u.e.p. has been lacking for quite a period of time.

This paper reviews recent advances in the development of analytical results for direct methods and provides insight into the underlying concepts and properties of various direct methods. Aside from a coherent view of the field, new material is presented. The exposition emphasizes fundamentals of direct methods rather than the heuristics which most of direct methods are based on. In order to demonstrate one of the advantages of employing the analytical approach instead of the heuristic approach to develop direct methods for transient stability analysis, these fundamentals are then applied to a recently developed direct method, called the boundary of stability region controlling unstable equilibrium point method (BCU method) for direct analysis of power system transient stability [14], [15]. The BCU method has been compared favorably with other methods on large-scale power systems, according to a recent EPRI report [16]. One implication from the development of the BCU method is that analytical results can sometimes lead to the development of reliable yet fast solution algorithms for solving problems.

II. Power System Model

We consider the following model which represents a large class of power system models for stability analysis

$$
\begin{aligned}
\dot{x}_1 &= y_1 \qquad\qquad\qquad\qquad (3)\\
M\dot{x}_1 &= -D_1 y_1 - f_1(x_1,\ x_2,\ y_2)\\
\dot{x}_1 &= -D_2 x_2 - f_2(x_1,\ x_2,\ y_2)\\
0 &= -g(x_1,\ x_2,\ y_2)
\end{aligned}
$$

where x_1, y_1, x_2, y_2 are vectors with appropriate dimension, M, D_1, D_2 are positive-definite matrix. The vector $f_1(\cdot,\cdot,\cdot)$, $f_2(\cdot,\cdot,\cdot)$ are bounded, the vector $f_1(x_1,\cdot,\cdot)$ is periodic in x_1 and the matrix $\frac{\partial f_1}{\partial x_1}$ is bounded. Most of currently available models for transient stability analysis can be put into this general form; including the classical model (in this model $x_2 = 0$, $y_2 = 0$, see section II.A) and the structure-preserving model with the effect of flux linkage (see section II.B), the classical model with control devices [17] and so on.

It will be shown in this section that there exists an energy function V(x) for this general power system model such that the following three conditions are satisfied:

(i) the derivative of the energy function V(x) along any system trajectory x(t) is non-positive, i.e.

$$
\dot{V}(x(t)) \leq 0
$$

(ii) If x(t) is a non-trivial trajectory (i.e. x(t) is not an equilibrium point (e.p.)), then there does not exist a time interval, say $[t_1, t_2]$, $t_2 > t_1$, such that $\dot{V}(x(t)) = 0$ for $t \in [t_1, t_2]$. Mathematically, this can be expressed as follows: along any non-trivial trajectory x(t) the set

$$
\{t \in R \ : \ \dot{V}(x(t)) = 0\}
$$

has measure zero in R

(iii) If a trajectory $x(t)$ has a bounded value of $V(x(t))$ for $t \in R^+$, then this trajectory $x(t)$ is also bounded. Stating this in brief :

$$\text{if } V(x(t)) \text{ is bounded, then } x(t) \text{ is also bounded.}$$

Property (i) indicates that the energy is non-increasing along its trajectory, but does not imply that the energy is strictly decreasing along its trajectory. There may exist a time interval $[t_1, t_2]$ such that $\dot{V}(x(t)) = 0$ for $t \in [t_1, t_2]$. Properties (i) and (ii) imply that the energy is strictly decreasing along any system trajectory. Property (iii) states that, along any system trajectory, the energy function is a proper map but its energy need not be a proper map for the entire state space.

Several power system models intended for transient stability analysis have been proposed in literature. There exists an energy function for each of these models. For the purpose of illustration, we will examine in the rest of this section two power system models in detail, namely (1) the classical model and (2) the structure-preserving model with flux decay, and show that an energy function indeed exist for each of them.

A. The Classical Model

Consider a power system consisting of n generators. Let the loads be modeled as constant impedances. Under the assumption that the transfer conductance of the reduced network after eliminating all load buses is zero, the dynamics of the i-th generator can be represented by the equations [18] :

$$\dot{\delta}_i = \omega_i \tag{4}$$

$$M_i \dot{\omega}_i = P_i - D_i \omega_i - \sum_{j=1, j \neq i}^{n+1} V_i V_j B_{ij} sin(\delta_i - \delta_j)$$

where the voltage at node n+1 is served as the reference, i.e., $\delta_{n+1} = 0$.

Consider the following function

$$V_1(\delta, \omega) = \frac{1}{2} \sum_{i=1}^{n} M_i \omega_i^2 - \sum_{i=1}^{n} P_i(\delta_i - \delta_i^s) \tag{5}$$

$$- \sum_{i=1}^{n} \sum_{j=i+1}^{n+1} V_i V_j B_{ij} \{cos(\delta_i - \delta_j) - cos(\delta_i^s - \delta_j^s)\} \tag{6}$$

where $x^s = (\delta^s, 0)$ is the stable equilibrium point under consideration.

We now show that $V_1(\cdot)$ is an energy function for the system (3). Differentiating $V_1(\cdot)$ along system trajectories gives

$$\dot{V}_1(\delta(t), \omega(t)) = \sum_{i=1}^{n} \left[\frac{\partial V_1}{\partial \delta_i} \dot{\delta}_i + \frac{\partial V_1}{\partial \omega_i} \dot{\omega}_i \right]$$

$$= - \sum_{i=1}^{n} D_i \omega_i^2 \neq 0$$

Suppose that there is an interval $t \in [T_1, T_2]$ such that

$$\dot{V}_1(\delta(t), \omega(t)) = 0$$

Hence, $\omega(t) = 0$ follows from (13). But this implies that $\omega(t) = 0$ and $\delta(t) = $ constant for $t \in [T_1, T_2]$. It then follows that

$$P_i - \sum_{j=1, j \neq i}^{n+1} V_i V_j B_{ij} \sin(\delta_i - \delta_j) = 0$$

which is precisely the equation for the equilibrium point of the system. Therefore, $(\delta(t), \omega(t)), t \in [T_1, T_2]$ must be on an equilibrium point, and we have shown that the function $V_1(\cdot)$ satisfies the properties (i) and (ii) of the energy function. The function $V_1(\cdot)$ is shown in [19] to possess the property (iii) of the energy function.

The classical model for transient stability analysis makes unwarranted simplifications in the formulation of generators and loads. This model is known to have the following shortcomings [7]: (1) It precludes consideration of reactive power demand and voltage variation at load buses; (2) Reduction of the network leads to loss of network topology and hence precludes study of transient energy shifts among different components of the network; (3) Neglect of transfer conductances leads to a bias in stability estimates of unknown magnitude and direction; and (4) Neglect of flux decay and exciter control may be unacceptable simplifications.

Structure-preserving models have been recently proposed with an aim to overcome some of the shortcomings of the classical model. These models also enable one to study the "voltage dip" phenomenon of power systems during a disturbance [65].

B. The Structure-Preserving Model

The first structure-preserving model was developed by Bergen and Hill [20] , who assumed frequency dependent real power demands and constant reactive power demands. Narasimhamurthy and Musavi [22] moved a step further by considering constant real power and voltage dependent reactive power loads. Padiyar and Sastry [23] have included nonlinear voltage dependent loads for both real and reactive powers. Tsolas, Araposthasis and Varaiya [21] developed a structure-preserving model with the effects of flux decay and constant real and reactive power loads. Without loss of generality, the Tsolas-Araposthasis-Varaiya model is used in the following discussion.

Consider a power system consisting of n generators and m load buses. For notation convenience, let $J_I := \{1, 2, \ldots, n\}$ generator internal bus, $n+1$ as the slack bus, $J_T := \{1, 2, \ldots, n\}$ generator terminal buses and $J_L := \{n + 2, n + 3, \ldots, n + m + 1\}$ load buses. Let the ith generator's terminal voltage be denoted as $V_i e^{i\theta_i}$ and its internal voltage be denoted as $E'_{iq} e^{i\delta_i}$. The model for internal bus i is :

$$\dot{\delta}_i = \omega_i \tag{7}$$

$$M_i \dot{\omega}_i = P_i - D_i \omega_i - P_i^e \tag{8}$$

$$T'_{doi}\dot{E}'_{qi} = -\frac{x_{di}}{d'_{di}}E'_{qi} + \frac{(x_{di}-x'_{di})}{x'_{di}}V_i cos(\delta_i - \theta_i) + E_{Fi} \tag{9}$$

$$P_i^e = \frac{E'_{qi}V_i \sin(\delta_i - \theta_i)}{x'_{di}} + \frac{V_i^2 \sin[2(\delta_i - \theta_i)](x'_{di} - x_{qi})}{2x_{qi}x'_{di}} \tag{10}$$

Equation (7) and (8) describe the dynamics of rotor in generator i. Equation (9) is the flux decay equation for a one-axis model. Equation (10) gives the electric power P_i^e injected into the network.

The real power demand at bus $k \in j_l$ is assumed to be $P_k^d + D_k\dot{\phi}_k$, and the reactive power demand is assumed to be a function of voltage magnitude. The models for the load bus and the generator terminal bus are expressed by the power flow equations. The real power flow equations are

$$P_i^e = \sum_{j \in J_T} V_i V_j B_{ij} \sin(\theta_i - \theta_j) + \sum_{l \in J_L} V_i V_l B_{il} \sin(\theta_i - \phi_l) \quad i \in J_T \tag{11}$$

$$0 = P_k^d - D_k\dot{\phi}_k - \sum_{j \in J_T} V_k V_j B_{kj} \sin(\phi_k - \theta_j) - \sum_{l \in J_L} V_k V_l B_{kl} \sin(\phi_k - \phi_l) \quad k \in J_L$$

The reactive power flow equations are

$$0 = 2a\frac{x'_{di} + x_{qi}}{2x_{qi}x'_{di}}V_i^2 - \frac{E'_{qi}V_i \cos(\theta_i - \delta_j)}{x'_{di}} \quad for \ i \in J_T \tag{12}$$

$$-\frac{V_i^2 \cos[2(\theta_i - \delta_j)](x'_{di} - x_{qi})}{2x_{qi}x'_{di}}$$

$$-\sum_{j \in j_T} V_i V_j B_{ij} \cos(\theta_i - \theta_j) - \sum_{l \in J_l} V_i V_j B_{il} \cos(\theta_i - \phi_l)$$

$$Q_k^d(V_k) = -\sum_{j \in J_T} V_k V_j \cos(\phi_k - \theta_j) \quad -\sum_{l \in J_L} V_k V_l \cos(\phi_k - \phi_l) \quad for \ k \in J_L$$

It can be shown that the following function $V_2(\cdot)$ is an energy function for the above structure-preserving model, under the assumption of the existence and uniqueness of trajectory in the system.

$$V_2(\delta, \omega, E'_q, \theta, V, \phi) := 1/2\sum_{i=1}^{n} M_i\omega_i^2 - \sum_{i=1}^{n} P_i\delta_i - \sum_{k \in J_l} P_k^d\phi_k$$

$$-\sum_{i=1}^{n+1}\sum_{j=i+1}^{n+1} V_i V_j B_{ij} \cos(\theta_i - \theta_j) - \sum_{i=1}^{n+1}\sum_{k \in J_l} V_i V_k B_{ik} \cos(\theta_i - \phi_k)$$

$$-\sum_{k \in J_l}\sum_{l \in J_l, l \neq k} V_k V_l B_{kl} \cos(\phi_k - \phi_l) - 1/2\sum_{k \in J_l} V_k^2 B_{kk}$$

$$-\sum_{k \in J_l}\int_{1}^{v_k} \frac{Q_k^d(v)}{v}dV - 1/2\sum_{i=1}^{n+1} V_i^2\left\{B_{ii} + \frac{(x'_{di} - x_{qi})[\cos(2\delta_i - \theta_i) - 1]}{2x_{qi}x'_{di}}\right\}$$

$$-\sum_{i=1}^{n+1} \frac{E'_{qi}V_i \cos(\delta_i - \theta_i)}{x'_{di}} + 1/2\sum_{i=1}^{n+1} E'^2_{qi}\frac{x_{di}}{x'_{di}(x_{di} - x'_{di})} - \sum_{i=1}^{n+1} \frac{E'_{qi}E_{Fi}}{x_{di} - x'_{di}}$$

Obeserve that the energy function $V_2(\cdot)$ is a summation of kinetic energy and potential energy. To show $V_2(\cdot)$ is an energy function, one differentiates $V_2(\cdot)$ along the trajectories of system (7)

$$
\begin{aligned}
\dot{V_2} &= \frac{\partial V_2}{\partial \delta}\dot{\delta} + \frac{\partial V_2}{\partial \omega}\dot{\omega} + \frac{\partial V_2}{\partial E_q'}\dot{E_q'} + \frac{\partial V_2}{\partial \phi}\dot{\phi} \\
&= -\sum_{i=1}^{n+1} D_i \omega_i^2 - \sum_{i=1}^{n+1} \frac{T_{doi}'}{x_{di} - x_{di}'}\dot{E_{qi}'}\dot{E_{qi}'} - \sum_{k=n+2}^{n+m+1} D_k \dot{\phi}_k^2 \\
&\leq 0
\end{aligned}
$$

From the above equation, we have that $\dot{V_2}(\cdot) = 0$ if and only if $\omega_i = 0$, $\dot{E_{qi}'} = 0$ and $\dot{\phi}_k = 0$. Substituting this into (7) gives that if $\dot{V_2}(\cdot) = 0$ for the interval $t \in [T_1, T_2]$, then $(\delta(t), \omega(t), E_q'(t), \phi(t))$ must be on an equilibrium point. Thus, we have shown that the function $V_2(\cdot)$ has the properties (i) and (ii) of the energy function. Also, the function $V_2(\cdot)$ is shown in [7] to possess the property (iii) of the energy function.

III. Theory of Stability Boundary

Power system transient stability analysis is essentially the problem of determining whether the fault-on trajectory at clearing time is lying inside the stability region of a desired s.e.p. of its post-fault system or not. Hence, the problem is closely related to the determination of a desired stability region. This section is devoted to the theory of stability regions of nonlinear dynamical systems. In Section 2 we have seen that the energy functions for different power system models have three properties. In this section we shall explore these three properties to develop a theory of stability boundary for general power system models. This theory enables us to establish a framework to examine the closest u.e.p. method and the controlling u.e.p. method for transient stability analysis.

A. Preliminaries

To unify our notation, let

$$
\dot{x}(t) = f(x(t)) \tag{13}
$$

be the power system model under study, where the state vector x(t) belongs to the Euclidean space R^n, and the function $f : R^n \rightarrow R^n$ satisfies the sufficient condition for the existence and uniqueness of solutions. The solution curve of (13) starting from x at t=0 is called a (system) trajectory, denoted by $\Phi(x, \cdot) : R \rightarrow R^n$. Note that $\Phi(x, 0) = x$.

The concepts of *equilibrium point* (e.p.), *invariant set, limit set* including α-*limit set* and ω-*limit set*, *stable* and *unstable manifolds* are important in dynamical system theory. Each of these concepts is defined next. A detailed discussion of these concepts and implication may be found in [24] - [29].

A state vector $\hat{x}$ is called an equilibrium point of system (13) if $f(\hat{x}) = 0$. We denote E to be the set of equilibrium points of the system. A state vector x is called a *regular point* if it is not an equilibrium point. We say that an equilibrium point of (13) is *hyperbolic* if the Jacobian of $f(\cdot)$ at $\hat{x}$, denoted $J_f(x)$, has no eigenvalues

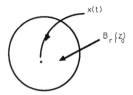

Figure 1: Case A: the trajectory x(t) stays in the ball $B_{r_1}(z_0)$ after some finite time of not being in the ball $B_{r_1}(z_0)$

with a zero real part. For a hyperbolic equilibrium point, it is a *(asymptotically) stable equilibrium point* if all the eigenvalues of its corresponding Jacobian have negative real parts; otherwise it is an *unstable equilibrium point*. If the Jacobian of the equilibrium point $\hat{x}$ has exactly one eigenvalue with positive real part, we call it a *type-one equilibrium point*. Likewise, $\hat{x}$ is called a *type-k equilibrium point* if its corresponding Jacobian has exactly k eigenvalues with positive real part.

A set $M \in R^n$ is called an *invariant set* of (13) if every trajectory of (13) starting in M remains in M for all t. A point p is said to be in the ω-limit set (respectively, α-limit set) of x if corresponding to each $\epsilon > 0$ and each $T > 0$ (respectively, $T < 0$) there is a $t > T$ (respectively, $t < T$) with the property that $\|\Phi(x,t) - p\| < \epsilon$. This is equivalent to saying that there is a sequence t_i in R, $t_i \to \infty$ (or $t_i \to -\infty$), with the property that $p = lim_{i\to\infty}\Phi(x,t_i)$. Thus, the ω-limit set of a trajectory captures the asymptotic behavior of the trajectory in positive time while the α-limit set of a trajectory captures the asymptotic behavior of the trajectory in negative time. The limit set of a trajectory includes its α-limit set and ω-limit set. One of the fundamental properties of the limit set is as follows:

Theorem 3-1 [27]: (Topological and dynamical properties of limit sets)

If $\Phi(x,t)$ is bounded for $t \geq 0$ (or $t \leq 0$), then its ω-limit set (or α-limit set) is a non-empty, compact, connected, and invariant set.

Usually, determining the limit set of a nonlinear dynamical system is not an easy task. Next, we show that if there exists an energy function for the system, then the structure of the limit set is very simple: it consists solely of equilibrium points.

Theorem 3-2 : (Structure of limit sets)

If there exists an energy function for system (13), then the ω-limit set of any bounded trajectory consists only of equilibrium points.

Proof: Assume z_0 with $\dot{V}(z_0) < 0$ is a ω-limit point of some trajectory. It follows from the differentiability of $V(\cdot)$ and the continuity of $f(\cdot)$ that there exists a $r > 0$ such that for all the points inside the ball $B_r(z_0)$, we have

$$\dot{V}(z) < -\delta \text{ for all } z \in B_r(z_0)$$

There are two cases to be considered: case A and case B.

Case A: the trajectory x(t) stays in the ball $B_{r_1}(z_0)$ after some finite time of not being in the ball $B_{r_1}(z_0)$. In this case, the closure of x(t), $t \geq 0$ is bounded.

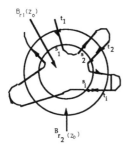

Figure 2: Case B: the trajectory x(t) keeps on coming back to the ball $B_{r_1}(z_0)$ at arbitrary large times.

Hence, V(x(t)) is bounded. Now, since along the system trajectory x(t), it follows that $lim_{t\to\infty}$ V(x(t)) exists and equals a constant, say $lim_{t\to\infty}$ V(x(t))=B. Then, the contradiction is achieved in the following

$$\int_0^\infty \dot{V}(x(t))dt \; < \; \int_T^\infty \dot{V}(x(t))dt \; < \; -\int_T^\infty \delta dt \qquad (14)$$

or

$$\int_0^\infty \dot{V}(x(t))dt \;\; = \;\; lim_{t\to\infty}V(x(t)) - V(x_0), \; x(0) = x_0 \qquad (15)$$

$$= \;\; B - V(x_0) \; < \; -\infty \qquad (16)$$

which is a contradiction.

<u>Case B</u>: the trajectory x(t) keeps on coming back to the ball $B_{r_1}(z_0)$ at arbitrary large times. Note that there exists a positive constant C such that $| V(z_0) - V(x) | < C$ for all x $\in B_{r_1}(z_0)$. We define another ball $B_{r_2}(z_0)$, with $r_2 > r_1$, and two increasing sequences $\{t_i\}$ and $\{s_i\}$, where t_i is the time that the trajectory x(t) enters the ball $B_{r_2}(z_0)$ and s_i is the time that the trajectory x(t) enters the ball $B_{r_1}(z_0)$ with x(t)$\in B_{r_2}(z_0)$ for $t_i < s_i$. It is clear that the sequence $\{s_i - t_i\}$ is positive and bounded below and the sequence $\{t_i\}$ exists; otherwise, the case is the same as case A.

Since

$$| x(s_i) - x(t_i) | \;\; = \;\; | \int_{t_i}^{s_i} \dot{x}(t)dt | \leq \int_{t_i}^{s_i} | \dot{x}(t) | \, dt \qquad (17)$$

$$= \;\; \int_{t_i}^{s_i} | f(x) | \, dt \; \leq \; m(s_i - t_i) \qquad (18)$$

$$\qquad (19)$$

where $m = max_{x\in(B_{r_1}(z_0)-B_{r_2}(z_0))}f(x).$

But $| x(s_i) - x(t_i) | = r_2 - r_1$, hence

$$(s_i - t_i) \; \geq \; \frac{r_2 - r_1}{m} \qquad (20)$$

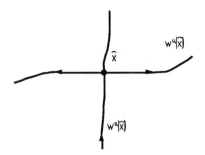

Figure 3: The stable and unstable manifolds of $\hat{x}$.

Observe that

$$V(x(t)) - V(x_0) \quad = \quad \int_0^t \dot{V}(x(\tau))d\tau \tag{21}$$

$$< \quad \sum_{i=1}^{n} \int_{t_i}^{s_i} \dot{V}(x(\tau))d\tau \; < \; -n\delta(s_i - t_i) \tag{22}$$

$$< \quad -n\delta \frac{r_2 - r_1}{m} \tag{23}$$

Since for any time $T \in R$, there exists a $t > T$ such that $x(t) \in B_{r_1}(z_0)$. i.e. n can be so large that (10) is violated, which is a contradiction.

Let $\hat{x}$ be a hyperbolic equilibrium point. Its stable and unstable manifolds, $W^s(\hat{x})$ and $W^u(\hat{x})$, are defined as follows:

$$W^s(\hat{x}) \; := \; \{ \, x \in R^n \; : \; \Phi(x,t) \to \; \hat{x} \quad as \; t \; \to \; \infty \, \}$$

$$W^u(\hat{x}) \; := \; \{ \, x \in R^n \; : \; \Phi(x,t) \to \; \hat{x} \quad as \; t \; \to \; -\infty \, \}$$

These two sets are invariant sets (see Fig. 3). Clearly, $\hat{x}$ is the ω-limit of every point in $W^s(\hat{x})$ and $\hat{x}$ is the α-limit of every point in $W^u(\hat{x})$.

The idea of *transversality* is basic in the study of dynamical systems. If A and B are two manifolds, we say that they satisfy the *transversality condition* if either (i) at every point of intersection $x \in A \cap B$, the tangent spaces of A and B span the tangent spaces of M at x, i.e.

$$T_x(A) \bigoplus T_x(B) = T_x(M) \quad for \; x \in A \cap B$$

or (ii) they do not intersect at all. The transversal intersection is important because it persists under perturbation of the vector field.

For a stable equilibrium point, it can be shown that there exists a number $\delta > 0$ such that $\|x_0 - \hat{x}\| < \delta$ implies that $\Phi(x_0,t) \to \hat{x}$ as $t \to \infty$. If δ is arbitrarily large, then $\hat{x}$ is called a *global stable equilibrium point*. There are many physical systems containing stable equilibrium points but not global stable equilibrium points. A useful concept for this kind of systems is that of *stability region* (or *region of attraction*). The stability region of a stable equilibrium point x_s is defined as

$$A(x_s) \; := \; \{x \in R^n \; : \; lim_{t \to \infty} \Phi(x,t) \; = \; x_s\}$$

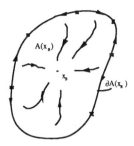

Figure 4: As time increases, every trajectory in the stability region $A(x_s)$ converges to the s.e.p. x_s.

From a topological point of view, the stability region $A(x_s)$ is an open, invariant and connected set. The boundary of stability region $A(x_s)$ is called the *stability boundary* (or *separatrix*) of x_s and will be denoted by $\partial A(x_s)$ (see Fig.4). The stability boundary is topologically an (n-1)-dimensional closed, invariant set.

B. Energy Function and Stability Boundary

We will analyze the stability boundary of power system (13) in this section. In particular, a characterization of the stability boundary of the system will be given. Throughout this section, it will be assumed that all the equilibrium points of system (13) are hyperbolic.

Theorem 3-3 : (A characterization of the stability boundary)

If there exists an energy function for the system (13), then the stability boundary $\partial A(x_s)$ is contained in the set which is the union of the stable manifolds of the u.e.p.'s on the stability boundary $\partial A(x_s)$, i.e.

$$\partial A(x_s) \subseteq \cup_{x_i \in \{E \cap \partial A(x_s)\}} W^s(x_i)$$

Proof: We prove this result by showing that every trajectory on the stability boundary $\partial A(x_s)$ converges to one of the equilibrium points. Consequently, the theorem follows. The following lemma is useful in the proof:

Lemma: Let p be a point on the stability boundary $\partial A(x_s)$ of system (13) and $V(\cdot)$ is an energy function of system (13). Then, $V(p) > V(x_s)$.

The proof of Lemma 3-4 is similar to the proof of Lemma 4-1 in [30]. Now, since $p \in \partial A(x_s)$, it follows from Lemma 3-4 that $V(\phi(p,t) > V(x_s)$ for $t \in R$. From this and property (iii) of energy function, we see that trajectory $\phi(p,t)$ is bounded in forward time. Next, we apply Theorem 3-1 and 3-2 to derive that the ω-limit set of $\phi(p,t)$ exists and is composed of equilibrium points. Since the equilibrium points are hyperbolic (hence they are isolated), we conclude that every trajectory on the stability boundary $\partial A(x_s)$ converges to one of the equilibrium points.

Remark:

Tsolas, Arapostathis and Varaiya [21] were the first to present the above characterization for the structure preserving model discussed in Sec. II. In the past two decades, several attempts to characterize (or approximate) the stability boundary of power system have been made, including the conventional Zubov method [32], [48]

and the Hyperplane method [35], but without much success. However, the result
of Theorem 3-1 confirms the spirit of the Hyperplane method that the hyperplane
tangential to the type-one equilibrium point x_i of the stability boundary $\partial A(x_s)$
can be used to locally approximate the stability boundary $\partial A(x_s)$ around x_i. A
complete characterization of the stability boundary for nonlinear systems having an
energy function will be detailed in the next section.

A common approach to estimate the stability boundary of system (13) is to use
an energy function approach. This is particularly true in the power system area.
In this approach, the set of the inverse of the energy function at a particular value,
which is of dimension n-1, is used to approximate the stability boundary which is
also of dimension n-1. In the rest of this section, we will explore the relationship
between the constant energy function surface and the stability boundary.

Next, we consider the following set

$$S_v(k) = \{x \in R^n : V(x) < k\} \tag{24}$$

where $V(\cdot) : R^n \rightarrow R$ is an energy function. Sometimes, we drop the subscript v of $S_v(k)$, simply writing S(k), if it is clear from the context. We shall
call the boundary of set (24), $\partial S(k) := \{x \in R^n : V(x) = k\}$ the *level
set* (or *constant energy surface*) and k the *level value*. If k is a regular value (i.e.
$\nabla V(x) \neq 0$, *for all* $x \in V^{-1}(k)$), then by the Inverse Function Theorem $\partial S(k)$
is a C^r (n-1)-dimensional submanifold of R^n. Moreover, if $r > n-1$, then by the
Morse-Sard theorem the set of regular values of V is residual; in other words 'almost
all' level values are regular. In particular, for almost all values of k, the level set
$\partial S(k)$ is a C^r (n-1)-dimensional submanifold.

Generally speaking, this set S_k can be very complicated with several different
components even for the 2-dimensional case. Let

$$S(k) = S^1(k) \cup S^2(k) \cup ... \cup S^m(k) \tag{25}$$

where $S^i(k) \cap S^j(k) = \phi$ when $i \neq j$. That is, each of these components is
connected and disjoint from each other. Since $V(\cdot)$ is continuous, $S(k)$ is an open
set. Because $S(k)$ is an open set, the level set $\partial S(k)$ is of (n-1) dimensions [20, p.46].
Furthermore, each component of $S(k)$ is an invariant set.

In spite of the possibility that a constant energy surface may contain several
disjoint connected components, there is an interesting relationship between the constant energy surface and the stability boundary. This relationship is that at most
one connected component of the constant energy surface $\partial S(r)$ has a non-empty
intersection with the stability region $A(x_s)$ (see Fig. 5). This relationship is established in Theorem 3-4.

Theorem 3-4: (Constant energy surface and stability region)

Let x_s be a s.e.p. of the system (13) and $A(x_s)$ be its stability region. Then, the
set S(r) contains only one connected component which has a non-empty intersection
with the stability region $A(x_s)$ if and only if $r > V(x_s)$.

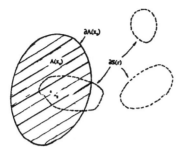

Figure 5: The constant energy surface $\partial S(r)$ may be composed of several disjoint connected components. Among them, only one connected component of $\partial S(r)$ has a non-empty intersection with the stability region $A(x_s)$.

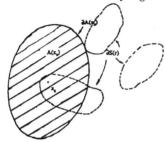

Figure 6: The constant energy surface $\partial S(r)$ contains only one connected component which has a non-empty intersection with the stability boundary $\partial A(x_s)$ if and only if $r > V(x_s)$. The situation shown in this figure cannot happen.

Proof : That the intersection of $S(r)$ and $A(x_s)$ is non-empty for $r > V(x_s)$ is obvious. The rest of the proof is similar to the proof of Theorem 4-1 in [36].

Theorem 3-4 can be extended to the relation between the constant energy surface $\partial S(r)$ and the stability boundary $\partial A(x_s)$. Let x_s be a s.e.p. of the system (13) and $\partial A(x_s)$ be its stability boundary. Then, the constant energy surface $\partial S(r)$ contains only one connected component which has a non-empty intersection with the stability boundary $\partial A(x_s)$. This result excludes the possibility of the situation shown in Fig. 6.

C. A Complete Characterization of Stability Boundary

For a fairly large class of nonlinear systems, the stability region of a stable equilibrium point can be completely characterized. This section reviews recent advances in the complete characterization of stability regions of such a class of nonlinear dynamical systems. Several necessary and sufficient conditions to determine whether a given equilibrium point (or limit cycle) is on the stability boundary will be presented, from which a complete characterization of the stability boundary will be derived.

Consider a nonlinear system described by equation (13) whose stability regions are to be analyzed. We first present conditions for an equilibrium point (or limit cycle) to be on the stability boundary $\partial A(x_s)$, which is a key step in the complete characterization of $\partial A(x_s)$.

Theorem 3-5 [37] : (Characterization of the critical elements on the stability boundary)

Let $A(x_s)$ be the stability region of x_s of the nonlinear dynamical system (13). Let $\hat{x} \neq x_s$ be a hyperbolic critical element of system (13). Then, we have the following

[1] $\hat{x}$ is on the stability boundary $\partial A(x_s)$ (i.e. $\hat{x} \in \partial A(x_s)$) if and only if $\{W^u(\hat{x}) - \hat{x}\} \cap \bar{A}(x_s) \neq \phi$

[2] Suppose $\{W^s(\hat{x}) - \hat{x}\} \neq \phi$., then, $\hat{x} \in \partial A(x_s)$ if and only if $\{W^s(\hat{x}) - \hat{x}\} \cap \partial A(x_s) \neq \phi$.

Theorem 3-5 only assumes that the critical elements on the stability boundary are hyperbolic. This is a generic property for nonlinear dynamical systems. Roughly speaking, let X be a complete metric space and P(x) a statement about points x in X. We say that P(x) is a generic property if the set of points where it holds contains a countable intersection of open dense sets. A formal definition is given in [38]. It has been shown in [39] that among the class of C^1 vector fields, the following properties are generic: (i) all critical elements are hyperbolic and (ii) the intersections of the stable and unstable manifolds of critical elements satisfy the transversality conditions.

A corollary to Theorem 3-5 is that if $W^u(\hat{x}) \cap A(x_s) \neq \phi$, then $\hat{x}$ is on the stability boundary $\partial A(x_s)$. In other words, a sufficient condition for $\hat{x}$ to be on the stability boundary $\partial A(x_s)$ is the existence of a trajectory in $W^u(\hat{x})$ which approaches $\hat{x}$. This condition can be checked numerically. From a practical viewpoint, it is fruitful to see when the condition is also necessary. The following result shows that the condition becomes necessary under the following assumptions on the vector field f(x).

(A1) All the critical elements on the stability boundary $\partial A(x_s)$ are hyperbolic

(A2) The stable and unstable manifolds of the critical elements on $\partial A(x_s)$ satisfy the transversality condition

(A3) Every trajectory on the stability boundary $\partial A(x_s)$ approaches one of the critical elements as $t \to \infty$

Assumption (A1) is a generic property of C^1 vector fields and can be checked numerically for a particular system by computing the corresponding eigenvalues. Assumption (A2) is also a generic property, but it is not easy to check numerically for a particular system. Assumption (A3) is a generic property only for planar systems. For higher dimensional systems, it may be verified by means of energy functions.

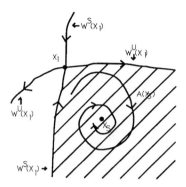

Figure 7: An example of a dynamical system whose trajectories on the stability boundary do not all converge to its critical elements.

Theorem 3-6 [37] : (Further characterization of the critical elements on the stability boundary)

Let $A(x_s)$ be the stability region of x_s of the nonlinear dynamical system (13). Let $\hat{x} \neq x_s$ be a hyperbolic critical element of system (13). If assumptions (A1) to (A3) are satisfied, then

[1] $\hat{x} \in \partial A(x_s)$ if and only if $W^u(\hat{x}) \cap A(x_s) \neq \phi$.

[2] $\hat{x} \in \partial A(x_s)$ if and only if $W^s(\hat{x}) \subseteq \partial A(x_s)$.

Remarks:

[1] Figure 7 shows a nonlinear system for which the assumption (A3) does not hold. In this system, the stability boundary $\partial A(x_s)$ is composed of a part of the unstable manifold and a part of the stable manifold of x_1. Note that the unstable manifold of x_1 does not intersect with the stability region $A(x_s)$ and a part of the stable manifold of x_1 is not on the stability boundary $\partial A(x_s)$ (See Theorem 3-6).

[2] To show that assumption (A2) is needed in Theorem 3-6, we consider an example taken from [21] as shown in Fig. 8. The intersection of the unstable manifold of x_1 and the stable manifold of x_2 does not satisfy the transversality condition because the tangent space at the intersection relative to both stable and unstable manifolds is of dimension 1. Note that the unstable manifold of x_1 intersects with the stability boundary (see Theorem 3-5), but not with the stability region (see Theorem 3-6). A part of the stable manifold of x_1 (upper part in the figure) is not on the stability boundary (see Theorem 3-6).

We are now in a position to characterize the stability boundary for a fairly large class of nonlinear dynamical system (13) whose stability boundary is non-empty.

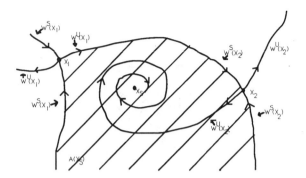

Figure 8: The intersection of the unstable manifold of x_1 and the stable manifold of x_2 does not satisfy the transversality condition. In this case, the unstable manifold of x_1 intersects with the stability boundary, but not with the stability region. A part of the stable manifold of x_1 (upper part in the figure) is not on the stability boundary.

Theorem 3-7 [37]: (Characterization of stability boundary)

Consider a nonlinear dynamical system described by equation (13) which satisfies assumptions (A1) to (A3). Let x_i, i = 1,2, ..., be the critical elements on the stability boundary $\partial A(x_s)$ of the stable equilibrium point x_s. Then,

$$\partial A(x_s) = \bigcup W^s(x_i) \tag{26}$$

This theorem completely characterizes the stability boundary for nonlinear dynamical systems satisfying assumptions (A1) to (A3) by asserting that the stability boundary is the union of the stable manifolds of all the critical elements on the stability boundary. This theorem gives an explicit description of the geometrical and dynamical structure of the stability boundary. To the best of our knowledge, the only other complete characterization of the stability boundary is the one via the Zubov method. The Zubov method characterizes the stability boundary by use of a Zubov function, which entails solving a set of nonlinear partial differential equations. In addition to the difficulty of solving a set of nonlinear partial differential equations, the Zubov method offers little information about the geometrical or dynamical structure of the stability boundary.

D. Stability Boundary of Power Systems

We have so far presented results on the characterization of the critical elements on the stability boundary as well as of the stability boundary itself for a fairly large class of nonlinear dynamical systems. Next, we present results pertaining to nonlinear dynamical systems having an energy function, such as several power system models for transient stability analysis. For those systems, as pointed out in Theorem 3-2, every trajectory on the stability boundary converges to one of the equilibrium points, thus satisfying the assumption (A3). And the assumptions (A1) and (A2) are modified as

(A1') All the equilibrium points on the stability boundary $\partial A(x_s)$ are hyperbolic.

(A2') The stable and unstable manifolds of the equilibrium points on $\partial A(x_s)$ satisfy the transversality condition.

Note that the above two assumptions are generic properties. So, for several power system models, we have the following general result, which is a corollary of Theorem 3-6 and 3-7.

Theorem 3-8 [19]: (Characterization of stability boundary for power systems)

Consider a nonlinear dynamical system described by equation (13) which has an energy function and satisfies assumptions (A1') and (A2'). Let E denote the set of equilibrium points and $x_i \in E$, i =1,2,... Then,

[1] $\hat{x} \in \partial A(x_s)$ if and only if $\{W^u(\hat{x}) - \hat{x}\} \cap A(x_s) \neq \phi$.

[2] $\hat{x} \in \partial A(x_s)$ if and only if $W^s(\hat{x}) \subseteq \partial A(x_s)$.

[3]

$$\partial A(x_s) = \bigcup_{x_j \in E \cap \partial A(x_s)} W^s(x_j) \tag{27}$$

E. Some Topological Properties of Stability Boundary

In this section we derive some topological properties of nonlinear dynamical system (13) using an energy function. More specifically, we derive a necessary condition for a stability boundary to be bounded, from which we also derive a sufficient condition for the stability boundary to be unbounded. Theorem 3-9 below presents a necessary condition for existence of certain types of equilibrium points on the stability boundary.

Theorem 3-9: (Structure of equilibrium points on the stability boundary)

Consider a nonlinear dynamical system described by equation (13) which has an energy function. If every equilibrium point on a stability boundary is hyperbolic, then the stability boundary must contain at least one type-one equilibrium point. If, furthermore, the stability boundary is bounded, then there exists at least a source and a type-one equilibrium point on the stability boundary.

Proof : Since the dimension of stability boundary of system (13) is n-1, Theorem 3-3 and the fact that the dimension of stable manifold of type-k is co-dimension k ensure that there must exist at least one type-one equilibrium point on the stability boundary so that its dimension is n-1. Next, under the assumption that the stability boundary is bounded and the fact that the stability boundary is a closed set, it follows that the stability boundary is (topologically) a closed and bounded set (compact set). In addition, the stability boundary is (dynamically) an invariant set. Since the maximum value and the minimum value of a continuous function exist over any compact set, it follows that both the maximum and the minimum values of $V(\cdot)$ over the stability boundary exist. Since the stability boundary is bounded, the α-limit set and ω-limit set of every trajectory on the stability boundary are non-empty. From the properties (i) & (ii) of energy functions, we notice that the point with the maximum value of $V(\cdot)$ on the stability boundary must belong to the α-limit set of some trajectory. Theorem 3-2 also states that the α-limit set consists entirely of equilibrium points. On the other hand, suppose $\hat{x}$ is a hyperbolic equilibrium point but not a source, then $\hat{x}$ is on the stability boundary $\partial A(x_s)$ if and only if $\{W^s(\hat{x}) - \hat{x}\} \cap \partial A(x_s) \neq \phi$(see Theorem 3-5). Next, we need the following result:

Lemma: Consider a nonlinear dynamical system described by equation (13) which has an energy function $V(\cdot)$. Let x be a regular point and $\omega(x)$ be its ω-limit point, then $V(x) > V(\omega(x))$.

The above Lemma implies that only the sources can be the points with the (local) maximum value of the energy function on the stability boundary. Consequently, there exists at least one source on the stability boundary.

Next, we make use of Theorem 3-9 to derive a sufficient condition for the stability regions to be unbounded. Opoitsev in [42] has derived a sufficient condition for the stability region of a C^1 nonlinear dynamical system to be unbounded, which entails checking the divergence of vector field f at every point in the state space; if the divergence of vector field f at every point is less than zero, then the stability region is unbounded. In the following, we derive another sufficient condition for the stability region to be unbounded. This condition only requires checking the types of equilibrium points instead of checking every point in the state space. If there are no sources in the nonlinear dynamical system (13), then the stability region of the system is unbounded.

Corollary 3-10: (Unboundedness of stability region)

Consider a nonlinear dynamical system described by equation (13) which has an energy function. If every equilibrium point on a stability boundary is hyperbolic and the stability boundary contains no sources, then the stability boundary is unbounded.

Proof : the negation of Theorem 3-9.

Corollary 3-10 can be used to show that the stability boundary of both the classical model and the structure-preserving model of power systems is unbounded.

IV. The Closest U.E.P. Method

The essence of the closest u.e.p. method is to employ the constant energy surface passing through the closest u.e.p. to approximate an interested stability boundary. This method is commonly used in the probabalistic dynamic security assessment [5], [34], [43], [44]. In this section we examine the closest u.e.p. method from both theoretical and computational points of view. The issue regarding the robustness of the closest u.e.p. relative to changes of different parameters in the system is also addressed.

A. The Closest U.E.P.

We begin the discussion of the closest u.e.p. with a graphical explanation. A formal definition of the closest u.e.p. will be presented later. In Fig. 9, the relation between the constant energy surfaces at different level values and the stability region $A(x_s)$ is shown. Recall that only one component of the constant energy surface may have a non-empty intersection with the stability region (see Theorem 3-4). It is

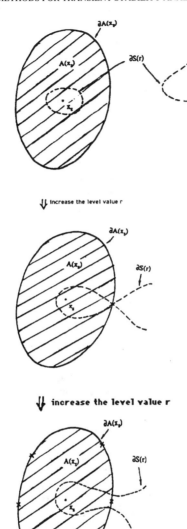

Figure 9: The relation between the constant energy surfaces at different level values and the stability region $A(x_s)$.

observed from this figure that the constant energy surface $\partial S(r)$ with small level value (critical value) r is very conservative in the approximation of the stability boundary $\partial A(x_s)$. As the constant energy surface $\partial S(r)$ is expanded by increasing the level value r, the approximation gets better until this constant energy surface hits the stability boundary $\partial A(x_s)$ at some point. This point will be shown to be an u.e.p. We shall call this point the closest u.e.p. of the s.e.p. x_s with respect to the energy function $V(\cdot)$.

Furthermore, as we increase the level value, the new constant energy surface would contain points which lie outside the stability region $A(x_s)$. It is therefore unduly to approximate the stability boundary $\partial A(x_s)$ by the connected constant energy surface (containing the s.e.p. x_s) with the level value higher than that of the lowest point on the stability boundary. From this figure, it appears that the connected constant energy surface containing the s.e.p. x_s is the best candidate to approximate the stability boundary $\partial A(x_s)$ among all the constant energy surfaces of $V(\cdot)$. These observations will be rigorously justified in the Theorem 4-1 below.

The concept of the closest u.e.p. has been introduced and applied to the estimation of power systems for a period of time. Conventionally, the closest u.e.p. of a stable equilibrium point x_s with respect to the energy function $V(\cdot)$ has been recognized to be the equilibrium point $\hat{x}$ with the following property

$$V(\hat{x}) = minimum_{V(x) > V(x_s), \, x \in E} \, V(x)$$

where E is the set of equilibrium points. Working with this definition, the classical closest u.e.p. method has been found to be very conservative in the estimation of stability regions. Recently (see for example [21], [45], [?], [?]), another concept of the closest u.e.p. has been recognized in the literature to be the equilibrium point, say $\hat{x}$, such that the following holds:

$$V(\hat{x}) = minimum_{x \, \in \, E \cap \partial A(x_s)} \, V(x)$$

In other words, the closest u.e.p. of the s.e.p. x_s with respect to the energy function $V(\cdot)$, is defined to be the u.e.p. on the stability boundary $\partial A(x_s)$ with the minimum value of $V(\cdot)$ among all the e.p.'s on $\partial A(x_s)$. The usefulness of this definition will be clear in the Theorem 4-1 below. It is noted that by Theorem 3-3 and the property of energy function, the closest u.e.p. of the s.e.p. x_s with respect to the energy function $V(\cdot)$ is also the point with the lowest value of $V(\cdot)$ among all the points on the stability boundary $\partial A(x_s)$, i.e. $V(p) = minimum_{x \in \partial A(x_s)} \, V(x)$. From now on, for the sake of easy exposition, sometimes we say an e.p. $\hat{x}$ is a closest u.e.p. without referring to its corresponding stable equilibrium point and energy function, when it is clear from the context.

We next examine the fundamental problems concerning the existence and uniqueness of the closest u.e.p. for system (13). The answer can be obtained by applying Theorem 3-5 and Theorem 3-6 to derive the following result :

Theorem 4-1:

Consider the nonlinear dynamical system (13) which has an energy function. Let x_s be a stable equilibrium point whose stability region $A(x_s)$ is not dense in R^n.

Then, the point with the minimum value of the energy function over the stability boundary $\partial A(x_s)$ exists and it must be an unstable equilibrium point. Moreover, this u.e.p. is generically unique.

Proof: The proof of the first part is similar to that of Theorem 3-1 in [46]. The proof of the second part is based on Thom's famous transversality theorem [47] and is quite lengthy, hence it is omitted here.

Two relevant questions arise: Given a nonlinear dynamical system with an energy function, can one find another energy function for the system ? If the answer is yes, then what are the relationships between the energy functions ? And do these energy functions give rise to different closest u.e.p.'s ? These two appear to be open questions.

B. Characterization of the Closest U.E.P.

After having shown the existence and uniqueness of the closest u.e.p., we next address the problem ; characterizations of the closest u.e.p. The characterization of the closest u.e.p. to be presented below is in terms of its unstable manifold. It will be shown that the necessary condition for $\hat{x}$ being the closest u.e.p. of x_s is that the unstable manifold of $\hat{x}$, denoted by $W^u(\hat{x})$, converges to x_s. This characterization of the closest u.e.p. will play an important role in the algorithm of finding the closest u.e.p.

Theorem 4-2: (Dynamical characterization of the closest u.e.p.)

Consider the nonlinear dynamical system (13), which has an energy function. Let $\hat{x}$ be a hyperbolic equilibrium point and have the minimum value of the energy function over the stability boundary $\partial A(x_s)$. Then, $W^u(\hat{x}) \cap A(x_s) \neq \phi$.

Proof : This proof is a slight extension of the proof of Theorem 3-3 in [46]. Since $\hat{x}$ is the closest u.e.p. of x_s, then (i) $\hat{x}$ is an equilibrium point on the stability boundary $\partial A(x_s)$ and (ii) $\hat{x}$ is the point with the lowest value of an energy function $V(\cdot)$ on the stability boundary $\partial A(x_s)$. According to Theorem 3-5, we have $W^u(\hat{x}) \cap \bar{A}(x_s) \neq \phi$. Next, it will be shown that $W^u(\hat{x}) \cap A(x_s) \neq \phi$. Consequently, we conclude that $W^u(\hat{x})$ converges to x_s. By contradiction, suppose $W^u(\hat{x}) \cap A(x_s) = \phi$, since $W^u(\hat{x}) \cap \bar{A}(x_s) \neq \phi$, this implies that $W^u(\hat{x}) \cap \partial A(x_s) \neq \phi$. In accordance with Theorem 3-3, which states that every trajectory on the stability boundary $\partial A(x_s)$ converges to one of the equilibrium points on the stability boundary, it is deduced that $W^u(\hat{x})$ converges to an equilibrium point, say $\hat{p} \in \partial A(x_s)$. Since the energy function $V(\cdot)$ strictly decreases along every trajectory of system (13), we have that $V(\hat{x}) > V(\hat{p})$. This is contradictory to part (ii) that $\hat{x}$ is the point with the lowest value of the energy function $V(\cdot)$ on the stability boundary $\partial A(x_s)$. Hence, $W^u(\hat{x}) \cap A(x_s) \neq \phi$.

C. The Closest U.E.P. Method

Given an energy function, the central point in the energy function approach to estimate the stability region of the stable equilibrium point x_s, is the determination of the critical level value. In this section, we discuss how to optimally determine the critical level value of an energy function for estimating the stability boundary $\partial A(x_s)$. We use the notation $\bar{A}^c(x_s)$ to denote the complement of the set $\bar{A}(x_s)$. ($\bar{A}(x_s)$ is the closure of the stability region.) Since the set $S(r)$ may contain several components, we use the notation $S_{x_s}(r)$ to denote the component of $S(r)$ containing the s.e.p. x_s.

Theorem 4-3 : (Optimal estimation of the stability region by an energy function)

Consider the nonlinear dynamical system (13) which has an energy function. Let x_s be a stable equilibrium point and let E be the set of equilibrium points. Let

$$\hat{c} = min_{x_i \in \partial A(x_s) \cap E} \quad V(x_i),$$

then

[1] the component $S_{x_s}(\hat{c}) \subset A(x_s)$, and

[2] the set $\{S_{x_s}(b) \cap A^c(x_s)\}$ is non-empty for any number $b > \hat{c}$.

The proof for part [1] of Theorem 4-3 can be found in [21], [37], [19] for different power system models. The proof of part [2] can be found in [37], [19]. Part (2) of Theorem 4-3 asserts that the scheme of choosing $\hat{c} = min_{x_i \in \partial A(x_s) \cap E} \quad V(x_i)$ to estimate the stability region $A(x_s)$ is optimal because the estimated stability region, characterized by the corresponding energy function, is the largest one within the entire stability region.

Theorem 4-3 asserts that using the set $\partial S(r)$, with $r = V(\hat{x})$, to approximate the stability boundary is optimal in the sense that the estimated stability region enclosed by the set $\partial S(r)$ is the largest one within the true stability region - $A(x_s)$, and any further increase of the level value r will make the set $\partial S(r)$ to contain some points which lie outside the stability region $A(x_s)$. Hence, this theorem indicates that it is very important to use the energy function at the precise closest u.e.p. as the level value (or critical value) when the closest u.e.p. method is used to approximate an interested stability boundary. Furthermore, Theorem 4-1 claims the uniqueness of the closest u.e.p., and (almost always) excludes the possibility that some other u.e.p. has the same value of energy function as that of the closest u.e.p. Therefore, the task of finding the correct closest u.e.p. is important in applying the closest u.e.p. method.

Theorem 4-1 together with the Theorem 4-3 can be used to show that the previous closest u.e.p. methods [48], [49], could give either rather conservative estimations or over estimations of the stability boundary (i.e. mis-classify some points, which are outside the stability region, to be the stability boundary). In correspondence to the importance of finding the precise closest u.e.p. for the closest u.e.p. method, our next objective is to propose a scheme to find the precise closest u.e.p. In doing so, we now summarize the characterizations of the closest u.e.p. of the s.e.p. x_s with respect to the energy function $V(\cdot)$:

[i] its unstable manifold converges to the s.e.p. x_s.

[ii] its energy function value is greater than $V(x_s)$.

[iii] it has the lowest value of the energy function $V(\cdot)$ among all the equilibrium points on the stability boundary $\partial A(x_s)$.

The procedure of the closest u.e.p. method to approximate the stability boundary $\partial A(x_s)$ using an energy function $V(\cdot)$ is as follows :

1. Determining the critical level value of the energy function.

   (a) Find all of the equilibrium points.

   (b) Order these equilibrium points whose corresponding values $V(\cdot)$ are greater than $V(x_s)$.

   (c) Of these, identify the one with the lowest value of the energy function and whose unstable manifold converges to the stable equilibrium point x_s (let this one be $\hat{x}$).

   (d) The value of the energy function at $\hat{x}$ gives the critical level value of this energy function (i.e. $V(\hat{x})$).

2. Estimating the stability region $A(x_s)$.

   (a) The connected component of $\{x \ : \ V(x) \ < \ V(\hat{x})\}$ containing the stable equilibrium point x_s, gives the estimated stability region.

Remark:
The computation associated with Step 1(a) could be very involved. Efficient numerical methods in conjunction with utilizing special properties of the system under study are needed to implement this step. A procedure of computing the unstable manifolds of equilibrium points is suggested in [37]. The computation efficiency of this scheme however can be much improved if the following conjecture holds:
Conjecture:

the equilibrium point with the minimum value of an energy function over the stability boundary is generically of type-one.

If this conjecture is true, then the computation associated with the above scheme can be considerably simplified as shown in the following:

1. Determining the critical level value of the energy function.

   (a) Find all the type-one equilibrium points.

   (b) Order these equilibrium points whose corresponding values $V(\cdot)$ are greater than $V(x_s)$.

   (c) Of these, identify the one with the lowest value of the energy function and whose unstable manifold converges to the stable equilibrium point x_s (let this one be $\hat{x}$).

   (d) The value of the energy function at $\hat{x}$ gives the critical level value of this energy function (i.e. $V(\hat{x})$).

2. Estimating the stability region $A(x_s)$.

   (a) The connected component of $\{x \ : \ V(x) \ < \ V(\hat{x})\}$ containing the stable equilibrium point x_s, gives the estimated stability region.

A procedure to check whether the unstable manifold of a type-one e.p. converges to the s.e.p. x_s is [37]:

[1] Find the Jacobian at the type-one equilibrium point (say, $\hat{x}$).

[2] Find the normalized unstable eigenvector of the Jacobian (say, y), (Note that it has only one unstable eigenvector for the type-one e.p.).

[3] Find the intersections of this unstable eigenvector with the boundary of an ϵ -ball of the equilibrium point. (the intersection points are $\hat{x} + \epsilon y$ and $\hat{x} - \epsilon y$).

[4] Integrate the vector field backward (reverse time) from each of these intersection points after some specified time. If the trajectory remains inside this ϵ-ball, then go to the next step. Otherwise, we replace the value ϵ by $\alpha \times \epsilon$ and also the intersection points $\hat{x} \pm \epsilon y$ by $\hat{x} \pm \alpha \times \epsilon y$, where $0 < \alpha < 1$. Repeat this step.

[5] Numerically integrate the vector field starting from these intersection points.

[6] If any of these trajectories approaches x_s, then the equilibrium point is on the stability boundary $\partial A(x_s)$.

Further investigation is needed to confirm this conjecture.

D. Examples

In order to illustrate the closest u.e.p. method developed in Sec. IV.C in a simple context, we consider the following example, which nearly represents a three-machine system, with machine number 3 as the reference machine. Two cases are studied.

Case 1 : (light-loaded case)

Consider the following system:

$$
\begin{aligned}
\dot{\delta}_1 &= \omega_1 \\
\dot{\omega}_1 &= -sin\delta_1 - 0.5sin(\delta_1 - \delta_2) - 0.4\delta_1 \\
\dot{\delta}_2 &= \omega_2 \\
\dot{\omega}_2 &= -0.5sin\delta_2 - 0.5sin(\delta_2 - \delta_1) - 0.5\delta_2 + 0.05
\end{aligned}
$$

It is easy to show that the following function is an energy function for this system.

$$
V(\delta_1, \delta_2, \omega_1, \omega_2) = \omega_1{}^2 + \omega_2{}^2 - 2cos\delta_1 - cos\delta_2 - cos(\delta_1 - \delta_2) - 0.1\delta_2 \quad (28)
$$

The point $x^s = (\delta_1{}^s, \omega_1{}^s, \delta_2{}^s, \omega_2{}^s) = (0.02001, 0, 0.06003, 0)$, is a stable equilibrium point whose stability boundary we are interested in. Applying the closest u.e.p. method in Sec. IV.C to approximate the stability boundary $\partial A(x^s)$, we have

Step 1: There are three type-one equilibrium points within the region $\{(\delta_1, \delta_2) : \delta_1{}^s - \pi < \delta_1 < \delta_1{}^s + \pi, \delta_2{}^s - \pi < \delta_2 < \delta_2{}^s + \pi\}$.

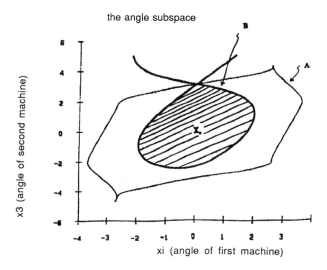

the angle subspace

Figure 10: Curve A is the intersection between the exact stability boundary and the angle space and Curve B is the intersection between the approximated stability boundary by the closest u.e.p. method and the angle space.

type-one e.p.	δ_1	ω_1	δ_2	ω_2	$V(\cdot)$
1	0.03333	0	3.10823	0	-0.31249
2	-2.69489	0	1.58620	0	2.07859
3	-3.03807	0	0.31170	0	1.98472

Step 2: The type-one equilibrium point (0.03333, 0, 3.10823, 0) is the closest u.e.p. because its unstable manifold converges to the s.e.p. (0.02001, 0, 0.06003, 0) and is the one with the lowest value of energy function among all the u.e.p.'s on the stability boundary.

Step 3: The constant energy surface of $V(\cdot)$, with the level value -0.31249 containing the s.e.p. x^s, is used to approximate the desired stability boundary as shown in Fig. 10. Curve A in this figure is the intersection between the exact stability boundary and the angle space $\{\delta_1, \delta_2 : \delta_1 \in R, \delta_2 \in R\}$. Curve B is the intersection between the approximated stability boundary by the closest u.e.p. method and the angle space.

To show the optimality of the method, the approximated stability boundary by the level surfaces with different level values are shown in Fig. 11, (level value = -0.2) and Fig. 12, (level value = -0.4). It can be seen from this figure that the approximated stability boundary by the level surface with a level value lower than -0.31249 gives more conservative than the closest u.e.p. method whereas the

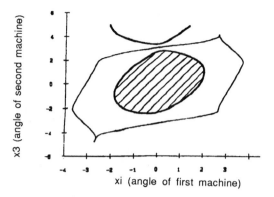

Figure 11: The approximated stability boundary by the level surface with a level value lower than the critical value, -0.31249, gives more conservative estimation results.

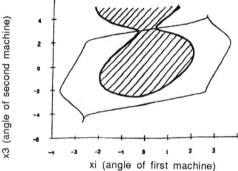

Figure 12: The approximated stability boundary by the level surface with a level value higher than the critical value, -0.31249, gives an inaccurate estimation because it contains points which are outside the exact stability boundary.

approximated stability boundary by the level surface with a level value higher than -0.31249 gives an inaccurate estimation because it contains points which are outside the exact stability boundary.

It is interesting to note that if the following scheme suggested in [49] is applied to this system

$$critical\ value\ =\ min\ \{V(\delta_{s1},\ \pi\ -\ \delta_{s2}, 0, 0),\ V(\pi\ -\ \delta_{s1},\ \delta_{s2}, 0, 0)\} \qquad (29)$$

then the critical value of $V(\cdot)$ would be -0.31276. This value is smaller than that obtained by the proposed method. This scheme however has an advantage in computation.

Case 2 : (heavy-loaded case)

Consider the following system

$$\dot{\delta}_1 = \omega_1$$
$$\dot{\omega}_1 = -sin\delta_1 - 0.5sin(\delta_1 - \delta_2) - 0.4\delta_1$$
$$\dot{\delta}_2 = \omega_2$$
$$\dot{\omega}_2 = -0.5sin\delta_2 - 0.5sin(\delta_2 - \delta_1) - 0.5\delta_2 + 0.3$$

It can be shown that the following function is an energy function for this system.

$$V(\delta_1, \delta_2, \omega_1, \omega_2) = \omega_1{}^2 + \omega_2{}^2 - 2cos\delta_1 - cos\delta_2 - cos(\delta_1 - \delta_2) - 0.6\delta_2 \quad (30)$$

We are interested in the stability boundary of the s.e.p. $(0.12137, 0, 0.36596, 0)$. The simulation results by application of the closest u.e.p. method is summarized in the following.

Step 1: There are two type-one equilibrium points lying in the region $\{(\delta_1, \delta_2) : \delta_1{}^s - \pi < \delta_1 < \delta_1{}^s + \pi, \delta_2{}^s - \pi < \delta_2 < \delta_2{}^s + \pi\}$.

type-one e.p.	δ_1	ω_1	δ_2	ω_2	$V(\cdot)$
1	0.19855	0	2.93462	0	-1.823925
2	0.19855	0	-3.34857	0	1.9459861

Step 2: The type-one equilibrium point $(0.19855, 0, 2.93462, 0, -1.823925)$ is the closest u.e.p..

Step 3: Curve A shown in Fig. 13 is the intersection between the exact stability boundary and angle space $\{\delta_1, \delta_2 : \delta_1 \in R, \delta_2 \in R\}$. Curve B is the intersection between the approximated stability boundary by the proposed close u.e.p. method and the angle space.

The critical value of this system by applying the scheme in [49] is -1.833300 (c.f. -1.823925).

E. An Improved Method

This section discusses an improved closest u.e.p. method [36] which requires less computation efforts than, yet yields the same estimated stability region as, the one discussed in Sec. IV B. The improved method explores the relationship between the closest u.e.p. of the underlying system and that of a reduced system, and finds the closest u.e.p. via the reduced system. The improved method is applicable to both the classical model and the structure preserving model. Without loss of generality, the classical model is discussed in this section.

Consider a power system consisting of n generators. Let the loads be modeled as constant impedances. Under the assumption that the transfer conductances are zero, the dynamics of the i-th generator can be represented by the following equations:

$$\dot{\delta}_i = \omega_i \qquad i = 1, 2, ..., n \qquad (31)$$
$$M_i\dot{\omega}_i = P_i - D_i\omega_i - \sum_{j \neq i}^{n+1} E_i E_j Y_{ij} sin(\delta_i - \delta_j)$$

the angle subspace

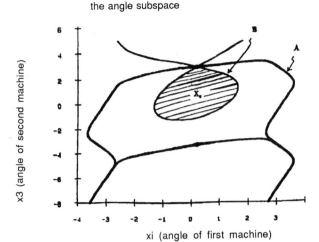

Figure 13: Curve A is the intersection between the exact stability boundary and angle space. Curve B is the intersection between the approximated stability boundary by the closest u.e.p. method and the angle space.

where node n+1 serves as the reference node, i.e. $E_{n+1} = 1$ and $\delta_{n+1}=0$. The damping constant D_i is assumed to be positive.

Let (δ^s, ω^s) be a stable equilibrium point of (31). We define the following function $V(\delta, \omega)$, which is an energy function for system (31).

$$\begin{aligned}
V(\delta, \omega) &= 0.5 \sum_{i=1}^{n} M_i \omega_i^2 - \sum_{i=1}^{n} P_i (\delta_i - \delta_i^s) \\
&\quad - \sum_{i=1}^{n} \sum_{j=i+1}^{n+1} E_i E_j Y_{ij} (\cos(\delta_i - \delta_j) - \cos(\delta_i^s - \delta_j^s)) \\
&= V_k(\omega) + V_p(\delta)
\end{aligned} \qquad (32)$$

where $V_k = 0.5 \sum_{i=1}^{n} M_i \omega_i^2$ is the kinetic energy and $V_p(\cdot)$ is the potential energy.

Notice that

$$\frac{\partial V_p(\delta)}{\partial \delta_i} = -P_i + \sum_{\substack{j \neq i}}^{n+1} E_i E_j Y_{ij} \sin(\delta_i - \delta_j) \qquad i = 1, 2, ..., n \qquad (33)$$

and (31) can be expressed in vector form as

$$\dot{\delta} = \omega \qquad (34)$$

$$M\dot{\omega} = -D\omega - \frac{\partial V_p(\delta)}{\partial \delta}$$

Next, we consider the following reduced system which is a gradient system

$$\dot\delta = -\frac{\partial V_p(\delta)}{\partial \delta} \tag{35}$$

$V_p(\delta)$ is an energy function for the reduced system (35), which is exactly the potential energy of the energy function for the original system (34). For ease of exposition, we use the notation d(M,D) to denote the original system (34) and the notation d(I)to denote system (35).

Now, we explore the relationship between the closest u.e.p. of system d(M,D) with respect to the energy function $V(\cdot)$ in Eq. (32) and the closest u.e.p. of system d(I) with respect to the energy function $V_p(\cdot)$. This viewpoint is motivated by the fact that the energy value $V_p(\cdot)$ at the equilibrium point (δ) of system d(I) is equal to the energy value $V(\cdot)$ at the equilibrium point $(\delta,0)$ of system d(M,D). We begin with a relationship between the equilibrium points of the system $d(I)$ and that of the original system $d(M, D)$.

Theorem 4-4 [36]: (Relationship between the equilibrium points of d(I) and that of d(M,D))
If zero is a regular value of $\frac{\partial V_p(\delta)}{\partial \delta}$, then

1. (δ_s) is a stable equilibrium point of $d(I)$ if and only if $(\delta_s,0)$ is a stable equilibrium point of d(M,D), and furthermore,

2. (δ) is a type-k equilibrium point of d(I) if and only if $(\delta,0)$ is a type-k equilibrium point of d(M,D).

The relationship between the closest u.e.p. of system d(M,D) and that of system d(I) is established below, upon which the improved closest u.e.p. method is based.
Theorem 4-5 [36]: (Relationship between the closest u.e.p. of d(I) and that of d(M,D))
$(\hat\delta, 0)$ is the closest u.e.p. of the s.e.p. $(\delta_s,0)$ of system d(M,D) with respect to the energy function $V(\cdot)$ in Eq. (32) if and only if $(\hat\delta)$ is the closest u.e.p. of the s.e.p. (δ_s) of system d(I) with respect to the energy function $V_p(\cdot)$ in Eq. (35).

Theorem 4-5 furnishes an approach resembling the model-reduction approach to find the closest u.e.p. of system d(M,D). It suggests that, in order to find the closest u.e.p. of the original system d(M,D), we only need to find the closest u.e.p. of the reduced system d(I), whose state-space dimension is n instead of 2 × n, thus reducing the computational burden dramatically. Theorem 4-5 leads to the following improved closest u.e.p. method:

1. Determining the critical level value of the energy function.

   (a) Find all of the equilibrium points of system d(I).

   (b) Order the equilibrium points whose corresponding values $V_p(\cdot)$ are greater than $V_p(\delta_s)$.

   (c) Of these, identify the one with the lowest value of the energy function and whose unstable manifold converges to the stable equilibrium point δ_s (let this one be $\hat\delta$).

(d) The value of the energy function at $\hat{\delta}$ gives the critical level value of this energy function (i.e. $V_p(\hat{\delta})$).

2. Estimating the stability region $A(\delta_s, 0)$.

(a) The connected component of $\{(\delta, 0 \ : \ V(\delta, 0) \ < \ V_p(\hat{\delta})\}$ containing the stable equilibrium point $(\delta_s, 0)$, gives the estimated stability region.

F. Robustness of Closest U.E.P.

The problem concerning the robustness of the closest u.e.p. relative to changes in matrices M and D is studied in this section. A more general study can be found in [36].

We consider the case that, due to the error in modelling or to the changes of parameter settings in control devices of generators, the machine inertia matrix and damping matrix of system d(M,D) are changed into another machine matrix $\hat{M}$ and damping matrix $\hat{D}$, which results in another system $d(\hat{M}, \hat{D})$. Note that the positions and types of the equilibrium points are independent of matrices M and D (Theorem 4-4). It can easily be shown that the following function is an energy function for the new system.

$$
\begin{aligned}
\hat{V}(\delta, \omega) \ = \ & 0.5 \sum_{i=1}^{n} \hat{M}_i \omega_i^2 - \sum_{i=1}^{n} P_i(\delta_i - \delta_i^s) \\
& - \sum_{i=1}^{n} \sum_{j=i+1}^{n+1} E_i E_j Y_{ij}(cos(\delta_i - \delta_j) - cos(\delta_i^s - \delta_j^s))
\end{aligned}
\tag{36}
$$

The following theorem shows that the closest u.e.p. of system d(M,D) relative to an energy function, which is a summation of kinetic energy and potential energy, is also independent of matrices M and D as long as they are positive definite.

Theorem 4-6 : (Invariant property of the closest u.e.p.)
Let $(\hat{\delta}, \ 0)$ be the closest u.e.p. of the s.e.p. $(\delta_s, 0)$ of system d(M,D) with respect to an energy function , which is a summation of $V_1(\delta)$ and $V_2(M, \omega)$. If the system d(M,D) is changed into $d(\hat{M}, \hat{D})$, then

[1] $(\hat{\delta}, \ 0)$ is on the stability boundary $(\delta_s, 0)$ of system $d(\hat{M}, \hat{D})$, and

[2] $(\hat{\delta}, \ 0)$ is the closest u.e.p. of the s.e.p. $(\delta_s, 0)$ of system $d(\hat{M}, \hat{D})$ with respect to the energy function $V_1(\delta) + V_2(\hat{M}, \omega)$.

Proof: The proof mimics the proof of theorem 5-3 in [36], hence it is omitted .

Theorem 4-6 illustrates another application of energy functions in the understanding of power system dynamical behaviors. Using an appropriate energy function, one is able to show that a particular equilibrium point (i.e. the closest u.e.p.) remains on the stability boundary of the new system $d(\hat{M}, \hat{D})$ which is derived from d(M,D) under large variations of both machine inertia matrix and damping matrix. By theorem 4-2, the unstable manifold of the closest u.e.p. always converges to the s.e.p. $(\delta_s, 0)$ even without the assumption of the transversality condition (see Fig.

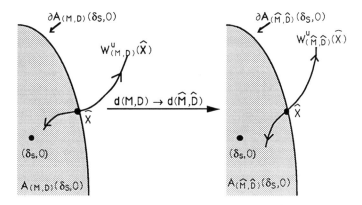

Figure 14: The closest u.e.p. $\hat{x}$ remains on the stability boundary of the new system $d(\hat{M}, \hat{D})$. The unstable manifold of the closest u.e.p. always converges to the s.e.p. $(\delta_s, 0)$.

14).

V. The Controlling U.E.P. Method

A power system is said to be transiently stable for a particular (pre-fault) steady-state operating condition and for a particular disturbance, if following that disturbance, it reaches an acceptable steady-state operating condition (i.e. a desired stable equilibrium point of a post-fault system). From this definition, transient stability analysis is essentially the problem of determining whether or not the fault-on trajectory at clearing time is lying inside the stability region of a desired s.e.p. of its post-fault system. Hence, the main point in transient stability analysis is not to estimate the whole stability boundary of the post-fault system. Instead, only the relevant part of stability boundary toward which the fault-on trajectory is heading is of main concern.

When the closest u.e.p. method is applied to power system transient stability analysis, this method has been found to yield conservative results. In fact, in the context of transient stability analysis, the closest u.e.p. method provides an approximated stability boundary for the post-fault system, and is independent of the fault-on trajectory. Thus, the closest u.e.p. method gives conservative results for transient stability analysis.

A desirable method for transient stability analysis would be the one which can provide more accurate approximation of the relevant part of stability boundary to which the fault-on trajectory is heading, even though it might provide a very poor estimate on the other part of stability boundary. To this end, the controlling u.e.p. method uses the (connected) constant energy surface passing through the controlling u.e.p. to approximate the relevant part of stability boundary of the post-fault system to which the fault-on trajectory is heading. This is the essence of the controlling u.e.p. method.

The evolution of the concept of controlling u.e.p. can be traced back to the mid-1970s. But a formal definition of the controlling u.e.p. and theoretical justification of the controlling u.e.p. method have been lacking. In [48], Prabhakara and El-Abiad argued that the controlling u.e.p. is the u.e.p. which is closest to the fault-on trajectory. Athay, et al. in [13] suggested that the controlling u.e.p. is the u.e.p. 'in the direction' of the fault-on trajectory. Ribbens-Pavella et al. [50] relate the controlling u.e.p. to the machine (or groups of machine) which first go out of synchronism if the fault is sustained. Fouad et al. [51], associated the controlling u.e.p. with the 'mode of instability' of machines.

In this section, we attempt to build a framework to analyze the controlling u.e.p. method. A fundamental theorem will be presented to interpret and justify the controlling u.e.p. method. Several characterizations for the controlling u.e.p. will be derived. Based on these characterizations of the controlling u.e.p., we will propose a controlling u.e.p. method. A simple example is studied to illustrate the proposed controlling u.e.p. method.

A. The Controlling U.E.P.

The controlling u.e.p. method for transient stability analysis proceeds as follows :

1. Determination of the critical energy

Step 1.1: Find the controlling u.e.p. x_{co} for a given fault-on trajectory $x_f(t)$.

Step 1.2: The critical energy v_c is the value of energy function $V(\cdot)$ at the controlling u.e.p., i.e.

$$v_c = V(x_{co})$$

2. Determination of Stability

Step 2.1: Calculate the value of the energy function $V(\cdot)$ at the time of fault clearance (say, t_{cl}) using the fault-on trajectory

$$v_f = V(x_f(t_{cl})).$$

Step 2.2: If $v_f < v_c$, then the post-fault system is stable. Otherwise, it is unstable.

In fact, this method is computationally equivalent to the following method. And it is from this viewpoint that we build a framework to analyze the controlling u.e.p. method.

Another Viewpoint of the Controlling U.E.P. Method.
1. Determination of the Critical Energy

Step 1.1: Find the controlling u.e.p. x_{co} for a given fault-on trajectory $x_f(t)$.

Step 1.2: The critical energy v_c is the value of energy function $V(\cdot)$ at the controlling u.e.p., i.e.

$$v_c = V(x_{co})$$

2. Approximation of the relevant part of stability boundary

Step 2.1: Use the connected constant energy surface of $V(\cdot)$ passing through the controlling u.e.p. x_{co} (containing the s.e.p. x_s) to approximate the relevant part of stability boundary for the fault-on trajectory $x_f(t)$.

3. Determination of Stability

Check whether the fault-on trajectory at the fault clearing time (t_{cl}) is located inside the stability boundary characterized in step 2.1. This is done as follows :

Step 3.1: Calculate the value of the energy function $V(\cdot)$ at the time of fault clearance (t_{cl}) using the fault-on trajectory $x_f(t)$

$$v_f = V(x_f(t_{cl})).$$

Step 3.2: If $v_f < v_c$, then the point $x_f(t_{cl})$ is located inside the stability boundary and the post-fault system is stable. Otherwise, it is unstable.

From this viewpoint, the aim of the controlling u.e.p. method is essentially to yield an approximation of the relevant part of the stability boundary of the post-fault system to which the fault-on trajectory is heading. More precisely, the controlling u.e.p. method uses the (connected) constant energy surface passing through the controlling u.e.p. to approximate the relevant part of stability boundary.

Before we further interpret the controlling u.e.p. method from the above angle, the following definition will be useful in our analysis:

Definition 5-1:

The point from which a given fault-on trajectory exits the stability boundary of the post-fault system is called the *exit point* of the fault-on trajectory.

Since the stability boundary $\partial A(x_s)$ is contained in the set which is the union of the stable manifolds of the equilibrium points on the stability boundary (under the assumption of the existence of energy function), the exit point of a fault-on trajectory must be lying on the stable manifold of some e.p. on the stability boundary. This implies the existence and uniqueness of the controlling u.e.p. relative to a fault-on trajectory.

In order to explain the concept of controlling u.e.p. clearly, we start with an example. Let us observe Fig. 15 which depicts a fault-on trajectory $x_f(t)$ being moving toward the stability boundary $\partial A(x_s)$ of the s.e.p. x_s of a post-fault system. The exit point of this fault-on trajectory is the point x_e which lies on the stable manifold of $\hat{x}$. The set $\partial S(r_1)$ is the constant energy surface passing through the closest u.e.p. x_{cl} and the set $\partial S(r)$ is the constant energy surface passing through the u.e.p. $\hat{x}$. Apparently, the stable manifold of $\hat{x}$ is the relevant part of stability boundary for the fault-on trajectory $x_f(t)$; if the fault is cleared before the fault-on trajectory $x_f(t)$ passes through the stable manifold of $\hat{x}$, then the post-fault trajectory will converge to the s.e.p. x_s and the post-fault system is stable. In order to approximate this relevant part of stability boundary for this fault-on trajectory $x_f(t)$, we may use the constant energy surface passing through $\hat{x}$ (i.e. $\partial S(r)$) to

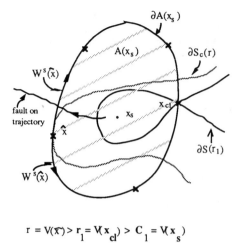

$$r = V(\hat{x}) > r_1 = V(x_{cl}) > C_1 = V(x_s)$$

Figure 15: The fault-on trajectory $x_f(t)$ moves toward the stability boundary $\partial A(x_s)$ of the s.e.p. x_s of a post-fault system. The exit point of this fault-on trajectory is the point x_e which lies on the stable manifold of $\hat{x}$. The equilibrium point $\hat{x}$ will be referred to the controlling u.e.p. of the fault-on trajectory $x_f(t)$.

approximate it (As a matter of fact, only part of the set $\partial S(r)$ is used to approximate the relevant stability boundary). The important point here is that the fault-on trajectory $x_f(t)$ passes through the set $\partial S(r)$ before it exits the stability boundary $\partial A(x_s)$ at x_e.

The above arguments suggest that the energy value at the equilibrium point $\hat{x}$ be used as the critical energy for the fault-on trajectory $x_f(t)$ and the constant energy surface passing through the u.e.p. $\hat{x}$ be used to approximate the relevant part of stability boundary. This is the essence of the controlling u.e.p. method. The equilibrium point $\hat{x}$ will be referred to the controlling u.e.p. of the fault-on trajectory $x_f(t)$. It is intuitively clear that the critical energy value given by the controlling u.e.p. method is no less than the critical energy value given by the closest u.e.p. method, and the only case that these two values are equal is that the exit point of the fault-on trajectory lies on the manifold of closest u.e.p. This observation of the controlling u.e.p. method will be rigorously justified in Theorem 5-1 below.

Theorem 5-1 : (Fundamental theorem for the controlling u.e.p. method)
Consider a nonlinear system described by system (13) which has an energy function $V(\cdot) : R^n \rightarrow R$. Let $\hat{x}$ be an equilibrium point on the stability boundary $\partial A(x_s)$ of this system. Let

- $S(r) :=$ the connected component of the set $\{ x : V(x) < r \}$ containing x_s, and

- $\partial S(r) :=$ the connected component of the set $\{ x : V(x) = r \}$ containing x_s.

Then,

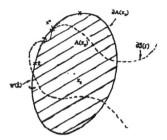

Figure 16: The set $S(V(x^u))$ contains only part of the stable manifold $W^s(\hat{x})$.

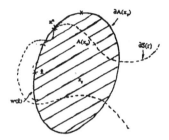

Figure 17: The set $S(V(x^u))$ contains the whole stable manifold $W^s(\hat{x})$

[i] the connected constant energy surface $\partial S(V(\hat{x}))$ intersects with the stable manifold $W^s(\hat{x})$ only at point $\hat{x}$; moreover, the set $S(V(\hat{x}))$ has an empty intersection with the stable manifold $W^s(\hat{x})$.

[ii] suppose $x_u \neq \hat{x}$, an unstable equilibrium point and $V(x_u) > V(\hat{x})$. The set $S(V(\hat{x}))$ has a non-empty intersection with the set $\{ W^s(\hat{x}) - \hat{x} \}$.

Part [1] of Theorem 5-1 implies that for any fault-on trajectory $x_f(t)$ starting from a point p with $p \in A(x_s)$ and $V(p) < V(\hat{x})$, if the exit point of this fault-on trajectory $x_f(t)$ lies on the stable manifold of $\hat{x}$, then this fault-on trajectory $x_f(t)$ must pass through the connected constant energy surface $\partial S(V(\hat{x}))$ before it passes through the stable manifold of $\hat{x}$ (thus exits the stability boundary $\partial A(x_s)$). This suggests that the connected constant energy surface $\partial S(V(\hat{x}))$ be used to approximate the relevant part of the stability boundary $\partial A(x_s)$ for the fault-on trajectory $x_f(t)$. More general, part [1] of Theorem 5-1 recommends that the connected constant energy surface $\partial S(V(\hat{x}))$ be used to approximate the relevant part of the stability boundary $\partial A(x_s)$ for all of the fault-on trajectories whose exit points are in the stable manifold of $\hat{x}$.

On the other hand, part [2] of Theorem 5-1 encompasses two possible cases; namely,

case (i): the set $S(V(x^u))$ contains only part of the stable manifold $W^s(\hat{x})$ (see Fig. 16).

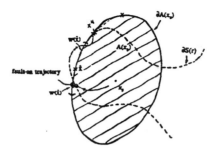

Figure 18: The fault-on trajectory $x_f(t)$ passes through the connected constant energy surface $\partial S(V(\hat{x}))$ before it passes through stable manifold $W^s(\hat{x})$.

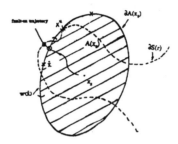

Figure 19: The fault-on trajectory $x_f(t)$ passes through the connected constant energy surface $\partial S(V(\hat{x}))$ after it passes through stable manifold $W^s(\hat{x})$.

case (ii): the set $S(V(x^u))$ contains the whole stable manifold $W^s(\hat{x})$ (see Fig. 17).

In case (i), the fault-on trajectory $x_f(t)$ may pass through the connected constant energy surface $\partial S(V(\hat{x}))$ before it passes through stable manifold $W^s(\hat{x})$ (see Fig. 18). In this situation, the controlling u.e.p. method using x^u as the controlling u.e.p. still gives an accurate stability assessment. However, the fault-on trajectory $x_f(t)$ may pass through the connected constant energy surface $\partial S(V(\hat{x}))$ after it passes through stable manifold $W^s(\hat{x})$ (see Fig. 19). In this situation, the controlling u.e.p. method using x^u as the controlling u.e.p. gives an inaccurate stability assessment.

In case (ii), the fault-on trajectory $x_f(t)$ always passes through the connected constant energy surface $\partial S(V(\hat{x}))$ after it passes through stable manifold $W^s(\hat{x})$ (see Fig. 20). Thus, under this situation the controlling u.e.p. method using x_u as the controlling u.e.p. always gives an inaccurate stability assessment. From these arguments, we have shown that for a given fault-on trajectory $x_f(t)$, if the exit point of this fault-on trajectory $x_f(t)$ lies on the stable manifold of an u.e.p. $\hat{x}$, then the controlling u.e.p. method, without using $\hat{x}$ as the controlling u.e.p., may give an incorrect stability assessment.

Motivated by this theorem, a formal definition of the controlling u.e.p. is given below:

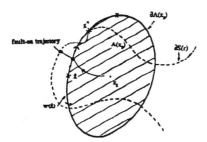

Figure 20: The fault-on trajectory $x_f(t)$ always passes through the connected constant energy surface $\partial S(V(\hat{x}))$ after it passes through stable manifold $W^s(\hat{x})$.

Definition 5-2: [19]
The *controlling u.e.p.* of a fault-on trajectory x_f is the u.e.p. whose stable manifold contains the exit point of x_f.

In summary, the implication of Theorem 5-1 is that the task to find the precise controlling u.e.p. of a fault-on trajectory is essential in the controlling u.e.p. method for transient stability analysis. The controlling u.e.p. method without using the precise controlling u.e.p. is likely to give either an incorrect or conservative stability assessment. Due to the importance of finding the precise controlling u.e.p. in the controlling u.e.p. method, our next objective is to derive characterizations for the controlling u.e.p.

B. Characterizations of the Controlling U.E.P.

In this section, we will derive two characterizations for controlling u.e.p.'s. The Proposition 5-2 below gives a characterization of the controlling u.e.p. in terms of its topological property, while the Proposition 5-3 gives a characterization of the controlling u.e.p. in terms of its dynamical property.

Proposition 5-2 : (Topological characterizations of controlling u.e.p.)

The controlling u.e.p. for any given fault-on trajectory is almost always a type-one equilibrium point.
Proof : Theorem 3-5 states that the stability boundary is contained in the union of the stable manifolds of the equilibrium points on the boundary. This proposition follows from the fact that the dimension of stable manifold of a type-one e.p. is n-1 and the dimension of other types of e.p.'s is n-2 or lower.

Remarks:

[1] A good conjecture, which is close (but not equal) to the conclusion of Theorem 5-2, was made by Yee and Spalding [35], claiming that the controlling u.e.p. must be a type-one equilibrium point.

[2] It may happen that the precise controlling u.e.p. is not a type-one equilibrium point; however, the measure (possibility) for this situation to occur is zero. Moreover, any error in the computation of the fault-on trajectory (say due to round-off error or truncation error in computation) will end up with a type-one equilibrium point.

Proposition 5-3 : (Dynamical characterization of controlling u.e.p.)

The controlling u.e.p. is an u.e.p. on the stability boundary $\partial A(x_s)$ and its unstable manifold almost always converges to the s.e.p. x_s.

Proof :Follows from Theorem 3-6.

These characterizations represent the necessary conditions for an u.e.p. to be the controlling u.e.p. of a fault-on trajectory. These results can be used to confirm whether or not the controlling u.e.p. is being obtained by any computational method.

C. The Controlling U.E.P. Method

Up to present, a number of methods based on physical reasonings to find the critical energy for a given fault-on trajectory have been proposed. These methods are classified into the following : (i) The transient energy approach [13], [52], (ii) The acceleration approach [50], [53], (iii) The potential energy boundary surface approach [11], [54], (iv) The system separation approach [9], [51], [59], and (v) The stability boundary approach [14], [15].

Based on Theorem 5-1, Proposition 5-2 and 5-3, the following conceptual control u.e.p. method is proposed. The term 'conceptual' is to indicate that the computational problem encountered in step 3 of the following method is enormous. It should be stressed that this method has the ability to consistently find the precise controlling u.e.p. for a given fault-on trajectory when the computation is feasible. In practice, however, this method is not suitable for large-scale power systems.

em A conceptual method to find the controlling u.e.p. :

Step 1: Find all the type-one equilibrium points of the post-fault system.

Step 2: Identify those type-one equilibrium points whose unstable manifolds (a one-dimensional curve) converge to the interested s.e.p. x_s.

Step 3: Derive the stable manifold of those type-one e.p.'s identified in step 2.

Step 4: Compute the fault-on trajectory $x_f(t)$ and detect the first type-one e.p. among those identified in step 2 such that $x_f(t)$ pass through the ϵ-neighborhood of its stable manifold, where ϵ is a small positive number. This type-one e.p. is the controlling u.e.p. of the fault-on trajectory $x_f(t)$.

Remarks :

[1] In step 1, we may limit our search for type-one u.e.p.'s to the region

$$- n\pi + \delta_{si} < \delta_i < n\pi + \delta_{si} \qquad (37)$$

where δ_i is the angle at bus i, the number n is dependent on the system under study. It was argued by the physical-based methods that the controlling u.e.p. is lie in the region Eq. (37) with n = 1.

[2] It should be pointed out that the main computational problem of this controlling u.e.p. method is the computation involved in step 3. The exact series

representation of the stable manifold can be obtained [62]. However, no numerical test has been reported. Two schemes can be used to approximate the stable manifold; namely, the hyperplane scheme and the hypersurface scheme. The hyperplane method proposed in [35] can be used as the first-order approximation of the stable manifold. The hypersurface method suggested in [63] can be used as the second-order approximation of the stable manifold.

[3] In general, the task of finding controlling u.e.p.'s for general power system models is very difficult. By exploiting special properties of the underlying power system model, and the insights derived in this section to develop a solution algorithm for finding controlling u.e.p.'s, may prove fruitful. In section VII, we will detail a solution algorithm for finding controlling u.e.p.'s for the classical model.

D. Example

In this section, we consider the simple system in Sec. IV.C to illustrate the concepts of controlling u.e.p. and to clarify the controlling u.e.p. method. Only the un-stressed case is studied.

Case 1 : (light-loaded case)

The point $x_1^s = (0.02001, 0, 0.06003, 0)$ is a stable equilibrium point that we are interested in. There are eight type-one equilibrium points within the region $\{(\delta_1, \delta_2) : \delta_1^s - 1.5\pi < \delta_1 < \delta_1^s + 1.5\pi, \delta_2^s - 1.5\pi < \delta_2 < \delta_2^s + 1.5\pi\}$.

type-one e.p.	δ_1	ω_1	δ_2	ω_2
1	0.03333	0	-3.17496	0
2	3.24512	0	0.31170	0
3	3.04037	0	3.24307	0
4	0.03333	0	3.10823	0
5	-2.69489	0	1.58620	0
6	-3.03807	0	1.58620	0
7	-3.24282	0	3.24387	0
8	3.04037	0	-3.03931	0

Among these eight type-one e.p.'s, the first six type-one e.p.'s are on the stability boundary $\partial A(x_1^s)$ because their unstable manifold converge to the s.e.p. x_1^s.

There are four type-two equilibrium points on the stability boundary $\partial A(x_1^s)$.

type-one e.p.	δ_1	ω_1	δ_2	ω_2
1	-2.67489	0	1.58620	0
2	-3.66392	0	-2.02684	0
3	2.61926	0	-2.02684	0
4	3.60829	0	1.58620	0

The stability boundary $\partial A(x_1^s)$ is contained in the set which is the union of the stable manifolds of these six type-one e.p.'s and four type-two e.p.'s. Fig. 21 shows the intersection between the exact stability boundary and the angle space. Fig. 22 marks

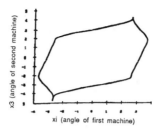

Figure 21: The intersection between the exact stability boundary and the angle space.

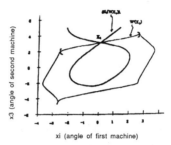

Figure 22: The boundary of the stable manifold of the type-one e.p. $x_1 = (0.03333, 0, 3.10823, 0)$ and the relationship between this stable manifold and the constant energy surface passing through this type-one e.p.

the boundary of the stable manifold of the type-one e.p. $x_1 = (0.03333, 0, 3.10823, 0)$, and shows the relationship between this stable manifold and the constant energy surface passing through this type-one e.p. Note that this relationship satisfies part [1] of Theorem 5-1.

Fig. 23 shows a given fault-on trajectory x_f^1. Among the stable manifolds of these six type-one e.p.'s on the stability boundary, the fault-on trajectory x_f^1, first passes through the stable manifold of the type-one e.p. x_1, thus x_1 is the controlling u.e.p. of this fault-on trajectory x_f^1. It is noted that this fault-on trajectory x_f^1 passes through the constant energy surface $\partial S(V(x_1))$, before it passes through the stable manifold of x_1. Fig. 24 (resp. Fig. 25) shows and delimits the stable manifold of the type-one e.p. $(0.03333, 0.0, 3.10823, 0.0)$ (resp. $(3.04237, 0.0, 3.24307, 0.0)$, and demonstrates its relationship with the constant energy surface passing through the associated type-one e.p. All of these simulations confirm the conclusion of Theorem 5-1 that the fault-on trajectory must pass through the constant energy surface of its controlling u.e.p. before it passes through the stable manifold of its controlling u.e.p. (thus exits the stability region), therefore providing a justification of the controlling u.e.p. method.

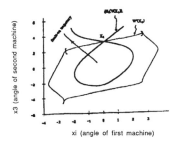

Figure 23: The fault-on trajectory x_f^1, first passes through the stable manifold of the type-one e.p. x_1, thus x_1 is the controlling u.e.p. of this fault-on trajectory x_f^1.

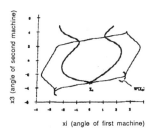

Figure 24: The stable manifold of the type-one e.p. $(0.03333, 0.0, 3.10823, 0.0)$ and the constant energy surface passing through it.

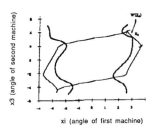

Figure 25: The stable manifold of the type-one e.p. $(3.04237, 0.0, 3.24307, 0.0)$ and the constant energy surface passing through it.

VI. The Potential Energy Boundary Surface Method

The Potential Energy Boundary Surface (PEBS) method, first proposed by Kakimoto et al. [11] was intended to circumvent the problem of determining the controlling u.e.p. relative to a fault-on trajectory. For a given fault-on trajectory, the method finds a 'local' approximation of the stability boundary via the PEBS. The process of finding this local approximation rests on the determination of the stability boundary of a lower dimensional system and, hence is computationally rather efficient. Despite its efficiency in computation, the PEBS method has been found to be not very reliable - it could give conservative estimates as well as over-estimates in critical clearing time. This may be attributed to the fact that the derivation of the PEBS method is based on heuristic arguments and has little theoretical basis.

The purpose of this section is to provide a theoretical foundation for the PEBS method based on the work by Chiang, Wu and Varaiya [64]. Sufficient conditions under which the PEBS method gives accurate stability estimates will be discussed. A method based on the theoretical analysis of the PEBS method has been developed for direct analysis of transient stability [14] and demonstrated in large-scale power systems [15]. The details of this method and its several simulation results on large scale power systems will be discussed in section VII.

A. Preliminary

Consider a power system consisting of n generators. Let the loads be modeled as constant impedances. Under the assumption that the transfer conductances are zero, the dynamics of the ith generator is represented by the following vector equations (see Eq. (31)):

$$\dot{\delta} = \omega \tag{38}$$
$$M\dot{\omega} = -D\omega - \frac{\partial V_p(\delta)}{\partial \delta}$$

where

$$\frac{\partial V_p(\delta)}{\partial \delta_i} = -P_i + \sum_{\substack{j \neq i}}^{n+1} E_i E_j Y_{ij} sin(\delta_i - \delta_j) \qquad i = 1, 2, ..., n \tag{39}$$

Let (δ^s, ω^s) be a stable equilibrium point of (38). Recall from Eq. (34) that the following function is an energy function for system (38)

$$V(\delta, \omega) = 0.5 \sum_{i=1}^{n} M_i \omega_i^2 + V_p(\delta)$$
$$= V_k(\omega) + V_p(\delta) \tag{40}$$

where $V_k = 0.5 \sum_{i=1}^{n} M_i \omega_i^2$ is the kinetic energy.

The PEBS method essentially provides a local approximation of the stability boundary of the system (38). The application of the PEBS as a local approximation of the stability boundary proceeds in two steps:

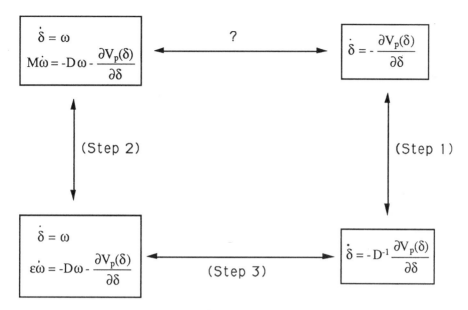

Figure 26: Three steps to determine the relationship between the stability boundary of the original system and the PEBS

Step1: From the fault-on trajectory $(\delta(t), \omega(t))$, detect the point δ^* at which the potential energy $V_p(\delta)$ reaches its local maximum. Let the value of $V_p(\cdot)$ at δ^* be v, i.e. $v = V_p(\delta^*)$.

Step2: Use the connected constant energy surface of the set $\{(\delta, \omega) : V(\delta, \omega) = V_p(\delta^*)\}$ containing the stable equilibrium point $(\delta_s, 0)$ as the local approximation of the stability boundary $\partial A(\delta_s, 0)$.

Next, the local approximation for the stability boundary provided by the PEBS method will be evaluated.

B. Analysis of the PEBS Method

Varaiya, Wu and Chen suggested [7] viewing the PEBS as the stability boundary of the associated gradient system:

$$\dot{\delta} = -\frac{\partial V_p(\delta)}{\partial \delta} \tag{41}$$

of the original system (38). This viewpoint is partly due to two facts: (i) all the equilibrium points of the original system (38) lie on the subspace $\{(\delta, \omega : \delta \in R^n, \omega = 0\}$ and (ii) the energy function $V(\delta, \omega)$ at an equilibrium point, say (δ_e, ω_e), is of the form $V(\delta_e, \omega_e) = V_p(\delta_e)$. This viewpoint further suggests the study of the relationship between the stability boundary of the original system (38) and that of the gradient system (41) (i.e. the PEBS).

The relationship between the stability boundaries of (38) and (41) can be established through the following three steps:

Step 1: Determine the relationship between the stability boundaries of the systems

$$\dot{\delta} = -\frac{\partial V_p(\delta)}{\partial \delta} \tag{42}$$

and

$$\dot{\delta} = -D^{-1} \frac{\partial V_p(\delta)}{\partial \delta} \tag{43}$$

where D^{-1}, the inverse of the damping matrix D, is also a positive diagonal matrix.

Step 2: Determine the relationship between the stability boundaries of the systems

$$\dot{\delta} = \omega \tag{44}$$
$$M\dot{\omega} = -D\omega - \frac{\partial V_p(\delta)}{\partial \delta}$$

and

$$\dot{\delta} = \omega \tag{45}$$
$$\epsilon I \dot{\omega} = -D\omega - \frac{\partial V_p(\delta)}{\partial \delta}$$

where ϵ is a small positive number and I denotes the Identity matrix.

Step 3: Determine the relationship between the stability boundaries of the systems

$$\dot{\delta} = -D^{-1} \frac{\partial V_p(\delta)}{\partial \delta} \tag{46}$$

and

$$\dot{\delta} = \omega \tag{47}$$
$$\epsilon I \dot{\omega} = -D\omega - \frac{\partial V_p(\delta)}{\partial \delta}$$

Thus, the relationship between the stability boundaries of (38) and (41) is established by connecting the relationship between the stability boundaries of (42) and (43), (44) and (45), and (46) and (47) (See Fig. 26).

The reasoning behind the three steps determining the relationship between the stability boundary of the original system and the PEBS is that we want to compare and study the relationship between the stability boundaries of each pair of systems. The original system can be rewritten as

$$M\ddot{\delta} + D\dot{\delta} + \frac{\partial V_p}{\partial \delta} = 0 \tag{48}$$

and the associated gradient system can be rewritten as

$$\dot{\delta} + \frac{\partial V_p(\delta)}{\partial \delta} = 0 \qquad (49)$$

If we can delete the term $(M\ddot{\delta})$ in (48), we can simply compare

$$D^{-1}\dot{\delta} + \frac{\partial V_p(\delta)}{\partial \delta} = 0$$

and

$$\dot{\delta} + \frac{\partial V_p(\delta)}{\partial \delta} = 0$$

To delete the term $(M\ddot{\delta})$, we consider an intermediate step where the coefficient matrix of that term is small, i.e. $\epsilon I\ddot{\delta} + D\dot{\delta} + \frac{\partial V_p(\delta)}{\partial \delta} = 0$.

For ease of exposition, we use the notation $d(M,D)$ to denote the original system (41) and the notation $d(D)$ to denote system (45). With these notations, recall that we are interested in the relationship between the stability boundaries of $d(M,D)$ and $d(I)$. We first study the relationship between the equilibrium points of the system $d(I)$ and that of the original system $d(M, D)$. Theorem 4-4 states that (δ) is a stable equilibrium point of $d(I)$ if and only if $(\delta,0)$ is a stable equilibrium point of $d(M,D)$ (hence, it makes sense to find the relationship between the stability boundaries of them). Furthermore, (δ) is a type-k equilibrium point of $d(I)$ if and only if $(\delta,0)$ is a type-k equilibrium point of $d(M,D)$.

We next study the relationship between the stability boundary of the original system $d(M,D)$ and the PEBS. The relationship is established in Theorem 6-1 below. Let $M_\lambda = \lambda M + (1 - \lambda) \epsilon I$, where ϵ is a small positive number, and $\lambda \in [0,1]$. Let $\bar{D}_\lambda = \lambda\bar{D} + (1 - \lambda)I$, where $\bar{D}$ is the inverse matrix of D, and $\lambda \in [0,1]$.

Theorem 6-1 [64]: (Relationship between the stability boundary of the original system and the PEBS)

There exists a $\hat{\epsilon} > 0$ such that if the dynamical system $d(M_\lambda, D)$ for $\epsilon < \hat{\epsilon}$ and the dynamical system $d(\bar{D}_\lambda)$ satisfy the following two assumptions

[C1] the intersections of the stable and unstable manifolds of the equilibrium points on the stability boundary satisfy the transversality condition

[C2] the number of equilibrium points on the stability boundary is finite

then

[i] the equilibrium point (δ_i) is on the PEBS if and only if equilibrium point $(\delta_i, 0)$ is on $\partial A(M, D)$

[ii]

$$\partial A(M, D) = \cup_{(\delta_i,0)\, \in\, \partial A(M,D)}\, W^s_{d(M,D)}(\delta_i,\, 0) \qquad (50)$$

$$PEBS = \cup\, W^s_{d(I)}(\delta_i) \qquad (51)$$

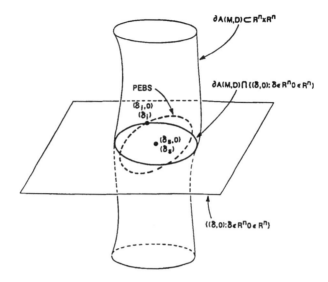

Figure 27: Illustration of the relationship between the PEBS and the stability boundary of the original system d(M,D)

Theorem 6-1 asserts that if the one-parameter transversality condition is satisfied and the number of equilibrium points on the stability boundary is finite, then the equilibrium points on the PEBS correspond to the equilibrium points on the stability boundary of the original system d(M,D). Moreover, the PEBS is the union of the stable manifolds of the equilibrium points, say (δ_i), $i = 1,2,...,n$, of the reduced system d(I) and the stability boundary of the original system d(M,D) is the union of the stable manifolds of the equilibrium points $(\delta_i, 0)$, $i = 1,2,...,n$. Fig. 27 illustrates the relationship between the stability boundary of the original system d(M,D) and the PEBS.

We next discuss under what conditions the PEBS method gives a good approximation of the relevant stability boundary of the original system d(M,D). Recall that the point at which the fault-on trajectory leaves the stability boundary (of the post-fault system) is called the *exit point* and what we mean by the relevant stability boundary is the stable manifold on which the exit point of the fault-on trajectory lies.

Given a fault-on trajectory $x_f(t) = (\delta(t), \omega(t))$, let $(\bar{\delta}_t, \bar{\omega}_t)$ be the corresponding exit point on the stability boundary of the original system d(M,D), and let $(\delta_{\hat{t}})$ denote the exit point relative to the projected trajectory $\delta(t)$ on the PEBS. Let $\partial S(V(x))$ denote the connected constant energy surface of $V(\cdot)$ containing the stable equilibrium point $(\delta_s, 0)$ passing through the point x; i.e. $\partial S(V(x)) :=$ the connected component of the set $\{(\delta, \omega) : V(\delta, \omega) = V(x)\}$ containing the stable equilibrium point $(\delta_s, 0)$, provided $V(x) > V(\delta, 0)$. The PEBS method uses the set $\partial S(V(\delta_{\hat{t}}, 0))$

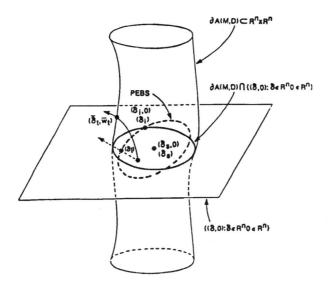

Figure 28: Illustration of the relationship between the exit point of the projected trajectory on the PEBS and the exit point of the original trajectory

to approximate the relevant stability boundary of the original system d(M,D).

To see how well the set $\partial V(\delta_{\hat{i}}, 0)$ approximates the relevant stability boundary of the original system d(M,D), we proceed as follows. Recall that the exit point $(\bar{\delta}_t, \bar{\omega}_t)$ must lie on the stable manifold of the controlling u.e.p., say $(\hat{\delta}, 0)$, relative to the fault-on trajectory $x_f(t)$. We propose to employ the controlling u.e.p. method as the framework to assess the PEBS method. One important issue then is how to relate the point $(\delta_{\hat{i}}, 0)$ to the controlling u.e.p. $(\hat{\delta}, 0)$. Note that the point $(\delta_{\hat{i}}, 0)$ cannot be an equilibrium point. Two cases can happen: (1) $V(\delta_{\hat{i}}, 0) < V(\hat{\delta}, 0)$ and (2) $V(\delta_{\hat{i}}, 0) > V(\hat{\delta}, 0)$. In the former case, the PEBS method gives relatively conservative results (more conservative than the controlling u.e.p. method). In the latter case, two subcases can happen. In the first case, the fault-on trajectory $x_f(t)$ passes through the constant energy surface $\partial S(V(\delta_{\hat{i}}, 0))$ before it passes through the stable manifold $W^s(\hat{\delta}, 0)$, the PEBS method gives a good approximation of the relevant stability boundary for this fault-on trajectory $x_f(t)$ (See Fig. 29). In this situation, the PEBS method provides less conservative results than the controlling u.e.p. method. In the second case, the fault-on trajectory $x_f(t)$ passes through the constant energy surface $\partial S(V(\delta_{\hat{i}}, 0))$ after it passes through the stable manifold $W^s(\hat{\delta}, 0)$. In this situation, the PEBS method gives incorrect stability assessment because the fault-on trajectory $x_f(t)$ after passing through the stable manifold $W^s(\hat{\delta}, 0)$ becomes unstable relative to the post-fault system, but is still classified to be stable by the PEBS method (See Fig. 30).

We are now in a position to derive a set of sufficient conditions under which the

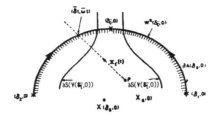

Figure 29: The fault-on trajectory $x_f(t)$ passes through the constant energy surface $\partial S(V(\delta_i, 0))$ before it passes through the stable manifold $W^s(\hat{\delta}, 0)$. In this case, the PEBS method gives a good approximation of the relevant stability boundary for this fault-on trajectory $x_f(t)$.

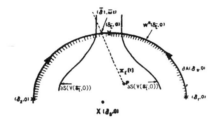

Figure 30: The fault-on trajectory $x_f(t)$ passes through the constant energy surface $\partial S(V(\delta_i, 0))$ after it passes through the stable manifold $W^s(\hat{\delta}, 0)$. In this case, the PEBS method gives an inaccurate stability assessment.

PEBS method gives accurate stability assessment. Based on the above arguments, it follows that if the following two conditions are satisfied

(i) (δ_i) on the stable manifold of ($\hat{\delta}$) (of the reduced system d(I)) corresponds to $(\bar{\delta}_t, \bar{\omega}_t)$ on the stable manifold of ($\hat{\delta}$, 0) (of the original system d(M,D))

(ii) the fault-on trajectory $x_f(t)$ passes through the constant energy surface $\partial V(\delta_i, 0)$ before it passes through the stable manifold $W^s(\hat{\delta}, 0)$

then the PEBS method gives a good approximation (better than the controlling u.e.p. method) of the relevant stability boundary of the original system d(M,D) relative to the given fault-on trajectory $(\delta(t), \omega(t))$.

Condition (i) in fact contains the following two subconditions : (ia) the equilibrium point $(\hat{\delta})$ is on the PEBS if and only if the equilibrium point $(\hat{\delta}, 0)$ is on the stability boundary of the original system d(M,D) and (ib) the exit point (δ_i) is on the stable manifold $W^s_{d(I)}(\hat{\delta})$ if and only if $(\bar{\delta}_t, \bar{\omega}_t)$ is on the stable manifold $W^s_{d(M,D)}(\hat{\delta}, 0)$. Theorem 6-1 has provided a basis for the subcondition (ia) to be true.

It is difficult to check whether or not the condition (ii) above is satisfied mainly because finding the exit point $(\bar{\delta}_t, \bar{\omega}_t)$ for the fault-on trajectory $(\delta(t), \omega(t))$ is hard. However, it can be enforced that condition (ii) is always satisfied if the PEBS method is modified in the following way :

Step 1: From the fault-on trajectory $(\delta(t), \omega(t))$, detect the point δ^* at which the projected trajectory $\delta(t)$ crosses the PEBS.

Step 2: Find the equilibrium point of system d(I) whose stable manifold contains the point δ^*, say $\hat{\delta}$. Let the value of $V_p(\cdot)$ at $\hat{\delta}$ be v, i.e. v $= V_p(\hat{\delta})$.

Step 3: Use the connected constant energy surface of the set $\{(\delta, \omega) : V(\delta, \omega) = v \}$ containing the stable equilibrium point $(\delta_s, 0)$ as a local approximation of the stability boundary $\partial A(\delta_s, 0)$.

The above procedure forms the backbone of the BCU method for direct analysis of transient stability, which is detailed in the next section.

VII. A BCU Method

We will discuss the boundary of stability region controlling unstable equilibrium point method (BCU method) for direct analysis of power system transient stability [14], [15]. This method is applicable to the classical power system model with transfer conductances. This method is based on the relationship between the boundary of stability region of a power system and that of a reduced system. One feature that distinguishes the BCU method from the existing direct methods is that it consistently finds the exact controlling u.e.p. relative to a fault-on trajectory and it has a sound theoretical basis. This method has been tested on several power systems with very promising results.

A. The Classical Model With Transfer Conductance

Consider a power system consisting of n generators. Let the loads be modeled as constant impedances. The dynamics of the ith generator can be represented by the following equations

$$\dot{\delta}_i = \omega_i \qquad i = 1, 2, ..., n \qquad (52)$$
$$M_i \dot{\omega}_i = P_i - D_i \omega_i - P_{ei}$$

where node n+1 serves as the reference node, i.e. $E_{n+1} = 1$ and $\delta_{n+1} = 0$. E_i is the constant voltage behind direct axis transient reactance. $P_{ei} = \sum_{j \neq i}^{n+1} E_i E_j B_{ij} sin(\delta_i - \delta_j) + \sum_{j \neq i}^{n+1} E_i E_j G_{ij} cos(\delta_i - \delta_j)$. M_i is the generator's moment of inertia. The damping constant D_i is assumed to be positive. G_{ij} represents the transfer conductance of the i-j element in the reduced admittance matrix of the system. $P_i = P_{mi} - E_i^2 G_{ii}$, where P_{mi} is the mechanic power. We assume uniform damping $\frac{D_i}{M_i} = \lambda$, i $= 1,2,...,n$.

By using one machine as reference, Eq. (53) can be transformed into the following:

$$\dot{\delta}_{in} = \omega_{in} \qquad i = 1, 2, ..., n - 1$$

$$\dot{\omega}_{in} = \frac{1}{M_i}(P_i - P_{ei}) - \frac{1}{M_n}(P_n - P_{en}) - \lambda\omega_{in} \qquad (53)$$

$\delta_{in} = \delta_i - \delta_n, \omega_{in} = \omega_i - \omega_n$. One (numerical) energy function for system (53) is of the following form [13] with $M_T = \sum_{i=1}^{n} M_i$:

$$V(\delta,\omega) = \sum_{i=1}^{n-1} \sum_{j=i+1}^{n} (\frac{1}{2M_T}M_i M_j(\omega_{in} - \omega_{jn})^2 \qquad (54)$$

$$- \frac{1}{M_T}(P_i M_j - P_j M_i)(\delta_{in} - \delta_{jn} - \delta_{in}^s + \delta_{jn}^s)$$

$$- V_i V_j B_{ij}(cos(\delta_{in} - \delta_{jn}) - cos(\delta_{in}^s - \delta_{jn}^s))$$

$$- V_i V_j G_{ij} \frac{\delta_{in} + \delta_{jn} - (\delta_{in}^s + \delta_{jn}^s)}{\delta_{in} - \delta_{jn} - (\delta_{in}^s - \delta_{jn}^s)}$$

$$(sin(\delta_{in} - \delta_{jn}) - sin(\delta_{in}^s - \delta_{jn}^s))$$

$$= V_p(\delta) + \frac{1}{2M_T} \sum_{i=1}^{n-1} \sum_{j=1}^{n} M_i M_j(\omega_{in} - \omega_{jn})^2$$

It should be pointed out that theoretically the controlling u.e.p. method should give stability estimates on the conservative side if an exact energy function exists. Recall that an 'exact' energy function is a well-defined function and its function value is strictly decreasing along any trajectory of the system. It has, however, been shown that there does not exist a general form of exact energy function for power systems with transfer conductances [30] (see also [55], [56], [57] for relevant issues). The existing numerical energy functions are derived on the assumption that the fault-on trajectory is close to a straight line. The assumption is accepted in the area of direct methods. Nonetheless, it should be cautioned that the validity of this assumption may affect the property of the controlling u.e.p. method: it always gives a stability estimate on the conservative side.

B. The BCU Method

The BCU method is based on the relationship between the stability boundary of the (post-fault) classical power system model (53) and the stability boundary of the following (post-fault) reduced system:

$$\dot{\delta}_{in} = (P_i - P_{ei}) - \frac{M_i}{M_n}(P_n - P_{en}) \qquad (55)$$

$$:= f_i(\delta) \qquad i = 1, 2, ..., n - 1$$

The reduced system (55) is a function of machine angles only with dimension of n-1 while the dimension of the original system (53) is of $2(n-1)$. It can be seen that $(\hat{\delta})$ is an equilibrium point of the reduced system (55) if and only if $(\hat{\delta},0)$ is an equilibrium point of the original system. It can be shown that, under the condition of small transfer conductances, the following results hold:

(R1) $(\hat{\delta}_s)$ is a stable equilibrium point of the reduced system (55) if and only if $(\hat{\delta}_s,0)$ is a stable equilibrium point of the original system (53).

(R2) $(\hat{\delta}_x)$ is a type-k equilibrium point of the reduced system (55) if and only if $(\hat{\delta}_s, 0)$ is a type-k equilibrium point of the original system (53).

(R3) If the one-parameter transversality condition is satisfied, then $(\bar{\delta})$ is on the stability boundary $\partial A(\hat{\delta}_s)$ of the reduced system (55) if and only $(\bar{\delta}, 0)$ is on the stability boundary $\partial A(\hat{\delta}_s, 0)$ of the original system (53).

Result (R1) indicates that it makes sense to compare the stability region $A(\hat{\delta}_s)$ of the reduced system (55) with the stability region $A(\hat{\delta}_s, 0)$ of the original system (53). Result (R3) establishes a relationship between the stability boundary $\partial A(\hat{\delta}_s)$ and the stability boundary $\partial A(\hat{\delta}_s, 0)$ and suggests the plausibility of finding the controlling u.e.p. of the original system (53) via finding the controlling u.e.p. of the reduced system (55).

The BCU method : find the controlling u.e.p. relative to a fault-on trajectory (version 1)

Step 1: From the fault-on trajectory $(\delta(t), \omega(t))$, detect the exit point δ^* at which the projected trajectory $\delta(t)$ exits the stability boundary of the reduced system (55). From the fault-on trajectory $(\delta(t), \omega(t))$, detect the exit point δ^* at which the projected trajectory $\delta(t)$ reaches the first local maximum of $V_p(\cdot)$.

Step 2: Use the point δ^* as initial condition and integrate the post-fault reduced system (55) to find the first local minimum of $\sum_{i=1}^{n} \| f_i(\delta) \|$, say at δ_o^*.

Step 3: Use the point δ_o^* as the initial guess to solve $\sum_{i=1}^{n} \| f_i(\delta) \| = 0$, say at δ_{co}^*.

Step 4: The controlling u.e.p. with respect to the fault-on trajectory is $(\delta_{co}^*, 0)$.

Once the controlling u.e.p. is found, the BCU method uses the same procedure as that of the conceptual controlling u.e.p. method presented in section 4 (i.e. step 2 - step 4) to perform stability assessment. The essence of the BCU method is that it finds the controlling u.e.p. via the controlling u.e.p. of the reduced system (55) which is only defined in the angle space and whose controlling u.e.p. is easier to compute. Steps 1 - 3 find the controlling u.e.p. of the reduced system and step 4 relates the controlling u.e.p. of the reduced system to the controlling u.e.p. of the original system.

Remarks:

[1] An effective computation scheme to implement Step 1 is : From the fault-on trajectory $(\delta(t), \omega(t))$, detect the exit point δ^* at which the projected trajectory $\delta(t)$ reaches the first local maximum of $V_p(\cdot)$.

[2] The reduced system (55) can be stiff. In such a case, a stiff differential equation solver is recommended to implement Step 2.

[3] Several existing direct methods can be viewed as finding the controlling u.e.p. in the angle space. For example, Fouad et al. finded the controlling u.e.p. via the 'mode of disturbance' of machines. Pavella et al. located the controlling

u.e.p. via the 'accelerating machines'. The BCU method finds the controlling u.e.p. of the reduced system (in the angle space) via two steps. The first step locates the exit point of the reduced system. Recall that the exit point must lie on the stable manifold of the controlling u.e.p. Hence, the second step integrates the reduced system starting from the exit point and the resulting trajectory will converge to the controlling u.e.p. In practical computation, the point calculated in Step 1 can not be exactly the exit point but lies in a neighborhood of the exit point. Consequently, the resulting trajectory will just pass by the controlling u.e.p.

C. Computational Considerations

There are basically three major computational tasks in the BCU method: (i) compute the point with the first local maximum of potential energy along the projected fault-on trajectory, (ii) compute the trajectory of the post-fault reduced system, and (iii) compute the controlling u.e.p. via solving the nonlinear algebraic equations related to the reduced system. It is very difficult to compute the exact point where the first local maximum of potential energy along the projected fault-on trajectory occurs because the simulation of the trajectory on a digital computer yields only a sequence of points and not the entire trajectory. Hence, step 1 of the BCU method usually gives a point which is close to the exit point. In order to achieve a speed-up in computation without sacrificing accuracy, the following method is recommended when a large integration step size is used in integrating the fault-on system:

The BCU method (version 2)

Step 1: From the fault-on trajectory $(\delta(t), \omega(t))$, detect the point δ^* at which the projected trajectory $\delta(t)$ reaches the first local maximum of $V_p(\cdot)$. Also, compute the point δ^+ that is one step after δ^*.

Step 2: Use the point δ^* as the initial condition and integrate the post-fault reduced system to find the first local minimum of $\sum_{i=1}^{n} \| f_i(\delta) \|$, say at δ_o^*.

Step 3: Use δ^+ as an initial condition and repeat Step 2 to find the corresponding points, say δ_o^+.

Step 4: Compare the values of $\|f(\delta)\|$ at δ_o^* and δ_o^+. The one with the smallest value is used as the initial guess to solve $f_i(\delta) = 0$, say the solution is δ_{co}.

Step 5: The controlling u.e.p. with respect to the fault-on trajectory is $(\delta_{co}, 0)$.

Using a large integration step size in step 1 may cause that the first local minimum in Step 2 cannot be found in the simulation. Should this happen, go back to Step 1 and take a smaller step size to detect the new points δ^* and δ^+ and repeat Step 2 to Step 5.

D. Test Results

The BCU method has been tested on several power systems. In this section we present test results on a 50-generator, 145-bus system. We have applied the proposed

Fault location at bus	Line clearing between buses	Critical Clearing Time		
		Step-by-Step simulation Unit: sec	The BCU Method Unit: sec	The BCU Method + corrected energy Unit: sec.
6	6 - 12	0.200	0.170	0.185
7	6 - 7	0.119	0.115	0.122
12	12 - 14	0.190	0.170	0.190
58	58 - 87	0.51	0.59	-
63	61 - 63	0.91	0.855	0.91
90	90 - 92	0.270	0.265	0.270
96	96 - 73	0.245	0.242	0.245
97	97 - 66	0.260	0.260	0.260
98	98 - 72	0.205	0.205	0.205
100	100 - 72	0.315	0.315	0.320
102	63 - 102	0.200	0.195	0.195
116	116 - 63	0.28	0.28	0.28

Table 1: The Critical Clearing Time estimated by the BCU method.

method to several three-phase faults with fault locations at both generator and load buses. The numerical energy function (55) is used in the simulations. Both severe and mild faults have been considered.

Table 1 displays the estimated critical clearing time of several faulted systems using three different methods: step-by-step numerical integration technique, the BCU method (version 2), and the BCU method (version 2) with corrected kinetic energy [51]. The results from the numerical integration technique are used as a benchmark. A few observations and comments on the simulation results follow.

The BCU method consistently finds the exact controlling u.e.p. In all but one case, the BCU method gives slightly conservative results in CCT (under-estimate). This is consistent with analytical results derived in Sec. V. The only case that the BCU method gives an over-estimate is when a fault occurs at bus 58. In this case, the CCT from benchmark is 0.51 sec. while the estimated CCT by the BCU method is 0.59 sec. After further studies, we found that the BCU method still finds the correct controlling u.e.p. However, a whole group of machines in this case tends to become unstable such that the assumption in deriving the energy function that the fault-on trajectory is a straight line does not hold. Thus, it is the accuracy of the energy function (i.e. the conditions required for energy functions are not satisfied) for this case that causes the over-estimate. It has been our experience that those cases causing a whole group of machines to become unstable usually damage the accuracy of energy functions.

The BCU method in conjunction with the corrected kinetic energy method gives on the average very good results except when they together give over-estimate results such as the faults occurring at bus #7 and bus #100. In all other cases, they reduce the conservativeness of the BCU method. This indicates that the corrected kinetic energy method is quite effective and deserves further (theoretical) investigation.

The BCU method is in fact an improvement of the PEBS method and appears to be fast and reliable; it consistently gives accurate and slightly conservative results of stability analysis. The essence of the BCU method is that it finds the controlling u.e.p. via the controlling u.e.p. of the reduced system which is defined in the angle space and whose controlling u.e.p. is easier to compute. It is believed that this essence can be extended to develop solution algorithms for transient stability analysis of detailed power system models. Future research along this line may prove fruitful.

VIII. Concluding Remarks

This paper has presented an overview of recent advances in the development of analytical results for direct methods. In addition to providing insight into the underlying concepts and properties of various direct methods, new material has been developed. The paper has demonstrated, through an exposition of the BCU method, that analytical results can sometimes lead to the development of reliable yet fast solution algorithms for solving problems. This further enhances the author's belief that solving practical problems efficiently can be accomplished through a thorough understanding of the underlying theory, in conjunction with exploring the features of the practical problem under study.

Several problems, in addition to those open problems posed in previous sections, require further investigations in the context of analytical results for direct methods. Among others, the development is desired of both analytical results and solution algorithms for stability regions of nonlinear systems described by ordinary differential equations coupled with algebraic equations, such as structure-preserving power system models. It is known that trajectories of such nonlinear systems may be non-existent and could be non-unique. The issues here include how to define, how to characterize and how to estimate stability regions for such systems. It has been suggested to estimate stability regions of such systems via estimating stability regions of singular-perturbed systems. This approach, though theoretically sound, may cause serious computation problems since the equations describing singular-perturbed systems are stiff. For several recent studies on such systems, see for example [58], [44], and [74].

Load characteristics are known to have a significant effect on power system dynamics. Inaccurate load modelling could lead a power system being operated in modes to result in actual collapse or separation [73]. Load models that can accurately capture load behaviors during dynamics are therefore necessary to allow more precise calculations of power system controls and stability limits. Most of the load models used in direct methods are limited to the so-called static load models, where loads are represented as constant impedance, constant current, constant MVA, or some combinations of these three. These static load models may not adequately capture load behaviors during transient. Hence, it may prove rewarding to show that direct methods can be extended to structure-preserving models with sufficiently detailed load models (more accurate static load models or even adequate dynamic load models). This problem seems challenging since the applicability of direct methods depends on the existence of energy functions for the underlying power system models.

In recent years, practical experience has shown that a power system may become unstable ten to thirty seconds after a disturbance, even if it is stable in transient states. The power system is short-term stable and mid-term unstable. One mechanism that may contribute to this phenomenon is that the initial state of the post-fault system lies inside the stability region of the short-term power system model but lies outside the stability region of the mid-term power system model. This points out the necessity of extending direct methods for mid-term power system models. In this regard, it may prove useful to develop general functions (more general than the energy function) such that for any nonlinear system having such a function (1) the ω-limit set of any of its bounded trajectories consists only of equilibrium points and limit cycles and (2) the function is non-increasing along its trajectories. Such a development will allow one to apply direct methods to detailed power system models whose stability boundaries contain equilibrium points and limit cycles. An example of such model can be found in [75]. Research efforts on extending direct methods to solve the 'voltage dip' problem [67], which is an emerging concern in power industry, may also prove fruitful.

References

[1] D.N. Ewart, 'Whys and wherefores of power system blackouts', *IEEE Spectrum*, Vol. 15, 1978.

[2] 'Proposed Terms and Definitions for Power System Stability', Task Force on Terms and Definition, System Dynamic Performance Subcommittee, Power System Engineering Committee, *IEEE Transaction on Power Apparatus and System*, Vol.PAS-101, July, 1982, pp. 1894-1898.

[3] G.E. Gless, 'Direct method of Lyapunov applied to transient power system stability,' *IEEE Trans. on Power Apparatus and Systems*, Vol.PAS-85, Feb., 1966, pp.159-168.

[4] A.H. El-Abiad and K. Nagappan, 'Transient stability regions for multi-machine power systems,' *IEEE Trans. on Power Apparatus and Systems*, Vol. PAS-85, Feb., 1966, pp.169-179.

[5] R.J. Kaye and F.F. Wu, 'Dynamic security regions for power systems,' *IEEE Trans. on Circuits and Systems* Vol. CAS-29, September, 1982, pp.612-623.

[6] M. Ribbens-Pavella and F.J. Evans,'Direct methods for studying dynamics of large-scale electric power systems - A Survey,' Automatica Vol-32, January, 1985, pp.1-21.

[7] P.P. Varaiya, F.F. Wu and R-L Chen, 'Direct methods for transient stability analysis of power systems : Recent results,' *Proceedings of the IEEE*, December 1985, pp. 1703-1715.

[8] F. Mercede, R. Fischl, F.F. Wu and H.D. Chiang, "A Comparison of Dynamical Security Indices Based on Direct Methods", *International Journal of Electrical Power and Energy Systems*, Vol. 10, No. 4, Oct. 1988, pp. 210-232.

[9] A.A. Fouad and V. Vittal, 'The transient energy function method', *International Journal of Electrical Power and Energy Systems*, Vol. 10, No. 4, Oct. 1988, pp. 233-246.

[10] M.A. Pai, *Energy Function Analysis for Power System Stability.* Kluwer Academic Publishers, 1989.

[11] N. Kakimoto, Y. Ohsawa and M. hayashi, ' Transient Stability Analysis Of Electric Power System via Lure-Type Lyapunov Function, Part I and II', *Trans. IEE of Japan*, Vol-98, pp.516, 1978.

[12] B. Toumi, R. Phifaoui, Th-Van cutsem and M. Ribbens-Pavella, 'Fast Transient Stability Assessment Revisited', *IEEE Trans. on Power Systems* Vol. 1, 1986, pp.211-220.

[13] T. Athay, R. Podmore and S. Virmani, 'A practical method for direct analysis of transient stability', *IEEE Transaction on Power Apparatus and System*, Vol. PAS-98, 1979, pp.573-584.

[14] H.D. Chiang, 'A theory-based controlling u.e.p. method for direct analysis of transient stability', IEEE 1989 International Symposium of Circuits and Systems, May 1989.

[15] H.D. Chiang, F.F. Wu and P.P Varaiya, 'A BCU Method for Direct Analysis of Power System Transient Stability', submitted for possible publication (in revised version).

[16] A. Rahimi, 'Evaluation of transient energy function method software for dynamic security analysis', EPRI Project RP 4000-18, Final Report, Dec. 1990.

[17] H. Miyagi and A.R. Bergen, 'Stability studies of multimachine power systems with the effect of automatic voltage regulators', *IEEE Trans. on Automatic Control* Vol. AC-31, March 1986, pp.210-215.

[18] M.A. Pai, *Power System Stability*, North Holland Publishing Co., New York, 1981.

[19] H.D. Chiang, F.F. Wu and P. P. Varaiya, 'Foundation of Direct Methods for Power System Transient Stability Analysis', *IEEE Trans. on Circuits and Systems* Vol. CAS-34, Feb. 1987, pp. 712- 728.

[20] A. R. Bergen and D. J. Hill, ' A structure preserving model for power system stability analysis,' *IEEE Trans on Power Apparatus and systems*, Vol. PAS-100, January, 1981, pp.25-35.

[21] N. Tsolas, A. Arapostathis and P.P. Varaiya, 'A structure preserving energy function for power system transient stability analysis,' *IEEE Trans. on Circuits and Systems* Vol. CAS-32, Oct. 1985, pp. 1041-1049.

[22] N. Narasimhamurthi and M. R. Musavi, ' A general energy function for transient stability analysis of power systems,' *IEEE Trans. on Circuits and Systems* Vol. CAS-31 1984, pp.637-645.

[23] K.R. Padiyar and K.K. Ghosh, 'Direct stability evaluation of power systems with detailed generator models using structure-preserving energy functions' *International Journal of Electrical Power and Systems*, Vol. 11, No. 1 , Jan. 1989, pp. 47-56.

[24] M. Vidyasagar, *Nonlinear Systems Analysis.* Prentice-Hall Inc., N.J., l978.

[25] R.K. Miller and A.N. Michel, *Ordinary Differential Equations*, Academic Press, 1982.

[26] J. Guckenheimer and P. Holmes, *Nonlinear Oscillations, Dynamical Systems, and Bifurcation Of Vector Fields,* Springer-Verlag, l983.

[27] M.W. Hirsch and S. Smale, *Differential Equations, Dynamical Systems and Linear Algebra*, Academic Press, 1974.

[28] P. Hartman, *Ordinary Differential Equations*, John Wiley, New York, 1973.

[29] Jack K. Hale, *Ordinary Differential Equations*, Robert E. Krieger Publishing Company, Huntington, New York, 1980.

[30] H.D. Chiang, 'Study of the existence of energy functions for power system with losses', *IEEE Trans. on Circuits and Systems* Vol. CAS-36, Nov. 1989, pp. 1423-1429.

[31] A. Arapostathis, S.S. Sastry, and P.P. Varaiya, 'Global analysis of swing dynamics,' *IEEE Trans. on Circuits and Systems* Vol. CAS-29, Oct. 1982, pp. 673-679.

[32] Y.N. Yu and K. Vongasuriya, 'Nonlinear power system stability study by Lyapunov function and Zubov's method', *IEEE Trans on Power Apparatus and systems,* Vol. PAS-86, pp.1480-1487.

[33] F.S. Prabhakara, A.H. El-Abiad and A.J. Kovio, 'Application of generalized Zubov's method to power system stability',*International Journal of Control,* Vol. 20, 1974, pp.203.

[34] F.F. Wu and Y-K. Tsai, 'Probabilistic dynamic security assessment of power systems, Part I: Basic model,' *IEEE Trans. on Circuit and Systems*, Vol. CAS-30, March ,1983, pp.148-159.

[35] H. Yee and B.D. Spalding, 'Transient stability analysis of multimachine systems by the method of hyperplanes',*IEEE Trans on Power Apparatus and systems,* Vol. PAS-96, 1977, pp. 276-284.

[36] H.D. Chiang and J.S. Thorp, 'The Closest Unstable Equilibrium Point Method for Power System Dynamic Security Assessment', *IEEE Trans. on Circuits and Systems* Vol. CAS-36, Sept. 1989, pp. 1187-1120.

[37] H.D. Chiang, M. Hirsch and F.F. Wu, 'Stability Regions of Nonlinear Autonomous Dynamical Systems', *IEEE Trans. on Automatic Control*, Vol. AC-33, pp. 14-27, Jan. 1988.

[38] S. Smale, 'Differentiable dynamical systems', *Bull. Amer. Math. Soc.*, Vol. 73, pp.747, 1967.

[39] M. Pexioto, 'On an approximation theorem of Kupka and Smale', *Journal of Differential Equations*, Vol. 3, 1967, pp. 214-227.

[40] N. Kopell and R.B. Washburn, Jr., 'Chaotic motion in two-degree-freedom swing equations,' *IEEE Trans. on Circuit and Systems, Vol. CAS-29, Sept., 1982, pp.612-623.*

[41] E.H. Abed and P.P. Varaiya, 'Nonlinear oscillations in power systems', *International Journal of Electric Power Energy and Systems*, Vol. 6, no. 1, pp.37-43.

[42] V.I. Opoitsev, 'Unbounded regions of asymptotic stability,' *Automatic and Remote Control*, 1981, pp.275-282.

[43] S.M. Shahidehpour and J. Qiu, 'Effect of random perturbations on dynamic behavior of power systems',*Electric Power System Reserch* Vol. 11, 1986, pp.117-127.

[44] C.L. De Marco and A.R. Bergen, 'Application of singular perturbation techniques to power system transient stability analysis,' Memo M84-7, Electronics Research Laboratory, University of California, Berkeley, CA., 1984.

[45] C.L. De Marco and A.R. Bergen, 'A security measure for random load disturbances in nonlinear power system models,' *IEEE Trans. on Circuits and Systems*, CAS-34, Dec. 1989, pp. 1229-1241.

[46] H.D. Chiang and J.S. Thorp, 'Stability regions of nonlinear dynamical systems: A constructive methodology' *IEEE Transactions on Automatic Control* Vol. 34, No. 12, Dec. 1989, 1229-1241.

[47] R. Thom, *Structural Stability and Morphogenests*, Reading, M.A: Benjamin, 1975.

[48] F.S. Prabhakara and A.H. El-Abiad, 'A simplified determination of stability regions for Lyapunov method', *IEEE Transaction on Power Apparatus and System*, Vol. PAS-94, 1975, pp. 672-689.

[49] M. Ribbens-Pavella and B. Lemal, 'Fast determination of stability regions for on-line transient power system studies', *Proceeding of IEE*, Vol. 123, 1976, pp.689-695.

[50] M. Ribbens-Pavella, P.G. Murthy, and J.L. Horward, 'The acceleration approach to practical transient stability domain estimation in power systems,' Proc. of the 20th IEEE Conference on Decision and Control, San Diego, CA, Dec. 16-18, 1981, pp. 471-477.

[51] A.A Fouad and S.E. Stanton, 'Transient stability of a multimachine power systems. Part I :Investigation of system trajectories' *IEEE Transaction on Power Apparatus and System*, Vol. PAS-100, 1981, pp. 3408-3414.

[52] ESCA Corp. 'Contribution to power system state estimation and transient stability analysis,' Final Report to U.S. Dept. of Energy, DOE/ET/29362-1, February, 1984.

[53] M. Ribbens-Pavella, Lj.T. Grujic, J. Sabatel and A. Bouffioux, 'Direct methods for stability analysis of large scale power systems,' Proc. of IFAC Symposium on Computer Applications in Large Scale Power Systems, New Delhi, India, Aug. 16-18, 1979, Pergamon Press.

[54] N. Kakimoto, Y. Ohsawa and M. Hayashi, 'Transient stability analysis of large-scale power systems by Lyapunov's direct method,' *IEEE Transactions on Power Apparatus and Systems*, Vol. PAS-103, January 1984, 160-167.

[55] N. Narasimhamurthi, 'On the existence of energy function for power systems with losses', *IEEE Trans. on Circuit and Systems* Vol. CAS-31, 1984, pp.637-645.

[56] H.G. Kwatny, L.Y. Bahar and A.K. Pasrija, 'Energy-like Lyapunov function for power systems stability analysis', *IEEE Trans. on Circuit and Systems* Vol. CAS-32, Nov. 1985, pp.1140-1148.

[57] U.D. Caprio, 'Accounting for transfer conductance effects in Lyapunov transient stability analysis of a multimachine power system', *Int. J. Electric Power and Energy Systems*, Vol. 8, Jan. 1986, pp. 27-41.

[58] S. Sastry and P.P. Varaiya, 'Hierarchical stability and alert steering control of interconnected power systems', *IEEE Trans. on Circuit and Systems* Vol. CAS-27, Nov. 1980, pp.1102-1112.

[59] A.N. Michel, A.A. Fouad and V. Vittal, 'Power system transient stability using individual machine energy functions,' *IEEE Trans. on Circuit and Systems*, Vol. CAS-30, May, 1983, pp.266-276.

[60] V. Vittal and A.N. Michel, 'Stability and security assessment of a class of systems governed by Lagrange's equation with application to multi-machine power systems', *IEEE Trans. on Circuit and Systems*, Vol. CAS-33, June, 1986, pp.623-636.

[61] R.J. Thomas and J.S. Thorp, 'Toward a direct test for large scale electric power system instabilities,' in IEEE Proc. of 24th Conf. on Decision and Control, Ft. Lauderdale, FL., 1985, pp. 65-69.

[62] F.M.A. Salam, A. Arapostathis and P. Varaiya, 'Analytic expressions for the unstable manifold at equilibrium points in dynamical systems of differential equation,' *Proc. of 22nd Conference on Decision and Control.*

[63] P.A. Cook and A.M. Eskicioglu, 'Transient stability analysis of electric power systems by the method of tangent hypersurfaces',*IEE Proc. C. Gen., Trans. & Distrib.*, Vol. 130, no. 4, 1983, pp. 183-193.

[64] H.D. Chiang, F.F. Wu and P. Varaiya, 'Foundation of PEBS Method for Power System Transient Stability Analysis', *IEEE Trans. on Circuits and Systems* Vol. CAS-35, June 1988, pp. 712- 728.

[65] A. Debs, 'Voltage dip at maximum swing in the context of direct stability analysis' *IEEE Trans. on Power Systems*, Vol. 5, Nov. 1990, pp. 1497-1502.

[66] P.W. Sauer, A.K. Behera, M.A. Pai, J.R. Winkelman and J.H. Chow, 'Trajectory approximations for direct energy methods that use sustained faults with detailed power system models', *IEEE Transaction on Power System*, Vol. 4, No. 2, May 1989, pp. 499-506.

[67] A.S. Debs and A.R. Benson, 'Security assessment of power systems', *System Engineering for Power : Status and prospects*, Proc. Eng. Foundation Conf., Ed. by L.H. Fink and K. Carlsen, Henniker, N.H., Aug. 1975, pp.144-176.

[68] P.W. Sauer, D.J. Lagesse, S. Ahmed-Zaid and M.A. Pai, 'Reduced order modeling of interconnected multimachine power systems using time-scale decomposition', *IEEE Transactions on Power Systems* Vol. PWRS-2, May, 1987, pp.310-320.

[69] S. Ahmed-Zaid, P.W. Sauer, M.A. Pai and M.K. Sarioglu, 'Reduced order modeling of synchronous machines using singular perturbation', *IEEE Trans. on Circuits and Systems*, Vol. CAS-29, Nov, 1982, pp. 782-786.

[70] J.H. Chow, *Time-Scale Modeling of Dynamic Network with Applications to Power Systems*, Vol. 46 in Lecture notes in Control and Information Sciences, Springer-Verlag, Berlin-Heidelburg-New York, 1982.

[71] M.A. Pai, 'Some mathematical aspects of power system stability by Lyapunov's method', presented at SIAM Int. Conf. on Electric Power Problems: The Mathematical Challenge, Seatle WA, March 18-20, 1980.

[72] G. Kwatny and X.-M. Yu, 'Energy analysis of load-induced flutter instability in classical models of electric power networks', *IEEE Trans. on Circuits and Systems*, Vol. CAS-36, Dec., 1989, pp. 1544-1557.

[73] R.H. Craven and M.R. Michael, 'Load characteristic measurements and representation of loads in the dynamic simulation of the Queensland power system', 1983 CIGRE and IFAC symposium, Florence 1983, paper S 39-83.

[74] D.J. Hill and M.Y. Iven, 'Stability theory for differential/algebraic systems with application to power systems' *IEEE Trans. on Circuits and Systems*, Vol. CAS-37, Nov, 1990, pp. 1416-1423.

[75] N. Tsolas, 'Stability and computer simulation of power systems' PH.D. dissertation, Dept. of Elec. Eng. and Comput. Sci., University of California, Berkeley, CA, 1983.

IMPROVED CONTROL AND PROTECTION OF POWER SYSTEMS THROUGH SYNCHRONIZED PHASOR MEASUREMENTS

ARUN G. PHADKE
Virginia Tech
Blacksburg, Virginia, 24061

JAMES S. THORP
Cornell University
Ithaca, New York, 14853

I. INTRODUCTION

A power system is one of the largest dynamic systems in existence. Generators of electric power are interconnected with loads by means of an extensive transmission and distribution network. The size of the power network is usually on the scale of continents. The secure and economic operation of the electric power network is a complex problem, involving a distributed, hierarchical, and multi–centered structure. Very sophisticated monitoring, protection, and control systems have been devised to achieve a system which provides electric power of high quality and reliability. Recent developments in the field of computer based substation functions, and satellite based dissemination of high accuracy time reference signals, has provided power system engineers with a new tool for achieving even better overall system performance more economically. This paper begins with an introduction to phasors and symmetrical components – two of the fundamental concepts used in

CONTROL AND DYNAMIC SYSTEMS, VOL. 43
Copyright © 1991 by Academic Press, Inc.

describing the power system, and then describes the techniques and uses of synchronizing phasor measurements for improved monitoring, protection, and control of a power network. Finally, some recent field trials of synchronized phasor measurements, and directions for future developments are described.

II. PHASORS

The phasor is a complex number, used to represent a sinusoidal function of time. The sinusoidal functions in question are ac voltages and currents, and for convenience in calculating the power in ac circuits from phasors, the phasor magnitude is set equal to the rms value of the sinusoidal waveform. A sinusoidal quantity and its phasor representation are shown in Figure 1, and are defined as follows:

Sinusoidal quantity   *Phasor*

$$x(t) = X_m \cos(\omega t + \varphi) \quad \Longleftrightarrow \quad X \equiv \frac{X_m}{\sqrt{2}} \epsilon^{j\varphi} \tag{1}$$

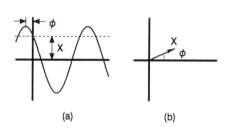

(a)   (b)

Figure 1. Phasor representation.

A phasor can only represent a single frequency sinusoid, and therefore phasor representation is not directly applicable under transient conditions. However, often the idea of a phasor can be used in the presence of more complex signals, by stipulating that the phasor represent the fundamental frequency component of a waveform observed over a finite window. In case of sampled data x_k, obtained from the signal $x(t)$ over a period of the waveform (with N samples taken over one period),

$$X = \frac{1}{\sqrt{2}} \frac{2}{N} \sum_{k=1}^{N} x_k \, \epsilon^{-jk\frac{2\pi}{N}} \tag{2}$$

or,

$$X = \frac{1}{\sqrt{2}} \frac{2}{N} \left\{ \sum_{k=1}^{N} x_k \, \cos k\frac{2\pi}{N} - j \sum_{k=1}^{N} x_k \, \sin k\frac{2\pi}{N} \right\} \tag{3}$$

Using θ for the sampling angle $\frac{2\pi}{N}$, it follows that

$$X = \frac{1}{\sqrt{2}} \frac{2}{N} (X_c - jX_s) \tag{4}$$

where

$$X_c = \sum_{k=1}^{N} x_k \, \cos k\theta \, , \text{ and}$$

$$X_s = \sum_{k=1}^{N} x_k \, \sin k\theta \tag{5}$$

Note that the sampling frequency is $(N\omega)$, and thus the input signal $x(t)$ must be band–limited to $N\omega/2$ to avoid aliasing errors. In the presence of white noise, the fundamental frequency component of the Discrete Fourier Transform (DFT) given by equations (2–5), is a least–squares estimate of the phasor. If the number of samples available for phasor computation is different from a multiple of $N/2$, the least–squares estimate is some other combination of X_c and X_s, and is no longer given by equation (4). Short window (less than one period) phasor computations are of interest in digital relaying applications. For the present, we will concentrate on data windows of one or more complete periods of the fundamental frequency component.

Figure 1 points out an important aspect of this definition. The data window begins at the instant when sample number 1 is obtained. Thus, returning to equation (1), the sample set x_k is given by

$$
\begin{aligned}
x_1 &= X_m \cos (\theta + \varphi) \\
x_2 &= X_m \cos (2\theta + \varphi) \\
&\cdot \\
&\cdot \\
x_N &= X_m \cos (N\theta + \varphi)
\end{aligned}
\tag{6}
$$

Substituting for x_k from equation (6) in equation (2),

$$
X = \frac{1}{\sqrt{2}} \frac{2}{N} \sum_{k=1}^{N} X_m \cos (k\theta + \varphi) \, \epsilon^{-Jk\theta}
\tag{7}
$$

If we substitute, in equation (7),

$$
\cos (k\theta + \varphi) = \frac{\epsilon^{J(k\theta + \varphi)} + \epsilon^{-J(k\theta + \varphi)}}{2}
\tag{8}
$$

we get

$$
X = \frac{1}{\sqrt{2}} \frac{2}{N} \frac{1}{2} \sum_{k=1}^{N} X_m \left\{ \epsilon^{J\varphi} + \epsilon^{-J(2k\theta + \varphi)} \right\}
$$

or,

$$
X = \frac{1}{\sqrt{2}} X_m \epsilon^{J\varphi} + \frac{1}{\sqrt{2}} \frac{1}{N} X_m \, \epsilon^{-J\varphi} \sum_{k=1}^{N} \epsilon^{-J2k\theta}
\tag{9}
$$

The sum $\displaystyle\sum_{k=1}^{N} \epsilon^{-J2k\theta} = \frac{\epsilon^{-J2\theta}(1 - \epsilon^{-J2N\theta})}{(1 - \epsilon^{-J2\theta})} = 0$, for $N\theta = 2\pi$. Thus,

$$X = \frac{1}{\sqrt{2}} X_m \epsilon^{J\varphi}$$

which is the familiar expression, equation (1), for the phasor representation of the sinusoid in equation (1). The instant at which the first data sample is obtained defines the orientation of the phasor in the complex plane. The reference axis for the phasor, i.e. the horizontal axis in Figure 1(b), is specified by the first sample in the data window.

Equations (4–5) define an algorithm for measuring the phasor from data samples obtained from an input signal. As such, a recursive form of the algorithm is far more useful for real–time measurements. Consider the phasors computed from two sample sets: $x_k\{k=1,\cdots,N\}$, and, $x'_k\{k=2,\cdots N+1\}$, and their corresponding phasors X^1 and $X^{2\prime}$ respectively:

$$X^1 = \frac{1}{\sqrt{2}} \frac{2}{N} \sum_{k=1}^{N} x_k \, \epsilon^{-Jk\theta} \tag{10}$$

$$X^{2\prime} = \frac{1}{\sqrt{2}} \frac{2}{N} \sum_{k=1}^{N} x_{k+1} \, \epsilon^{-Jk\theta} \tag{11}$$

We may modify equation (11) to produce a recursive phasor calculation as follows:

$$X^{2\prime} = \frac{1}{\sqrt{2}} \frac{2}{N} \epsilon^{J\theta} \sum_{k=1}^{N} x_{k+1} \, \epsilon^{-J(k+1)\theta} \tag{12}$$

or,

$$\epsilon^{-J\theta} X^{2\prime} = \frac{1}{\sqrt{2}} \frac{2}{N} \sum_{k=2}^{N+1} x_k \, \epsilon^{-Jk\theta}$$

$$= \frac{1}{\sqrt{2}}\frac{2}{N}\sum_{k=1}^{N}x_k\,\epsilon^{-jk\theta} + \frac{1}{\sqrt{2}}\frac{2}{N}\left\{x_{N+1}\,\epsilon^{-j(N+1)\theta} - x_1\epsilon^{-j\theta}\right\}$$

$$X^2 \equiv \epsilon^{-j\theta}X^{2\prime} = X^1 + \frac{1}{\sqrt{2}}\frac{2}{N}(x_{N+1} - x_1)\epsilon^{-j\theta} \tag{13}$$

Since the angle of the phasor $X^{2\prime}$ is greater than the angle of the phasor X^1 by the sampling angle θ, (this is clear if we consider the picture corresponding to Figure 1 for the data set x_k'), the phasor X^2 as defined in equation (13), having been obtained from $X^{2\prime}$ by reducing its angle by θ, has the same angle as the phasor X^1. Thus, when the input signal is a constant sinusoid, the recursive phasor calculated according to equation (13) is a constant complex number. This is illustrated in Figure 2. In general, the phasor X^r, corresponding to the data set $x_k\{k=r,r+1,\cdots,N+r-1\}$, is recursively modified into X^{r+1} according to the formula

$$X^{r+1} = X^r + \frac{1}{\sqrt{2}}\frac{2}{N}(x_{N+r} - x_r)\epsilon^{-jr\theta} \tag{14}$$

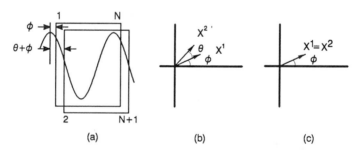

(a)   (b)   (c)

Figure 2.   Non-recursive and recursive phasors.

The recursive phasor calculation, as given by equation (13) is computationally very efficient. It regenerates the new phasor from the old one, and utilizes most of the computations performed for the phasor with the old data window. Furthermore, on a stationary input signal, the phasor produced by the recursive computation is a constant complex number. The factor $(\sqrt{2}/N)$ in equation (14) is generally omitted in the real–time computation, and the multiplier $\exp(-jr\theta)$ can be reduced to very few shift operations by a careful choice of the sampling angle θ. In many practical systems, a sampling angle of 30°, corresponding to N=12, (i.e. a sampling frequency of 720 Hz for a 60 Hz power system), has been found to be an optimum choice.

III. TIME SYNCHRONIZATION

Phasors, representing voltages and currents at various buses in a power system, define the state of the power system. If several phasors are to be measured, it is essential that they be measured with a common reference. The reference, as mentioned in the previous section, is determined by the instant at which the samples are taken. In order to achieve a common reference for the phasors, it is essential to achieve synchronization of the sampling pulses.

The precision with which the time synchronization must be achieved depends upon the uses one wishes to make of the phasor measurements. For example, one use of the phasor measurements is to estimate, or validate, the state of the power systems; so that crucial performance features of the network, such as the power flows in transmission lines could be determined with a degree of confidence. Many other important measures of power system performance, such as contingency evaluation, stability margins, etc. can be expressed in terms of the state of the power system, i.e. the phasors. Accuracy of time synchronization directly translates into the accuracy with phase angle differences between various phasors can be measured. Phase angles between the ends of transmission lines in a power network may vary between a few degrees, and may approach 180° during particularly violent stability oscillations. Under these circumstances, assuming that one may wish to measure angular differences as little as 1°, one would want the accuracy of measurement to be better than 0.1°. Fortunately, synchronization

accuracies of the order of 1 μsec are now achievable because of some
recent technological developments.   One micro second corresponds to
0.022° for a 60 Hz power system, which more than meets our needs.
In terms of technically and economically feasible alternatives, two
sources of synchronization are available at this time:

(a)   Transmission of synchronization pulses over fiber–optic
communication links, and

(b)   The use a common access time signal transmitted by the
Global Positioning System (GPS) satellites.[1]

Synchronization over fiber–optic links.

Of these two methods, the one using fiber–optic links depends upon
the availability of a fiber optic communications facility at every
substation where a measurement is to be made.   If such a link is
available, a pulse could be sent from a master transmitter as shown in
Figure 3.   The received pulse is delayed by an adjustable time delay τ,
set to compensate for the propagation delays between the master station
and the substation in question.   All delays can be adjusted, so that the
pulses produced at each station are coincident to within 1 μsec across
the entire power system.   Each of the time delays can be adjusted
individually to achieve the desired precision.

In order to maintain the synchronization to within 1 μsec, a
dedicated fiber is required.   The uncertainties introduced by framing
protocols in multiplexing schemes (such as the T1 protocol) are too
large, and can not be tolerated in the present application.   A dedicated
fiber, and an appropriate spare, is usually called for in such systems.
Furthermore, one must not switch the routing of the fiber being used
for synchronization, unless the delay for each route is compensated for
automatically.   The availability of fiber optic links to all measurement
sites is questionable on most power systems at the present time.   The

cost of fiber optic installation must be justified for some other, commercially viable reason.   Most electric utility companies have only a

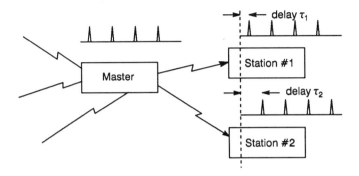

Figure 3. Synchronization with a master transmitter.

few experimental fiber–optic links available between major substations. Under these circumstances, the use of fiber–optic links for a system–wide phasor measurement system seems impractical.   Of course, when the fiber–optic links are available, it would be perfectly appropriate to use them for transmitting synchronizing pulses.

Synchronization with GPS satellite signals.

A system of satellites, known as NAVSTAR, was inaugurated in 1974, by the US Air Force.[1]   These satellites are elements of a Global Positioning System (GPS), which, in its final configuration will operate 24 satellites in 12–hour orbits at an altitude of 10,898 miles.   The planned orbits are as shown in Figure 4.   The final system will provide visibility of 6 to 11 satellites at 5 degrees or more above the horizon to users located anywhere in the world at any time.   Using the transmissions from these satellites, positions of objects can be determined with an accuracy of 10 meters in three dimensions. However, our interest is in the common–view time transmission provided by these satellites, which when decoded by appropriate receiver clocks, will provide 1 pulse–per–second (1 pps) at any location in the world with an accuracy of about 1 μsec. As the full complement of satellites is not yet deployed, the GPS receivers usually augment the satellite transmission with an internal oscillator, providing the same accuracy of trans– mission. Clearly, when the full complement of satellites is deployed in the coming years, the internal oscillators will no longer be needed. In addition to the 1 pps signal, the GPS satellites provide, along with other information, the time–stamp unique to the pulse.

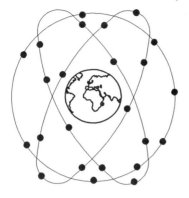

Figure 4.   NAVSTAR GPS phase III configuration for 24 satellites.

The GPS receiver clocks must also be designed to provide the sampling pulses needed by the phasor measurement units. These pulses, which are nominally at a multiple of the fundamental power system frequency, must be phase–locked with the 1 pps signal. Some commercially available GPS receivers, designed to produce sampling pulses at 12 times the 60 Hz power system frequency, are illustrated in

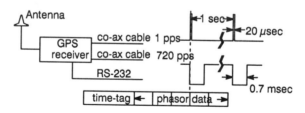

Figure 5. GPS receiver and its output signals.

Figure 5. The pulse waveforms, and the time–stamp format are not standardized as yet, but a recently formed Working Group in the Power System Relaying Committee of the IEEE Power Engineering Society has begun a preliminary investigation of the possibilities and opportunities available with this new technology.

Phasor Measurement Unit

The major subsystems of a phasor measurement unit are shown in Figure 6. The power system voltages and currents are converted to standard secondary levels by current and voltage transformers, and properly isolated and filtered   by the   signal   conditioning   unit.   The

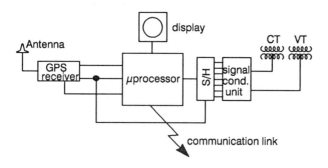

Figure 6. Phasor Measurement Unit block diagram.

filtering is necessary to avoid aliasing of the signal in the phasor calculation. The analog signals are strobed by the sampling clock pulse provided by the GPS receiver using sample–and–hold circuits, and then sampled by a multiplexed analog–to–digital converter. The data samples are stored in scratch–pad tables, and the recursive DFT algorithm executed to calculate the positive sequence (see below) voltage and current phasors from the samples data. The raw data samples, as well as the phasors are stored in event tables, following the occurrence of an abnormal power system condition. The time stamp provided by the GPS receiver clock on the serial line constitutes a part of the time–tag of the phasors. The other part is provided by the sample number corresponding to the first sample in the data window. For a sampling rate of 720 Hz, the sample number varies between 1 and 719. Sample number 1 coincides with the 1 pps signal at each measurement site. Thus, by matching the complete time–tag of a phasor measurement, simultaneous measurements can be assured to within 1 μsec. The time–tagged phasors are communicated to remote locations on a serial line. The channel speed available for this line determines the rate at which the phasors can be communicated to a central site. Typically, with a channel speed of 9600 baud, several phasors can be communicated to the center at a refresh rate of about five cycles of the fundamental frequency of the power system. This scan rate is more than adequate to track all electromechanical oscillations taking place in a power system. As the number of phasor measurement units installed on a system increases, a possible data bottle–neck may develop at the central receiving site. However, by incorporating a front–end communications handling processor at the center, this problem can be resolved. Several experimental systems utilizing these ideas are being installed in many power systems.

IV. SYMMETRICAL COMPONENTS

Symmetrical components are linear transformations on voltages and currents of a three phase network. The symmetrical component transformation matrix S transforms the phase quantities, taken here to be voltages E_φ, (although they could equally well be currents), into symmetrical components E_s:

$$E_s = \begin{bmatrix} E_0 \\ E_1 \\ E_2 \end{bmatrix} = SE_\varphi = \frac{1}{3} \begin{bmatrix} 1 & 1 & 1 \\ 1 & \alpha & \alpha^2 \\ 1 & \alpha^2 & \alpha \end{bmatrix} \begin{bmatrix} E_a \\ E_b \\ E_c \end{bmatrix} \tag{15}$$

where $(1,\alpha,\alpha^2)$ are the three cube–roots of unity. The symmetrical component transformation matrix S is a similarity transformation on the impedance matrices of balanced three phase circuits, which diagonalizes these matrices.   The symmetrical components, designated by the subscripts (0,1,2) are known as the zero, positive, and negative sequence components of the voltages (or currents).   The negative and zero sequence components are of importance in analyzing unbalanced three phase networks.   For our present discussion, we will concentrate on the positive sequence component E_1 (or I_1) only.   This component measures the balanced, or normal, voltages and currents that exist in a power system.   Dealing with positive sequence components only, allows the use of single phase circuits to model the three phase network, and provides a very good approximation for the state of a network in quasi–steady state.   All power generators generate positive sequence voltages, and all machines work best when energized by positive sequence currents and voltages.   The power system is specifically designed to produce and utilize almost pure positive sequence voltages and currents in the absence of faults or other abnormal unbalances.

It follows from equation (15) that the positive sequence component of the phase quantities X_φ is given by

$$X_1 = \frac{1}{3}(X_a + \alpha X_b + \alpha^2 X_c) \tag{16}$$

Or, using the recursive form of the phasors given by equation (14),

$$X_1^{r+1} = X_1^r + \frac{1}{\sqrt{2}}\frac{2}{N}\left\{(x_{a,N+r} - x_{a,r})\epsilon^{-jr\theta}\right.$$
$$\left. + \alpha(x_{b,N+r} - x_{b,r})\epsilon^{-jr\theta} + \alpha^2(x_{c,N+r} - x_{c,r})\epsilon^{-jr\theta}\right\} \tag{17}$$

Recognizing that for a sampling rate of 12 times per cycle – which is the rate most commonly being used at present – α and α^2 correspond to $\exp(j4\theta)$ and $\exp(j8\theta)$ respectively, it can be seen from equation (17) that

$$X_1^{r+1} = X_1^r + \frac{1}{\sqrt{2}} \frac{2}{N} \left\{ (x_{a,N+r} - x_{a,r})\epsilon^{-jr\theta} + (x_{b,N+r} - x_{b,r})\epsilon^{j(4-r)\theta} \right.$$
$$\left. + (x_{c,N+r} - x_{c,r})\epsilon^{j(8-r)\theta} \right\} \tag{18}$$

With a carefully chosen sampling rate – such as a multiple of 3 times the nominal power system frequency – very efficient symmetrical component calculations can be performed in real time. Equations similar to (18) hold for negative and zero sequence components also.

Operation at off–nominal frequency[2]

A normal power system is rarely operating at its nominal frequency. Usually, the power system frequency will vary randomly around the nominal value, with excursions of the order of less than 1 Hz on either side of the nominal value. It is interesting to see the effect of this power system frequency variation on the symmetrical component computation. It will be shown in this section that a purely positive sequence quantity at off–nominal frequency produces a small amount of negative sequence quantity, and *vice versa*. As a consequence, if the input is unbalanced, (i.e. it contains both positive and negative sequence components), each sequence component makes a small contribution to the other at off–nominal frequency, and must be taken into account when true measurements of positive and negative sequence components are required.

Consider three phase inputs of pure positive sequence at a frequency $\omega = \omega_0 + \Delta\omega$, where ω_0 is the nominal power system frequency. (120π radians per second for a 60Hz system). The three time functions for the phase quantities are given by

$$x_a(t) = X_m \cos (\omega t + \varphi) = \sqrt{2}\text{Re}\{ X \epsilon^{j\omega t} \}$$
$$x_b(t) = X_m \cos (\omega t + \varphi - \frac{2\pi}{3}) = \sqrt{2}\text{Re}\{ X \alpha^2 \epsilon^{j\omega t} \} \tag{19}$$
$$x_c(t) = X_m \cos (\omega t + \varphi - \frac{4\pi}{3}) = \sqrt{2}\text{Re}\{ X \alpha \epsilon^{j\omega t} \}$$

where X is the phasor representation of the phase a signal $x_a(t)$. The phase quantities can be expressed with the help of Euler's formula

$$x_a(t) = \sqrt{2}\, \frac{1}{2} \{ X\epsilon^{\jmath\omega t} + X^* \epsilon^{-\jmath\omega t} \}$$
$$x_b(t) = \sqrt{2}\, \frac{1}{2} \{ X\alpha^2 \epsilon^{\jmath\omega t} + X^* \alpha \epsilon^{-\jmath\omega t} \} \tag{20}$$
$$x_a(t) = \sqrt{2}\, \frac{1}{2} \{ X\alpha \epsilon^{\jmath\omega t} + X^* \alpha^2 \epsilon^{-\jmath\omega t} \}$$

since the complex conjugate of α is α^2. Now consider the r'th phasor computation of the phase a quantity obtained by generalizing equation (10) for the data window starting at the r'th sample

$$X_a^r = \frac{2}{N} \sum_{k=r}^{r+N-1} x_{a,k}\, \epsilon^{-\jmath k\omega_0 \Delta t}$$

$$= \frac{2}{N} \sum_{k=r}^{r+N-1} \frac{1}{2} \{ X\epsilon^{\jmath k\omega\Delta t} + X^* \epsilon^{-\jmath k\omega\Delta t} \} \epsilon^{-\jmath k\omega_0 \Delta t} \tag{21}$$

where the sampling angle θ has been replaced by $\omega_0 \Delta t$, corresponding to the *fixed* sampling frequency ω_0. Substituting expressions similar to (21) for phases b and c in the definition of the positive sequence component, i.e. in equation (16),

$$X_1^r = \frac{1}{3} \frac{2}{N} \sum_{k=r}^{r+N-1} \frac{1}{2} \{ 3X\epsilon^{\jmath k\omega\Delta t}\, \epsilon^{-\jmath k\omega_0 \Delta t} \}$$
$$+ \frac{1}{3} \frac{2}{N} \sum_{k=r}^{r+N-1} \frac{1}{2} X^* \{ 1 + \alpha\alpha^{2*} + \alpha^2\alpha^* \} \epsilon^{-\jmath k\omega\Delta t}\, \epsilon^{-\jmath k\omega_0 \Delta t} \} \tag{22}$$

Recognizing that $\{ 1 + \alpha\alpha^{2*} + \alpha^2\alpha^* \} = 0$, and $\omega = \omega_0 + \Delta\omega$,

$$X_1^r = \frac{1}{N} \sum_{k=r}^{r+N-1} X\epsilon^{\jmath k\Delta\omega\Delta t} \tag{23}$$

The geometric series in equation (23) can be summed to produce

$$X_1^r = X\epsilon^{jr\Delta\omega\Delta t}\left[\frac{\sin\frac{N\Delta\omega\Delta t}{2}}{N\sin\frac{\Delta\omega\Delta t}{2}}\right]\epsilon^{-j(N-1)\frac{\Delta\omega\Delta t}{2}} \tag{24}$$

We will return to the first term (i.e. $X\epsilon^{jr\Delta\omega\Delta t}$) in equation (24) in the next section. The remainder of the expression is independent of r, and is a complex attenuation factor which distorts the true phasor measurement. It is interesting to examine the magnitude and phase angle of this attenuation factor . Figure 7 shows the effect of a frequency variation between 55 and 65 Hz for a nominal system frequency of 60 Hz and a nominal sampling frequency of 720 Hz. This frequency deviation is certainly quite extreme – in

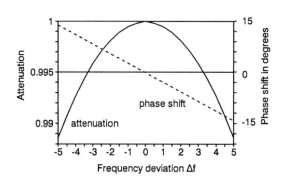

Figure 7. Effect of off-nominal frequency operation.

most cases frequency excursions would be limited to between 59 and 61 Hz. Even for such a large excursion, the attenuation lies between about 0.989 and 1.00, and the phase angle error varies between about −13.5° to about +13.5°. For the more common excursion of ± 1 Hz, the corresponding numbers are: attenuation between 0.999 and 1.0, and the phase angle error between ±2.7°. In most practical cases, both of these factors could be safely neglected. Where they can not be neglected, they can be taken into account if the actual frequency of the input quantity can be estimated (see the next section). It should be recognized that all inputs with the same off–nominal frequency will have equal attenuation and phase shifts. In many power system applications, for example in some relaying algorithms, equal errors in all phasors are of no consequence.

An interesting effect of operating at off–nominal frequency is to produce a false negative sequence component from a purely positive sequence input. The negative sequence component of the input quantities given by equation (20) can be expressed as

$$X_2 = \frac{1}{3} (X_a + \alpha^2 X_b + \alpha X_c) \tag{25}$$

Using expressions for the three phasors as in equation (21), it is seen that when the negative sequence component is computed, the corresponding expression is similar to equation (22), with the exception that the first sum becomes zero, while the second sum makes a contribution to the negative sequence quantity:

$$X_2^r = \frac{1}{3} \frac{2}{N} \sum_{k=r}^{r+N-1} \frac{1}{2} \{ 3X^* \epsilon^{-jk\omega\Delta t} \epsilon^{-jk\omega_0\Delta t} \} \tag{26}$$

This expression can also be expressed in a compact form by summing the geometric series:

$$X_2^r = X^* \epsilon^{-jr(\omega+\omega_0)\Delta t} \left[\frac{\sin N(\omega+\omega_0)\Delta t/2}{N \sin(\omega+\omega_0)\Delta t/2} \right] \epsilon^{-j(N-1)(\omega_2+\omega_0)\Delta t/2} \tag{27}$$

This is a false negative sequence component, as it is produced by a pure positive sequence input of magnitude X at an off–nominal frequency. Again, the part of equation (27) that is independent of r is plotted in Figure 8, and represents an attenuation and a phase shift in the false negative sequence component. It can be seen that even for a frequency deviation of 5 Hz, the attenuation factor is under 0.05. In other words, the false negative sequence component is about 1 % of the positive sequence component per 1 Hz deviation. In most cases, this is of no significance. However, when the input signal contains a negative sequence component, this false negative sequence will create an error in the negative sequence estimation. An easy method of eliminating this false negative sequence component from the measurement is to average three measurements which are made with data windows which are 120° apart from each other. The exponent $-jr(\omega+\omega_0)$ in the first term of equation (27) will cause the three false negative sequence components to cancel each other. Strictly speaking, one must

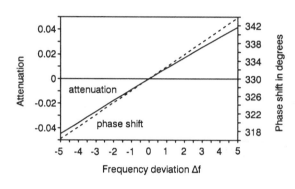

Figure 8. Effect of off-nominal frequency operation. False negative sequence from positive sequence.

also follow this procedure in computing the positive sequence component, as any negative sequence component in the input will create a small (but non–zero) false positive sequence component. As most power system signals have at most a small negative sequence component, this last correction is unnecessary, except in case of unbalanced faults. As a final note, it should be noted that the zero sequence component calculation is not affected by the off–nominal frequency operation.

It should be clear that equations similar to (24) can be derived for calculating harmonic components at off–nominal frequencies. Each of the angles in the formula are multiplied by the appropriate harmonic order. This results in different (somewhat smaller) attenuation factor for higher order harmonics. Nevertheless, if the frequency of the input signal is known, the attenuation factors can be computed, and accounted for, in the harmonic estimation process.

V. FREQUENCY MEASUREMENT[2]

We may disregard the attenuation and the phase shift factors (which are small, and moreover, can be taken into account if necessary), since they are independent of r in equation (24).   The recursive phasor formula then becomes

$$X_1^r = X_\epsilon j r \Delta \omega \Delta t \tag{28}$$

The phase angle of this phasor, denoted by φ_r, is given by

$$\varphi_r = r \Delta \omega \Delta t \tag{29}$$

If time is measured from some suitable value of r, we may replace $r\Delta t$ by the variable t, and then

$$\frac{d\varphi_r}{dt} = \Delta \omega \tag{30}$$

This method of measuring frequency deviation from the nominal frequency ω_0 is one of the most sensitive available for measuring local power system frequency.   Local frequency is an important parameter in many important relaying and control function implementations.   It should be emphasized that the frequency calculation is accurate only if the phase angle of the positive sequence voltage is used.

In practice, it is best to obtain a sequence of phase angle measurements φ_r { r=1,2,$\cdots$,n} to determine a least–square estimate of the frequency and the rate of change of frequency.   Assuming that each phase angle is obtained at multiples of Δt, we may assume a second degree polynomial for $\varphi_r(t)$:

$$\varphi_r(t) = a_0 + a_1 t + a_2 t^2 \tag{31}$$

The constants a_0, a_1, and a_2 can now be computed by the least–squares formula

$$\begin{bmatrix} a_0 \\ a_1 \\ a_2 \end{bmatrix} = [M^T M]^{-1} M^T \begin{bmatrix} \varphi_1 \\ \varphi_2 \\ \vdots \\ \varphi_n \end{bmatrix} \tag{32}$$

where

$$M = \begin{bmatrix} 1 & \Delta t & \Delta t^2 \\ 1 & 2\Delta t & 4\Delta t^2 \\ \vdots & \vdots & \vdots \\ 1 & n\Delta t & n^2\Delta t^2 \end{bmatrix} \tag{33}$$

The triple product $[M^T M]^{-1} M^T$ is calculated off–line once, and then re–used whenever the frequency and the rate of change of frequency are to be determined. These are expressed in terms of the constants a_i

$$\Delta \omega = a_1 + 2a_2 t = a_1 + 2a_2 r\Delta t$$
$$\dot{\omega} = 2a_2 \tag{34}$$

A somewhat simpler and approximate procedure results if one considers the first differences of the phase angles:

$$\Delta \varphi_r = \varphi_r - \varphi_{r-1} \tag{35}$$

And, from equation (34)

$$\Delta \omega_r \cong \frac{\Delta \varphi_r}{\Delta t} = a_1 + 2a_2 r\Delta t \tag{36}$$

Using $\Delta \omega_r$ as linear functions of $r\Delta t$, the constants a_1 and $2a_2$ can be calculated by the well–known regression formulas:

Let $\overline{\Delta\omega}$ and $\overline{r\Delta t}$ be the averages of the corresponding variables.   Then a_1 and $2a_2$ are given by

$$2a_2 = \frac{\Sigma(r\Delta t - \overline{r\Delta t})(\Delta\omega_r - \overline{\Delta\omega})}{\Sigma(r\Delta t - \overline{r\Delta t})^2}$$

(37)

$$a_1 = \overline{\Delta\omega} - 2a_2 \overline{r\Delta t}$$

Equations (37) are the least–squares solutions for the frequency deviation and the rate of change of frequency at the beginning of the measurement window.   The approximate procedure generally gives sufficiently accurate results.   It should be remembered that we have not accounted for bad data in this procedure.   The phasors are computed from data samples, and bad data are usually eliminated at that stage. Furthermore, one could assume that the local frequency of a power system could not change at rates above a certain upper bound.   One could thus invoke a bad data detection procedure based upon the magnitude of $\Delta\varphi_r$.   If the change in the phasor angle is detected to be above this upper limit, the measurement is rejected.   Indeed, one would be justified in restarting the entire estimation procedure from the next sample.   This will allow for any switching transients on the power system, which may change the phase angles by a step function, without affecting the local system frequency.

The data window used to determine the frequency and the rate–of– change of frequency has an important bearing upon the ability of the estimation procedure to track the power system dynamics.   A long window will, due to the form of the polynomial chosen, mask frequency components with periods smaller than twice the window length.   On the other hand, a too–short window will create an unacceptably high error in the frequency estimate.   One must have a good idea as to what frequencies are likely to occur on the power system, and then choose the data window accordingly.

VI. STATE ESTIMATION

Static State Estimation

The set of the positive sequence voltage phasors at all the buses of a power system constitutes the state vector of a power system in quasi-steady state. The knowledge of the current state of the power system is extremely useful in determining operational controls to be exercised on the power system in order to maintain it in a secure and economic mode of operation. The present state estimation procedures are based upon obtaining measurements from throughout the power system at fast as a scan rate as possible. The measurements obtained usually consist of line power flows, line currents, and bus voltage magnitudes. The scan rates may vary between a few seconds, to some minutes. Under these circumstances, the state estimation is a nonlinear process, and further, it does not represent a true snap–shot of the power system. The metering systems tend to have fairly long time constants, so that some averaging of the measurements is implicit in the process. The resulting state estimate is understood to be a quasi–steady state approximation – an averaged snap–shot over the scan period.

The synchronized phasor measurements offer a unique opportunity to improve the state–estimation function in a remarkable manner. Since the measurements are precisely synchronized, and are identified by a precise time–tag, measurements obtained at the control center by relatively slow communication channels do not affect the simultaneity of the measurements. It is only necessary to compare the time–tags, and a true synchronous snap–shot of the power system state is established at the center. If communication channels of adequate speed are available, these snap–shots can be obtained at a sufficiently fast rate so as to make it possible to follow the dynamics of the power system in real–time. More about this aspect of state estimation in the next section.

Another remarkable advantage of the synchronized measurement system is that, since the voltage and current phasors are used, the state estimation equations become linear.   The positive sequence voltage phasors are the state vector of the power system, and therefore the state estimation is a trivial procedure if all positive sequence voltages (and only these) are measured; i.e., the measurements themselves are the state estimate.   However, it is reasonable to expect that positive sequence currents also are measured in order to provide redundancy in measurements.   Thus, a complete measurement vector z is given by [E,I]^t, where E is the vector of all positive sequence bus voltages, and I is the vector of all measured positive sequence currents in transmission elements.

$$z = \begin{bmatrix} E \\ \hline I \end{bmatrix} = \begin{bmatrix} B \end{bmatrix} [\, E \,] = \begin{bmatrix} 1 \\ \hline C \end{bmatrix} [\, E \,] \tag{38}$$

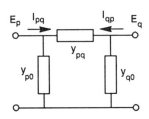

Figure 9.   Voltage and current measurements in a line.

The matrix C is constructed from a measurement–node connection matrix, and the primitive admittances of the connected branches.   Consider the π–section representation of a transmission line shown in Figure 9. If y_{pq}, y_{p0}, and y_{q0} are the three admittances of the π–section, the current measured at bus p of the line is given by

$$I_{pq} = y_{pq}(E_p - E_q) + y_{p0} E_p \tag{39}$$

Denoting the measurement–node incidence matrix by A, the current measurement vector I can be expressed as

$$I = [yA^T + y_s] \tag{40}$$

where **y** is the primitive (diagonal) matrix of the series admittances of all elements whose currents have been measured, and $\mathbf{y_s}$ is the primitive matrix of the shunt admittances of the elements.[2] If the measurement errors are assumed to have zero–mean and a covariance matrix $\mathbf{W}$, the weighted least–squares estimate of $\mathbf{E}$ obtained from the measurement vector $\mathbf{z}$ of equation (38) is given by

$$\mathbf{G}\hat{\mathbf{E}} = \mathbf{B}^{\dagger}\mathbf{W}^{-1}\,\mathbf{z} \tag{41}$$

where $\dagger$ denotes the complex conjugate transpose, and the gain matrix $\mathbf{G}$ is given by

$$\mathbf{G} = \mathbf{B}^{\dagger}\mathbf{W}^{-1}\mathbf{B} \tag{42}$$

It has been shown in reference [2] that the structure of the gain matrix is such that in most cases it leads to very simple solution algorithms for equation (41). It is a constant matrix in any case, so that an LU decomposition and Gaussian elimination technique can be used to find the state estimate. Furthermore, $\mathbf{G}$ is often real, and when it is not real, its imaginary part is so small that it can be neglected. However, it is the fact that $\mathbf{G}$ is constant which is all important. Unlike the presently used state estimation procedures (which are non–linear), the estimator based upon phasor measurements does not require iterative solution techniques, and consequently is much faster to execute. This fact, coupled with its precise synchronization at the source, makes it possible to develop dynamic estimates of the power system in real time. This has not been possible until now.

It is also possible to supplement existing state estimators with phasor measurement based state estimators. This can be done in one of two ways: one could put all measurements (including the linear and nonlinear ones) in the measurement vector $\mathbf{z}$, and then solve the entire problem as a nonlinear state estimation problem. However, a far more efficient procedure is to append the phasor based estimate of a part of the system upon the estimate based upon the nonlinear measurements obtained from another part of the power system. Both techniques have been shown to work well. It must be remembered however, that the overall response time of the hybrid measurement procedure is slower than that of the pure phasor measurement based procedure, due to the nonlinear measurements included in z in the former case.

Dynamic State Estimation

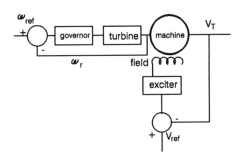

The real–time phasor measurements, and the computed frequency can be used to develop dynamic state estimates of generating stations. A generating station can be modeled as a nonlinear vector differential equation. For example, the equations representing the system

Figure 10. Functional block diagram of a turbine generator and its excitation system.

shown in Figure 10 are given by

$$\dot{x} = \begin{bmatrix} \dot{x}_1 \\ \dot{x}_2 \end{bmatrix} = \begin{bmatrix} A_{11} & 0 \\ 0 & A_{22} \end{bmatrix} \begin{bmatrix} x_1 \\ x_2 \end{bmatrix} + Bu \tag{43}$$

where the partitioned state vector is given by

$$x_1^T = [\lambda_d \ \lambda_f \ \lambda_{kd} \ \lambda_{kq} \ \lambda_q \ e_f]$$

$$x_2^T = [P_g \ P_m \ \omega_r \ \delta \ ] \tag{44}$$

The matrices A and B are functions of the constants describing the generating station model of Figure 10. The various λ's are the flux linkages of the different machine windings, P's are machine powers, and ω_r and δ are the machine rotor speed and angle respectively[3]. The input vector **u** is given by

$$\mathbf{u}^T = [\; e_d \;\; e_q \;\; \sqrt{(e_d^2 + e_q^2)} \;\; (e_d i_d + e_q i_q) \;\; 1] \tag{45}$$

The measured quantities, viz. the terminal voltage and current phasors, can be used to derive the rotor frequency ω_r and the rotor angle δ as explained earlier. The measurements can also be used to calculate the d– and q–axis voltages and currents of the stator. Treating all the quantities derived from the phasors (as well as the field voltage and current, which are assumed to be measured) as the measurement vector,

$$\mathbf{z}(k+1) = \mathbf{H}\mathbf{x}(k+1) + \epsilon(k+1) \tag{46}$$

an *extended Kalman filter* can be used to determine the complete state vector of the generating station. The details can be found in the reference cited.[3] The same reference also provides an observer based solution to the estimation problem.

The generating station models obtained from real time measurements of phasors are useful in studying the dynamic contingencies on the power system. The solution to the dynamic contingency problem – which attempts to determine whether or not the system is stable for several possible faults and outages – can be found for relatively simple systems which behave like a two machine system. However, for the more general case of a complex network system, no comparable solutions are not yet available. Work on the more general problem is in progress at many centers of research, and several promising approaches are being investigated. The dynamic equivalent of the generating systems is one of the key ingredients of the solution of the dynamic contingency problem.

VII. CONTROL

There are a number of controllable elements in a modern power system. For example, the generators have excitation and turbine governor controls. The network may be controlled by opening or closing circuit breakers. High voltage direct current converter terminals can be controlled by changing the firing angles of the converters. Dynamic brakes may be applied to a power system as the need arises. Series and shunt compensating elements can be controlled by switching, or through controlled thyristor bridges. And, finally, the loads themselves may be modulated if needed. All these elements are controlled to achieve some operational goal. For example, the objective of a control scheme may be to increase the loadability margin of certain transmission lines. Another objective may be to minimize the excursions of power system frequency from its nominal value. Yet another control objective may be to minimize the deviation of operating voltages on the network from a desirable operating range. These types of control objectives are operative during a disturbed state of the power system. On the other hand, the control objective may be simply to improve the operating economy of operation.

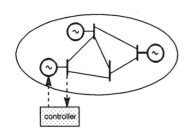

Figure 11. Power system control with local feedback.

Regardless of the nature of the control objectives, all presently used power system controllers use local feedback signals. This is shown schematically in Figure 11. Consider the excitation system of one of the generators in the system. Its feedback signal consists of the magnitude of the voltage measured at the terminals of the generator. In more advanced designs, the controller may use other signals derived from the locally measured power system frequency, rate of change of frequency, or some other appropriate quantity.

As the objective of the controller is to achieve some measure of improvement in the performance of the total system, consisting of the generator being controlled, and the rest of the power system, the controller design must be based upon some mathematical model of the total system.   Thus, implicit in the controller design is an assumed model of the complete dynamic system.   It is well known that the power system performance depends upon the system state vector in a highly nonlinear fashion.   This in turn implies that the controller design must be based upon linearization of the system behavior about an operating point.   It is known that controllers designed under these conditions may fail to control properly – or even become unstable – when the operating state of the power system is significantly different from the operating state for which the linearization was performed.

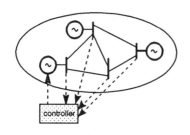

Figure 12. Power system control with remote phasor feedback.

The phasor measurements provide a unique opportunity to employ a state–vector feedback to improve the performance of the control system.   Let the state of the power system be denoted by χ. Depending on the level of modeling, and the components included in the formulation, the state could include generator angles and speeds, field flux linkages, exciter state variables, and governor state variables.   The control vector , $\mathbf{u}$, may include the exciter control signals, the governor control signals, and the control signals for any HVDC (High Voltage Direct Current) converters or SVC (Static Var Compensators) present in the system.   Given a concern with large disturbances, the state equation is nonlinear in general, and of the form

$$\dot{\chi}(t) = \mathbf{f}[\chi(t),\mathbf{u}(t)] \tag{47}$$

An output vector $\mathbf{y}(t)$ which is used to quantify system performance is associated with the control problem. The variables in $\mathbf{y}(t)$ could include: incremental changes in voltage magnitudes at generator terminals, rotor angles and generator speeds, and field voltages. To be completely general, $\mathbf{y}(t)$ can also be taken to be a nonlinear function of the state as

$$\mathbf{y}(t) = \mathbf{h}[\chi(t)] \tag{48}$$

For example, equation (48) would allow nonlinear functions of the state variables such as the generator terminal voltages to be considered as useful measurements.

It is assumed that the system has been moved away from equilibrium by some disturbance. The control problem is to steer the system along some desired trajectory – defined in terms of the desired output $\mathbf{y}_d(t)$ – to a desired state. Even with real–time measurements of all states it is not possible to solve such a nonlinear control problem in real time. Some approximations and simplifications are necessary. The use of real–time phasor measurements make it possible to treat the nonlinear system in an unusual way. Let χ_0 be the prefault equilibrium of the controlled system, i.e.

$$0 = \mathbf{f}(\chi_0, 0) \tag{49}$$

If we let $\chi(t)$ represent the incremental changes in the system state about the equilibrium point,

$$\chi(t) = \chi_0 + \mathbf{x}(t) \tag{50}$$

then, using the incremental (vector) variable $\mathbf{x}(t)$, we may write

$$\dot{\mathbf{x}}(t) = \mathbf{A}\mathbf{x} + \mathbf{B}\mathbf{u} + [\mathbf{f}(\chi_0 + \mathbf{x}, \mathbf{u}) - \mathbf{A}\mathbf{x} - \mathbf{B}\mathbf{u}] \tag{51}$$

and

$$y(t) = Cx + [h(\chi_0+x) - Cx] \tag{52}$$

It should be recognized that equations (51) and (52) are valid for any matrices A, B, and C, and are not approximations or Taylor expansions. On the other hand, it is clear that reasonable choices of the matrices are necessary. If we take the matrices to be the gradients of the appropriate functions (f or h) with respect to x or u (i.e. first terms of the corresponding Taylor expansions) evaluated at the stable equilibrium χ_0, then we can write a state equation for the incremental system:

$$\dot{x} = Ax + Bu + r(t) \tag{53}$$

$$y(t) = Cx + s(t) \tag{54}$$

where $r(t)$ and $s(t)$ are residuals given by

$$r(t) = [f(\chi_0+x,u) - Ax - Bu] = \dot{\chi} - Ax - Bu \tag{55}$$

and

$$s(t) = h(\chi_0+x) - Cx \tag{56}$$

Given the measurements of χ and $\dot{\chi}$, the residuals may be computed or measured in real time. The system described by equations (53) and (54) is a little unusual from a control system point of view. Rather than the nonlinear system described by equations (47) and (48), we have a stable (stable because χ_0 is stable) linear system with inputs $r(t)$ and $s(t)$, which can be measured in real time. Note that at an instant t_1, for example, $r(t)$, $t_0 < t < t_1$, is known, but the future values of $r(t)$ for $t > t_1$ are not known.

We take as a cost function a quadratic function of control $u(t)$ and differences between the actual $y(t)$ and the desired value $y_d(t)$, i.e.

$$J(u) = \tfrac{1}{2}[e^T(t_f)He(t_f)] + \frac{1}{2} \int_{t_0}^{t_f} [e^TQe + u^TRu]dt \tag{57}$$

with

$$e(t) = y(t) - y_d(t) \tag{58}$$

where Q and R are positive semi–definite and positive definite matrices respectively.   The minimization of the cost function is an attempt to make the system output follow a desired output while using little control effort.   The first term in equation (57) represents a penalty on the output error at the terminal time t_f.   The problem of minimizing $j(u)$ subject to equations (53) and (54), given the knowledge of $r(t)$ and $s(t)$ in the interval $t_0 < t < t_f$ is given by the well known equations

$$u(t) = R^{-1}B^T[g(t) - K(t)x(t)] \tag{59}$$

$$\dot{K}(t) = -A^TK - KA + KBR^{-1}B^TK - C^TQC \tag{60}$$

$$\dot{g}(t) = -[A^T - K(t)BR^{-1}B^T]g - C^TQ[y_d - s] + K(t)r \tag{61}$$

The difficulty with the measurement of $r(t)$ and $s(t)$ in real time is that equations (60) and (61) are solved in reverse time with terminal conditions at time t_f:

$$K(t_f) = C^THC, \text{ and } g(t_f) = C^TH[y_d(t_f) - s(t_f)] \tag{62}$$

That is, in order to determine the control at time t_1 from equation (59), we need the state at t_1, and the value of $g(t_1)$ from equation

(61). However, equation (61) is solved backward from t_f, so that the future values of $r(t)$ and $s(t)$ are needed to determine $g(t_1)$. Attempts have been made to predict $r(t)$ and $s(t)$ to overcome that difficulty. Another solution to the problem which has been shown by simulation to be quite acceptable, is to use K_s, the steady state state solution to equation (60), the so–called Algebraic Riccati Equation (ARE):

$$0 = -A^T K - KA + KBR^{-1}B^T K - C^T QC \tag{63}$$

Equation (63) can be thought of as equivalent to taking $t_f = \infty$ in equation (57). The steady state solution has an additional advantage in that K_s can be computed off–line, and stored for real–time use. Equation (60) would also be solved off–line, but would require considerably more storage. With a constant K_s in equation (61), the solution for $g(t)$ is not a large computational burden. Further simplification can be obtained by using the DC gain of the system described by equation (61) (i.e. assuming that the time varying inputs to the system are very slow). This gives

$$g_s(t) = -[A^T - K_s BR^{-1}B^T]^{-1}\{C^T Q[y_d - s] - K_s r(t)\} \tag{64}$$

Equation (64) only involves multiplication by known matrices and is an approximate but surprisingly effective control strategy.

The above procedure has been tried on controllers for HVDC converters and generator excitation systems on realistic power systems.[4] It has been shown that with reasonable rates of state vector data collection at a central site, the controller designed according to the procedure described above is able to limit excursions in the voltage magnitudes, and in generator rotor angles, following power system disturbances. These ideas await field trials, which may follow after considerable experience with the phasor measurement hardware has been gained. In the mean time, some work has begun in the field of adaptive relaying, which may be viewed as a special case of control. This development will be discussed next.

VIII. ADAPTIVE RELAYING[5]

Relaying is the branch of electric power engineering which deals with the protection of power systems and equipment against faults. Traditionally, protective devices, known as relays, are dedicated, pre–set devices, which operate in a predetermined manner in response to power system abnormalities, such as faults. In a sense, protection systems are rather simple control systems, which derive their input signals from measurements made at the site where they are located, and which control circuit breakers or other controllable elements in response to the abnormal state of the power system. The control law, embodied in the settings of the relays, is determined from an exhaustive study of the power system, and once established, the relay settings are fixed. The necessary speed of response of relays is such that only local inputs can be used in defining the relay logic, and in most cases this proves to be sufficient. In a power system, there are thousands of relays and protection systems, each of them carrying its own settings determined with the help of detailed system studies.

In the event that the primary protection system fails to operate as planned, there is a second (and often a third) line of defense, in what are known as backup protection systems. These backup systems are generally slower in responding to an abnormal system condition, since in all cases the preferred mode of operation is that the primary system should be given sufficient time to operate, if it is going to do so. The backup systems, as they are slower, may have more complex logic, and may use a greater variety of input signals in arriving at their decisions.

It has been found that most power systems are indeed well protected against the effects of any faults that may occur on the power system. However, in some rare cases, there is a combination of faults, followed by another contingency, which may initiate a chain of events which may lead to power system blackouts. In a significant number of cases, this second (un–anticipated) contingency is some hidden deficiency in a protection system. The protection system may thus prove to be inadequate to meet the requirements of the power system under the prevailing conditions.

Another aspect of relay settings is that they are often compromise settings, which are reasonable for many alternative states of the power system – but are not optimal for any particular state. Thus, many relay settings are poor in terms of sensitivity or speed of response for *every* fault, in order that they be adequate for *all* faults. The recently initiated field of adaptive relaying, attempts to correct this situation. It accepts the fact that many relay settings are system dependent, and no single setting may exist which is optimum for all states of the power system. A definition of adaptive relaying, which embodies this idea, is as follows:

"Adaptive protection is a protection philosophy which permits and seeks to make adjustments in various protection functions automatically in order to make them more attuned to prevailing power system conditions."

We will consider three examples of adaptive relaying which have received some attention in technical literature. It should be remembered that the concept of adaptive relaying has not yet been accepted by the power engineering community. The ideas expressed here are likely instances where the adaptive relaying may see its first manifestation in practice. The first example is actually not very dependent on the synchronized phasor measurements. But it will serve as an introduction to the subject which is less specific, and catches the philosophical flavor of the movement towards adaptive relaying. The next two examples are more specific, and will require some familiarity with the technical details of modern relaying practices.

Controllable dependability and security of protection:

Most modern protection systems have a built–in bias towards dependability. What this means is that protection systems are so arranged that if a fault should occur on the power system, it will be removed by some component of the overall protection system with a high degree of certainty. To achieve this high degree of certainty, often more than one relay may be arranged to provide an identical tripping function. Thus, even if some protection systems fail to operate correctly for a fault, some others – known as backup systems – will do so. Essentially, the tripping outputs of the systems are arranged to be in parallel with each other, and all of them are designed to energize the trip coils of the appropriate circuit breakers. This concept is illus–

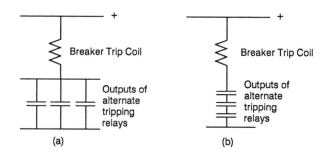

Figure 13.   Use of multiple relays to produce (a) highly dependable relaying, and (b) highly secure relaying.

trated by the highly simplified connection diagram in Figure 13(a). The bias towards dependability is justified, because the consequences of not removing a fault from a modern power system, when it is in a normal state, can be quite severe. On the other hand, if due to some misoperation of the protection system, a circuit element is tripped unnecessarily, the consequences of this over tripping are not very severe, because the power system offers many alternative paths to the flow of power if one element is lost.

However, the situation is quite different when the power system is in a stressed state. This may occur because of some unexpected outages, or because of some incomplete construction projects, or due to some unexpected increases in load on the power system. For any of these reasons, the power system may not be able to withstand an unnecessary loss of a power system element. Instead, one may be willing to take a risk that the relay system may not trip for a fault, provided that the risk of unnecessary tripping is significantly removed. Thus, under conditions of stress in the power system, it may be acceptable to have the power system somewhat less dependable, if it can be made more secure against false trips. Schematically, this is achieved by connecting in series the relay tripping outputs as shown in Figure 13(b). Thus, one may wish to alter the balance between dependability and security of a protection system for a network, based upon an estimate of the health (or otherwise) of the power system. Thus, the concept of adaptively controlling this dependability–security balance by intervention of a central computer seems like a good idea. This is illustrated in Figure 14. Under the control of a central computer, the outputs of alternative protections for the

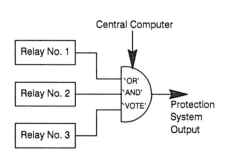

Figure 14. Adaptive adjustment between dependability and security of relaying.

same fault are used in an 'OR', 'AND', or 'VOTE' configuration. The 'OR' configuration is equivalent to the present bias towards increased dependability at the expense of some security. The 'AND' logic will provide an extremely secure protection system, while it may not be as dependable as our existing relaying practice. This option will be used when the power system is operating under extreme stress. Finally, the 'VOTE' option provides the two–out–of–three capability, and is a system with an intermediate level of the dependability–security balance. As with all adaptive relaying ideas, it is essential that any such intervention must be designed and implemented with appropriate safeguards and fall–back positions in case the central computer malfunctions, or communication to it is lost.

Adaptive Reclosing[6]

Reclosing of circuit breakers is usually attempted after a fault has been successfully cleared from a transmission line. The objective is to restore the line to the system as quickly as possible, in the event that the fault that produced the initial trip was of a temporary nature. In this context, we are only interested in automatic reclosing. This may be attempted in about one–half second after the first clearing, in which case it is known as High Speed Reclosing, or it could be attempted in a few seconds, in which case it is known as automatic reclosing. If the fault happened to be temporary, the reclosing is termed successful. Otherwise, the reclosing is unsuccessful, and the circuit breakers must trip the line again. This second exposure to a fault is often quite a severe shock to the power system, and may lead to equipment damage, or power system instability. The unsuccessful reclosing into a three phase fault is frequently used as a criterion to determine the loadability limit of a transmission line.

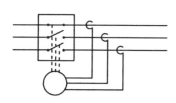

Figure 15. Adaptive reclosing relay for a 3-phase line.

It may be possible to eliminate the likelihood of ever reclosing into a multi–phase fault, if the reclosing relay logic is made adaptive to the prevailing power system conditions. Consider the reclosing relay shown in Figure 15. The relay must control each pole of the reclosing circuit breaker individually. The inputs to the relay are the line–side voltage and current phasors. The reclosing logic closes the breaker of that phase first, which is least likely to have a fault on it. For example, if the initial fault happened to be phase–a to ground, the first phase to reclose would be either phase b or c. One could go even further, and determine which of the two phases – b or c – is nearest to the originally faulted phase, in anticipation that the fault arc may have involved it in the original fault. With one phase successfully energized, the line–side voltages on the remaining phases provide a clear indication of whether those phases have a standing fault, or whether they are unfaulted. At least, it is possible to determine whether one more phase closure could be attempted. If the next phase closing is successful, the logic determines whether the remaining phase could be reclosed. If at any time it is determined that there is a standing fault on the transmission line (through an analysis of the currents and voltages on the transmission line), the entire reclosing procedure can be aborted. Adaptive reclosing is among the first adaptive relaying applications to be studied in some detail.

Detection of instability[7]

Detection of incipient instability of the power system, following the occurrence of a fault, is one of the key functions of many relaying and control functions. Of particular interest, is the out–of–step relaying function, which must determine appropriate control action to be taken, if a power system is on the verge of instability. Typically, as the machines in a power system begin to oscillate with respect to each other, many impedance relays on the power system may perceive the swings as evolutions into fault regimes. It would be wrong to permit the impedance relays to trip for a swing that is not going to lead to instability. On the other hand, if it is determined that a particular swing is going to lead to instability (out–of–step) for some machines, it would be appropriate to separate the power system along preferred boundaries, so that the remaining islands of the power system are afforded a good chance of recovering on their own. Thus the function of the out–of–step relays is two–fold: to block tripping of some relays if a condition is unlikely to lead to instability, and force tripping of some relays if the system is judged to be going towards an out–of–step condition.

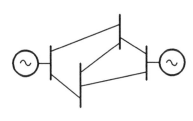

Figure 16. Two-machine power system for instability prediction.

As an example, consider a two–machine power system as shown in Figure 16. This is the simplest possible power system which exhibits the phenomenon of instability, and is the only configuration for which reasonable progress towards solving the problem of detecting instability in real time has been made at present. Although this system would appear to be extremely simple, it turns out that many practical power systems approach this configuration under certain operating conditions. Thus a discussion of such a system is of definite practical interest. The equation of motion of an equivalent rotor for the two machine case is given by the well–known swing equation:

$$\ddot{\delta} = c - k \sin\delta = c - P_e \tag{65}$$

where c and k are constants, and P_e is the electric power being transferred between the two machines. The solution to the differential equation (65) can be graphically illustrated with the help of a $P_e-\delta$ characteristic, and is generally known as the equal–area criterion. We must assume that the particulars of the sine curve – i.e. the parameter k – is known from a knowledge about the power system. The power level c is generally unknown, as it is determined by the post–disturbance equilibrium of the power system. If the power level c can be estimated, it becomes particularly simple to answer the question about the stability of the power system by comparing the area A_1 with the maximum available area A_2 (see Figure 17). If A_1 is smaller than A_2, the oscillation of the machines will be stable, otherwise it will be unstable. The observations made in real time consist of the phase angle difference between the two machines, δ_k $\{k=1,\cdots,n\}$. If it is assumed that the P–δ curve is piece–wise linear between the measurements, it has been shown[7] that the least–squares estimate of the parameter c is given by

$$\hat{c} = \frac{2}{\tau^2(\nu^T\nu)} [\nu^T\delta - \nu^T 1\delta(0)] + \frac{1}{(\nu^T\nu)} [\nu^T MP + \tfrac{1}{3} \nu^T\overline{P}] \tag{66}$$

where τ is the measurement interval,

$$\nu = \begin{bmatrix} 1 \\ 4 \\ 9 \\ \vdots \\ n^2 \end{bmatrix}, \ P = \begin{bmatrix} P_0 \\ P_1 \\ P_2 \\ \vdots \\ P_{n-1} \end{bmatrix}, \ \overline{P} = \begin{bmatrix} P_1-P_0 \\ P_2-P_0 \\ P_3-P_0 \\ \vdots \\ P_n-P_0 \end{bmatrix}, \ \delta = \begin{bmatrix} \delta_1 \\ \delta_2 \\ \delta_3 \\ \vdots \\ \delta_n \end{bmatrix}, \ 1 = \begin{bmatrix} 1 \\ 1 \\ 1 \\ \vdots \\ 1 \end{bmatrix} \tag{67}$$

and

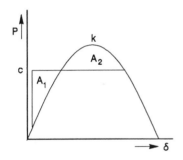

$$M = \begin{bmatrix} 1 & 0 & 0 & 0 & \cdots & 0 \\ 2 & 2 & 0 & 0 & \cdots & 0 \\ 3 & 4 & 2 & 0 & \cdots & 0 \\ 4 & 6 & 4 & 2 & \cdots & 0 \\ \cdot & \cdot & \cdot & \cdot & \cdots & \cdot \\ n & \cdot & \cdot & 6 & 4 & 2 \end{bmatrix} \tag{68}$$

Figure 17. Power angle curve for a two-machine system.

Given enough measurements, it is always possible to determine whether or not the power system is unstable. However, it is of utmost importance to be able to do this in about one–quarter of the electro–mechanical oscillation, so that appropriate control action can be taken to affect the outcome of the oscillation. For the simple two–machine system, it has been shown that it is possible to do this. The real–time computations involved are not too time consuming, and there is work in progress at this time to demonstrate that such on–line prediction and control of instability is practical.   Once the outcome of the oscillation is determined, it is a simple matter to decide the appropriate out–of–step relaying action for the distance relays on the system.

IX. FUTURE PROSPECTS

A number of field installations of the synchronized phasor measurement systems are being planned at this time.   About half a dozen such systems have already been installed.   At the same time, the GPS receiver technology is undergoing significant improvements and cost reductions.   It seems entirely likely that the complete synchronized phasor measurement computer, along with the GPS clock receiver can be packaged in one unit.   It is also expected that the cost of such a unit will be reasonable – comparable to that of a modern relay terminal.   As more experience with these systems is gained, it can be expected that more applications for this technique will be found. Certainly, in the context of an educational institution, this field of research is very attractive to the students, as it brings together the most modern developments in technology, and modern electric power engineering; and thereby makes attractive a field of engineering which has often been accused of being too old–fashioned and stodgy.   From this point of view, it has been a very rewarding activity to initiate and develop the field of synchronized phasor measurements in power systems.

X. REFERENCES

1.   'Global Positioning System', Volumes I, II, and III, Papers published in Navigation, reprinted by The Institute of Navigation, Washington, D.C., 1980.

2.   A.G. Phadke and J.S. Thorp, *Computer Relaying for Power Systems*, (Book), Research Studies Press Ltd., John Wiley & Sons Inc., Second Printing, April 1990.

3. P. Pillay, A.G. Phadke, D.K. Lindner, J.S. Thorp, "State Estimation for a Synchronous Machine: Observer and Kalman Filter Approach", Princeton Conference, 1987.

4. A.G. Phadke and J.S. Thorp, "Improved power system protection and control through synchronized phasor measurements", Proceedings of the 6th National Conference, Bombay, June 4–7, 1990, pp 339–346, McGraw–Hill Publishing Company, New Delhi, India.

5. A.G. Phadke and S.H. Horowitz, "Adaptive Relaying", IEEE Computer Applications in Power, Volume 3, Number 3, July 1990, pp 47–51.

6. A.G. Phadke, S.H. Horowitz, and A.K. McCabe, "Adaptive Automatic Reclosing", CIGRE 1990, Paper No. 34–204, Proceedings of the CIGRE General Assembly, September 1990, Paris.

7. J.S. Thorp, A.G. Phadke, S.H. Horowitz, and M.M. Begovic, "Some Applications of Phasor Measurements to Adaptive Protection", Proceedings of Power Industry Computer Applications (PICA), 1987, pp 467–474.

REAL TIME POWER SYSTEM CONTROL: ISSUES RELATED TO VARIABLE NONLINEAR TIE-LINE FREQUENCY BIAS FOR LOAD FREQUENCY CONTROL

RAYMOND R. SHOULTS
JESUS A. JATIVA
Energy Systems Research Center
The University of Texas at Arlington

I. INTRODUCTION

The interconnection of power systems has allowed various economic and technical advantages that were otherwise unavailable. Incremental increases in system reliability, ability to sell, buy or exchange energy, feasibility of installing larger power plants, incremental increases in system stability, sharing of spinning reserve capacities, and taking advantage of load diversity for economy of operation are the main achievements. However, the interconnection carries with it some difficulties and obligations such as incremental increases in operational complexity, reduced ability to control steady state power flows, propagation of the effect of faults through the entire system, propagation of steady state oscillations, responsibility of matching generation to load within each control area, and shared responsibility to maintain frequency and time error within certain established limits.

CONTROL AND DYNAMIC SYSTEMS, VOL. 43

The area secondary control, referred to as automatic generation control (AGC), is responsible for regulating the system frequency within acceptable error bounds, maintaining correct interchange schedules and distributing generation within each area according to minimum operating cost criteria. Even though tie-line frequency-bias control [1] has been in practice for many years, it still represents the state-of-the art in AGC. It is based on an area control error (ACE) defined as the generation change required to restore frequency and net interchange to desired values. The derivation of ACE is based upon the assumption of steady state conditions. A non-zero ACE represents the load-generation-net interchange unbalance within an area.

The current practice for determining ACE in many electric utilities is based in part on an assumed constant frequency bias. Some experts in the field of real-time power systems control have pointed out that the frequency bias should follow the variable and nonlinear natural frequency response of a given control area. However, others have argued in favor of maintaining a fixed frequency bias larger than any expected natural response in order that all control areas in a given interconnection would respond in the proper direction to restore scheduled values. This is viewed by some as the "good neighbor" policy by which each member of an interconnection should operate, i.e. each control area contributes to restore scheduled values regardless of which area is responsible for the deviation. This policy raises the question of unnecessary control action taken by those areas that were not responsible for the deviation. Is it possible to closely estimate a value for the frequency bias in a real-time environment in order to approximate the variable and nonlinear nature of a power system frequency response? If possible, this has the potential to improve area control performance without jeopardizing the integrity of the interconnection. Such is the case as reported [2] by engineers with Union Electric Company of St. Louis, Missouri. They have reported improvement in control performance, reduced unit regulation, and reduced production cost savings by means of implementing a variable, nonlinear frequency bias.

Exactly what are the benefits that can be realized by using a variable and nonlinear frequency bias? How can it contribute to improve control performance? The intent here is to clarify the role of the frequency bias parameter in the ACE calculation. The area control principle of an interconnected power system requires only the measurements of area frequency and area net interchange to calculate the ACE. This principle leads to a straightforward and efficient method of decentralized control. The input to ACE, which is an assumed steady state calculation, represents a continua of dynamic behavior. During dynamic conditions, ACE comprises the key input to load-frequency control (LFC). The final controller design for LFC may be carried out by either on-line tuning or some sort of control theory approach.

II. INTERCONNECTED POWER SYSTEMS FREQUENCY RESPONSE

The study of the natural frequency response of a given area in an interconnected system is expedited by considering such a system as being divided into two control areas. One is the area of interest, referred to as the internal area, and the remainder of the interconnected system forms the second. Assume that an isolated load change in the internal area, such as that shown in Fig. 1, causes the governors of all the on-line generators to respond, causing the system frequency to stabilize to a new value. This perturbation will be absorbed as a result of generator primary governor response and the frequency dependent portion of the load in both the internal and external areas.

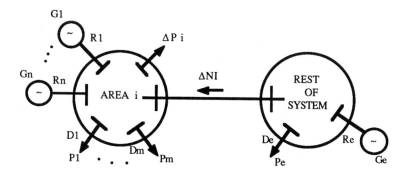

Fig. 1 Isolated load change in area i

Therefore,

$$\Delta Pi = \text{Area (i) Response} + \text{Rest of System Response}$$
$$\Delta Pi = \{\text{Area (i) Generators' Primary Regulation}\}$$
$$+ \{\text{Area (i) Load Damping}\}$$
$$+$$
$$\{\text{Rest of System Generators' Primary Regulation}\}$$
$$+ \{\text{Rest of System Load Damping}\}$$

A. Generator Primary Regulation and Load Damping

The generating unit response due to its governor primary action is approximated by a linear characteristic with a negative slope, often referred to as governor droop. This means that for a sustained increase in load, the governor will respond with increased output but stabilizing at a lower frequency. Load damping represents the self-regulating characteristic of certain types of loads sensitive to frequency changes. It is represented by a positive slope characteristic indicating that with a decrease in frequency a corresponding decrease in load will occur and vice-versa. Figure 2 illustrates these two approximated characteristics in an incremental plane.

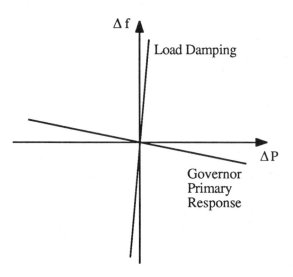

Fig. 2 Generator and Load Frequency Responses.

Thus for the isolated load change in area (i):

$$\Delta P_i = \{\Delta f \sum_{j=1}^{n} \frac{1}{R_j}\} + \{\Delta f \sum_{k=1}^{m} D_k\} + \{\Delta f \sum_{l=1}^{p} \frac{1}{R_l}\} + \{\Delta f \sum_{r=1}^{q} D_r\} \tag{1}$$

$$\Delta P_i = \Delta f \{\sum_{j=1}^{n} \frac{1}{R_j} + \sum_{k=1}^{m} D_k\} + \Delta f \{\frac{1}{R_{rest}} + D_{rest}\} \tag{2}$$

B. Power System Frequency Bias Calculation

The load disturbance in area (i) is made up from area (i) primary response and net interchange deviation. The net interchange deviation and frequency deviation measurements, and the control area's frequency response characteristic provide the information to determine whether or not the load-generation unbalance is inside or outside of the area (i) jurisdiction.

$$\Delta P_i = \beta_i \Delta f + \beta_{rest} \Delta f \tag{3}$$

where:

$$\beta_i = \sum_{j=1}^{n} \frac{1}{R_j} + \sum_{k=1}^{m} D_k \tag{4}$$

is the internal area frequency response

$$\beta_{rest} = \frac{1}{R_{rest}} + D_{rest} \tag{5}$$

is the rest of the system frequency response.
As can be seen,

$$\Delta NI = \beta_{rest} * \Delta f \tag{6}$$

then,

$$\Delta P_i = \beta_i \Delta f + \Delta NI_i \tag{7}$$

By knowing the net interchange deviation and the frequency deviation, the natural frequency response for area (i) may be evaluated based upon Eq. (7). Assuming that only primary response has taken place throughout the interconnection, the natural frequency response for the internal area can be determined by introducing an external load change:

$$\beta_i = -\frac{\Delta NI_i}{\Delta f} \tag{8}$$

since $\Delta P_i = 0$ when the disturbance occurs outside of area (i).

III. AREA CONTROL ERROR CALCULATION [3]

After a frequency deviation has been observed, the area in which the disturbance occurred must take action to return the system to scheduled values by adjusting its generation to meet its own load plus interchange agreements according to interconnection regulations. This supplementary control is realized by applying a reset function to the governor speed reference set point. The area control error (ACE) is the control performance parameter that represents the area generation change required to return to schedule. In essence it contains the information given in Eq. (7) :

$$ACE_i = \Delta NI_i + B_i \Delta f \tag{9}$$

where,

$$\Delta NI = NI_{actual} - NI_{scheduled} \tag{10}$$

NI is positive for exported power, and negative for imported power.

$$\Delta f = f_{actual} - f_{scheduled} \tag{11}$$

and B is called tie-line frequency-bias setting.

It should have the same value as β, the natural frequency response which changes as the system load and generation mix change. It represents a control area response to frequency deviation and may be used to distinguish internal from external load/generation unbalances in such a way that each area may regulate its own load as well as be part of the interconnection frequency regulation. The operating guide 1 of the North American Electric Reliability Council (NERC) recommends guidelines [4] for calculating B, which should closely match the area primary response.

For an area with over or under generation, ACE will be positive or negative, respectively, and the corrective action taken will be to reduce or increase area generation, respectively. An intrinsic negative sign for B has been widely used in the electric industry. Under this modification, with R entered as a positive number, the tie-line control parameters are calculated as:

$$B_i = -\frac{1}{R_i} - D_i \tag{12}$$

and

$$ACE_i = \Delta NI_i - B_i * \Delta f \tag{13}$$

If B is expressed in MW/0.1 Hz,

$$ACE_i = \Delta NI_i - 10*B_i*\Delta f \tag{14}$$

A. ACE Calculation with B Equal to Area Natural Frequency Response

In the following two-area example, both internal and external frequency bias values are set equal to their area frequency response. A 12% load increase occurs in the external area. Calculations represented in Table I show that (a) system frequency drops 0.374%, (b) generation increases in both areas, (c) load decreases in both areas, (d) the internal area helps the external area due to its increase in generation and decrease in load, (e) net interchange deviation indicates an outgoing power to the external area, since there is a lack of generation in the external area, (f) external ACE shows an undergeneration in the external area while internal ACE shows a load/generation balance i.e. ACE=0 in the internal area. Therefore, supplementary control will be executed in the external area only to return to schedule.

B. Analytical Determination of Unnecessary Generation Movements

If B does not closely match the natural frequency response, the following analysis reveals how unnecessary control action is taken by the area with no perturbation. The internal frequency bias in the above example is now set larger than its natural response characteristic.

Table I Two-Area System Steady-State Calculations for a +12% Load Change

System Characteristics	Internal Area	External Area	Total
Connected Generation [MW]	1750	8000	9750
Initial Connected Load [MW]	1400	6400	7800
Scheduled Net Interchange [MW]	-100	100	0
Initial Generation [MW]	1400-100=1300	6400+100=6500	7800
Load Change [MW]	0	+0.12*6400=768	+768
Final Connected Load [MW]	1400	6400+768=7168	8568
1/R [MW/Hz]	1750/(.05*60)=583.33	8000/(.05*60)=2666.667	3250
D [MW/Hz]	1.2*1400/60=28	1.2*7168/60=143.36	171.36
B [MW/Hz]	-583.33-28=-611.333	-2666.667-143.36= -2810.021	-3421.36

$$\Delta f = 768/-3421.36 = -0.22447 \text{ Hz}$$
$$f \text{ actual} = 60-0.22447 = 59.77553 \text{ Hz}$$

Governor Response [MW]	-583.33*Δf=130.942	-2666.667*Δf= 598.592	729.534
Load Response [MW]	28*Δf=-6.285	143.36*Δf=-32.180	-38.4656
Total Contribution [MW]	130.942+6.285= 137.227	598.592+32.180= 630.772	-768
New Generation [MW]	1300+130.942= 1430.942	6500+598.592= 7098.592	8529.534
New Load [MW]	1400-6.285= 1393.715	7168-32.180= 7135.82	8529.534
Actual Net Interchange [MW]	1430.942-1393.715= 37.227	7098.592-7135.82= -37.227	0
ACE=ΔNI-B Δf [MW]	[37.2-(-100)]-(-611.3)*Δf = 0	(-37.2-100)-(-2810.)*Δf= -768	

1. Primary System Response to An External Area Load Increase

Figure 2 illustrates that the operating point has moved from 0 to 1 after a 12% load increase in the external area. Calculations in Table II show that both ACE values call for an increase of generation. At this point an assumption needs to be made in order to illustrate a sequential analysis of generation control: the internal control area executes supplementary control first, followed by the external control area executing supplementary control and so on in an alternating sequence [5].

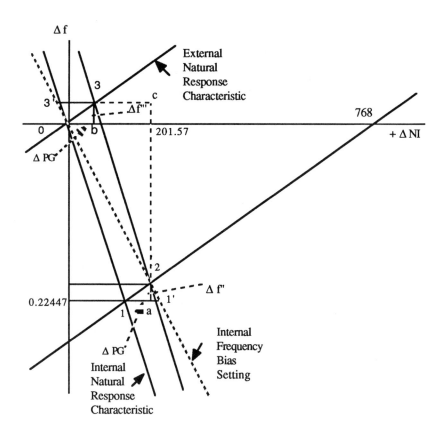

Fig. 2 Unnecessary Supplementary Control Analysis

2. First Internal Area Supplementary Action

The internal ACE from 1 to 1' is corrected by internal supplementary action after which the new operating point is shifted to 2, not to 1' because system elements will respond to the frequency change caused by the internal control action. Point 2 is found in the intersection of both frequency bias settings.

Table II Two-Area System Steady-State Calculations for a +12% Load Change with Internal
Area Frequency Bias Setting Different from its Natural Response Characteristic

System Characteristics	Internal Area	External Area	Total
Connected Generation [MW]	1750	8000	9750
Initial Connected Load [MW]	1400	6400	7800
Scheduled Net Interchange [MW]	-100	100	0
Initial Generation [MW]	1400-100=1300	6400+100=6500	7800
Load Change [MW]	0	+0.12*6400=768	+768
Final Connected Load [MW]	1400	6400+768=7168	8568
1/R [MW/Hz]	1750/(.05*60)=583.3	8000/(.05*60)=2666.6	3250
D [MW/Hz]	1.2*1400/60=28	1.2*7168/60=143.36	171.36
B [MW/H	-583.33-28=-611.3	-2666.667-143.36= -2810.021	-3421.36
Bias Setting [MW/Hz]	-1000	-2810.021	

$$\Delta f = 768/-3421.36 = -0.22447 \text{ Hz}$$
$$f \text{ actual} = 60-0.22447 = 59.77553 \text{ Hz}$$

Governor Response [MW]	-583.33*Δf=130.942	-2666.667*Δf= 598.592	729.534
Load Response [MW]	28*Δf=-6.285	143.36*Δf=-32.180	-38.4656
Total Contribution [MW]	130.942+6.285= 137.227	598.592+32.180= 630.772	-768
New Generation [MW]	1300+130.942= 1430.942	6500+598.592= 7098.592	8529.534
New Load [MW]	1400-6.285= 1393.715	7168-32.180= 7135.82	8529.534
Actual Net Interchange [MW]	1430.942-1393.715= 37.227	7098.592-7135.82= -37.227	0
ACE=ΔNI-B Δf [MW]	[37.2-(-100)]-(-1000)*Δf= -87.24484	(-37.2-100)-(-2810.027)*Δf= -768	

The new conditions are calculated in Table III. The net internal area
generation is made up of primary governor response, load response, and
supplementary control generation. The new internal generation is equal to
previous generation, primary governor response and supplementary
generation. Internal ACE calls for no further action since 2 lies on its
frequency bias setting, while external ACE requests a generation increase
equal to its load perturbation.

Table III Steady-State Calculations at Point 2 of Fig. 2

$$\Delta f'' = 0.0229 \text{ Hz}$$
$$\Delta f = -0.22447 + 0.0229 = -0.20157 \text{ Hz}$$

Governor Response [MW]	$-583.33*\Delta f''=-13.36$	$-2666.667*\Delta f''=$ -61.0667	-74.425
Load Response [MW]	$28*\Delta f''=0.6412$	$143.36*\Delta f''=3.283$	3.924
Total Contribution [MW]	$-13.36-0.6412=$ -13.9995	$-61.0667-3.28294=$ -64.34964	-78.35

Generation By Internal Area Supplementary Control	$\Delta PG2 = PGRC - PLR + PGSC$ $64.3461 = -13.3583 -0.6412 + PGSC$ $PGSC = 78.3456 \text{ MW}$

New Generation [MW]	$1430.942-13.36$ $+78.3456=1495.93$	$7098.592-61.0667=$ 7037.5253	8533.455
New Load [MW]	$1393.715+0.6412=$ 1394.3562	$7135.819+3.283=$ 7139.1019	8533.458

Actual Net Interchange [MW]	$1495.93-1394.35=$ 101.5731	$7037.52-7139.1=$ -101.5766	0

ACE=ΔNI-B Δf	$[101.6-(-100)]-(-1000)*\Delta f=$ 0	$(-101.58-100)-(-2810.027)*\Delta f=$ -768	

3. External Area Supplementary Action

Next, the external area executes supplementary control to shift the operating point to 3. As the shifting progresses, all of the system elements respond to the frequency change. Point 3 is found in the intersection of both natural response characteristics, whose new conditions are calculated in Table IV. The supplementary generation power is equal to the actual external load increase. The new external generation is composed of previous generation, governor primary response , and supplementary generation.

External ACE calls for no further action, whereas internal ACE requests a generation movement in order to reduce its overgeneration. At this point the external area has commanded its controls to supply its load increase.

Detailed calculations at points 2 and 3 are given in the appendix.

4. Second Internal Area Supplementary Action

The internal ACE has a value equal to that at point 1, but with opposite sign, which goes from 3 to 3'. To correct its ACE, the internal area must shift point 3 back to the schedule at point 0. Calculations in Table V show that the frequency deviation is zero and both ACEs are zero. The supplementary generation is equal to that calculated from 1 to 2 but with opposite direction.

Table IV Steady-State Calculations at Point 3 of Fig. 2

	$\Delta f = 0.0229$ Hz		
	$\Delta ft = 0.0229 - (-0.20157) = 0.22447$ Hz		
Governor Response [MW]	$-583.33*\Delta ft = -130.9$	$-2666.7*\Delta ft =$	-729.534
		-598.592	
Load Response [MW]	$28*\Delta ft = 6.285$	$143.36*\Delta ft = 32.180$	38.47
Total Contribution [MW]	$-130.942-6.285=$	$-598.592-32.18=$	-768
	-137.227	-630.772	
Generation By External Area	$\Delta PG3 = PGRC - PLR + PGSC$		
Supplementary Control	$201.5721 - 64.3467 = -598.592 - 32.18 + PGSC$		
	$PGSC = 768$ MW		
New Generation [MW]	$1495.93-130.942=$	$7037.5-598.6+$	8571.92
	1364.9873	$768 = 7206.9358$	
New Load [MW]	$1394.3562+6.285=$	$7139.1019+32.18=$	8571.92
	1400.6412	7171.2819	
Actual Net Interchange [MW]	$1364.99-1400.64=$	$7206.94-7171.3=$	0
	-35.653	35.653	
ACE=ΔNI-B Δf	$[-35.653-(-100)]-(-1000)*\Delta f=$	$(35.653-100)-(-2810.027)*\Delta f=$	
	87.2461	0	

Therefore, the area with no perturbation executes two unnecessary generation movements, the external area supplementary generation is moved for an amount equal to the load increase, ACE calculated for the area with erroneous frequency bias calls for a generation larger than that needed to make it zero.

Table V Steady-State Calculations at Point 0 of Fig. 2

	$\Delta f''' = -0.0229$ Hz		
	$\Delta f = 0.0229 - 0.0229 = 0$ Hz		
Governor Response [MW]	-583.33*$\Delta f'''$=13.36	-2666.7*$\Delta f'''$= 61.0667	74.425
Load Response [MW]	28*$\Delta f'''$=-0.6412	143.36*$\Delta f'''$=-3.3	-3.924
Total Contribution [MW]	13.36+0.6412= 13.9995	61.0667+3.3= 64.34964	78.34914
Generation By Internal Area Supplementary Control	ΔPG0= PGRC - PLR + PGSC -64.3461 = 13.3583 + 0.6412 + PGSC PGSC = -78.3456 MW		
New Generation [MW]	1364.99+13.36 -78.3456 = 1300	7206.9358+61.0667= 7268	8568
New Load [MW]	1400.6412-0.6412= 1400	7171.2819-3.2829= 7168	8568
Actual Net Interchange [MW]	1300-1400 = -100	7268-7168= 100	0
ACE=ΔNI-B Δf	[-100-(-100)]-(-1000)*Δf= 0	(100-100)-(-2810.021)*Δf= 0	

C. Consequences of Setting B Different from β

Many electric utilities set their frequency bias at a fixed value according to one of the NERC recommendations. A straight-line function of the tie-line deviation versus frequency deviation may be determined by observing and averaging the frequency response characteristic for several disturbances during on-peak hours. By using this approach, the largest value of β may be calculated since the greatest amount of generation regulation is on line during these peak hours.

1. Frequency Bias Larger Than Frequency Natural Response

If a fixed frequency bias larger than the natural frequency characteristic is used to calculate an area ACE, as illustrated in Fig. 3, it will help the interconnection by assisting with generation for an external unit loss or an external load increase. ACE will be negative as is calculated in the previous section, which calls for more generation resulting in smaller frequency deviations but larger inadvertent interchanges. It is the key point for maintaining this policy. However, there will be an excess of control in the nonperturbating area generating units.

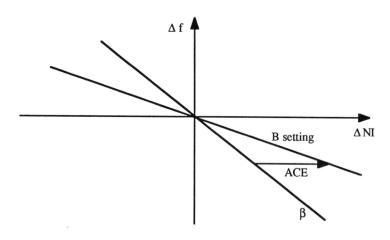

Fig. 3 B setting larger than β

2. Frequency Bias Smaller Than Frequency Natural Response

On the other hand, if a fixed frequency bias smaller than the natural frequency characteristic is used, as illustrated in Fig. 4, it will aggravate the interconnection condition by calling for a reduction of generation when an external unit loss or an external load increase exists. ACE will be positive and the control area will pull the system frequency down to a lower value but with smaller inadvertent interchange.

Because the number of generating units on line can change significantly with time, it can become difficult to guarantee that B will always be larger than β. As pointed out above, when B is smaller than β the control area could work against desired interconnection performance.

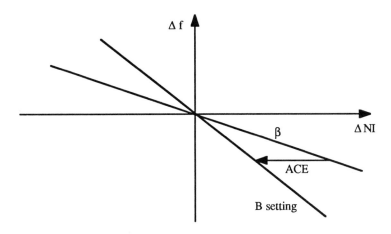

Fig. 4 B setting lower than β

IV. VARIABLE AND NONLINEAR FREQUENCY BIAS

The composition of a power system varies significantly depending upon the time of the day and the season of the year. Generating units are brought on/off line to meet the system load demand. Each unit may substantially affect the system frequency characteristic.

The load damping contribution to β will depend on the amount and type of load connected to the network at a particular period of time. During on-peak hours, the system load may be more frequency sensitive than at of off-peak hours. The transmission network composed of transmission lines, transformers and other shunt type elements may change from one hour to another affecting slightly the system response. Therefore, it can be concluded that the natural frequency response of a power control area varies as the connected load varies.

Control area frequency response nonlinearities are intrinsic to each component element. Time delays to input signals can be found mainly in the speed governor of all units under primary control mode, large frequency-sensitive loads and any other element which needs to update its output for a new input change.

An estimate of the frequency bias may be obtained based on the on-line regulating units and the total area generation, which accounts for area load plus network losses. Nonlinear effects should be accounted for and estimated based on the control area operation experience.

V. DETERMINATION OF β VIA DIGITAL COMPUTER SIMULATION

Basically, there are two system variables that can be monitored in order to evaluate the natural response characteristic β of a control area within a power pool. They are the frequency and net interchange deviations. If a particular power control area is being studied, each frequency deviation due to any external perturbation produces a change in the internal generation and load, which are directly reflected in the net interchange. Consequently, β may be calculated as the ratio of the net interchange deviation to frequency deviation caused by continuous changes of the external area condition.

An automatic generation control simulator (AGCS) [6] has been adapted to determine β of a power control area within a power pool. The AGCS main objective is to provide an appropriate tool for studying the dynamics of a multi-area power system in the time domain. Nine power plants, time-varying loads and energy control centers are modeled to take into account their actual dynamic performance. The dynamic characteristic of major interest in this long-term simulation is the exchange of energy among control areas and from generating units to loads.

A. Procedure to Determine β

Two-area system simulation studies provide the necessary information to evaluate β. All supplementary controls have been disabled to allow a free oscillation of frequency and interchange deviation. The internal area, for which β will be obtained, is composed of detailed models of generators and load. The external area, which is the rest of the power pool, is represented with a single lag equivalent generator and its load. Both areas include typical nonlinearities found in generators and load to resemble an actual power system.

β is obtained by changing the external area load in such a way that the complete non-linear effects be present in the system performance. The external load will be varied in the following manner: a 30-second steady state period is allowed, then a 12% load decrease is produced until the fluctuations are damped out, and the frequency stabilizes at a higher value. After that, a 12% load increase is produced to bring the load back to original conditions. Then, a 12% load increase is produced to cover the other side of the nonlinearities. This approach examines the complete effect of the nonlinear elements present in the system. In order to close this first excursion of external load perturbation in both sides, a 12% load decrease takes the system back again to its original load condition.

At this moment, two points of the characteristic frequency deviation - net interchange deviation have been obtained. To produce a continuous shape, the β-curve requires numerous small, external load changes in both directions. Thus, the external load is moved in steps of 2% following the strategy described above.

B. Natural Response Characteristic of a Power Control Area

By using the procedure described above, the characteristic frequency deviation versus net interchange deviation is obtained for a power control area with 1750 MW of maximum generation.

The natural frequency response characteristic has a non-linear and variable nature. It is non-linear since the power system physical elements contain intrinsic delayed behavior. Also, it is highly variable since the load demand varies in a random manner. Moreover, the load at distinct periods of the day has a very different composition, which requires the commitment of different numbers of units to be on line. Three load levels are considered for the present sample system: heavy, medium and light loads. A composite graph of β for the three load conditions is shown in Fig. 5.

For the heavy load, it can be noted that the internal system responds with higher nonlinear effects as the external area loses load and the internal area helps it by absorbing part of the over-generation. The generators will have less output demand, their primary response operating points will move up, and their nonlinear elements will be forced to move back in order to reduce the area generation. Meanwhile, on the other side, the area generators appear to have a better tendency to track the increasing load. Frequency deviation, interchange deviation, actual frequency bias and an error in linear setting are shown in Table VI.

Table VI Variable and Nonlinear Tie-Line Frequency Bias for Heavy Load.

External Load Change [%]	Frequency Deviation [Hz]	Interchange Deviation [MW]	Actual Freq. Bias [MW/Hz]	Linear Freq. Bias [MW/Hz]	Error in Linear Bias Setting [%]
-12	0.22829	-133.544	-584.98	-611.33	4.5
-10	0.19091	-108.934	-570.60	-611.33	7.1
-8	0.15325	-85.300	-556.61	-611.33	9.8
-6	0.11513	-63.138	-548.41	-611.33	13.5
-4	0.07692	-41.439	-538.73	-611.33	11.9
-2	0.03852	-20.450	-530.89	-611.33	15.2
2	-0.03799	21.721	-571.76	-611.33	6.9
4	-0.07522	45.409	-603.68	-611.33	1.3
6	-0.11214	69.734	-621.85	-611.33	1.7
8	-0.14906	93.907	-629.99	-611.33	3.0
10	-0.18630	116.951	-627.76	-611.33	2.6
12	-0.22296	141.466	-634.49	-611.33	3.6

For medium load, the internal area characteristic is found to be more affected by the non-linear responses. The phenomenon seen in the previous case appears more defined, i.e., the internal area generation is able to better track load increases than load decreases. Moreover, the greater the external load is reduced, the larger the frequency response deviates from an assumed linear response. On the other hand, for external load increases, the area response behaves approximately proportional to the external perturbation. Frequency deviation, interchange deviation, actual frequency bias and an error in linear setting are shown in Table VII.

Table VII Variable and Nonlinear Tie-Line Frequency Bias for Medium Load

External Load Change [%]	Frequency Deviation [Hz]	Interchange Deviation [MW]	Actual Freq. Bias [MW/Hz]	Linear Freq. Bias [MW/Hz]	Error in Linear Bias Setting [%]
-12	0.18000	-80.505	-447.25	-476.27	6.5
-10	0.15860	-67.688	-426.78	-476.27	11.6
-8	0.13033	-53.249	-408.57	-476.27	16.6
-6	0.09261	-39.403	-425.47	-476.27	11.9
-4	0.06047	-25.769	-426.15	-476.27	11.8
-2	0.03002	-13.060	-435.04	-476.27	9.5
2	-0.02970	13.882	-467.40	-476.27	1.9
4	-0.05900	28.760	-487.46	-476.27	2.3
6	-0.08850	42.960	-485.42	-476.27	1.9
8	-0.11804	56.938	-482.36	-476.27	1.3
10	-0.14758	70.813	-479.83	-476.27	0.7
12	-0.17718	84.463	-476.71	-476.27	0.1

For light load, the frequency response presents strong deviations from an assumed linear setting. Again, the area generation exhibits a large nonlinear presence under external load decreases. It may be established that for light load, β should definitely not be approximated by a constant linear function. Frequency deviation, interchange deviation, actual frequency bias and an error in linear setting are shown in Table VIII.

Table VIII Variable and Nonlinear Tie-Line Frequency Bias for Light Load.

External Load Change [%]	Frequency Deviation [Hz]	Interchange Deviation [MW]	Actual Freq. Bias [MW/Hz]	Linear Freq. Bias [MW/Hz]	Error in Linear Bias Setting [%]
-12	0.13903	-5.460	-39.27	-334.533	751.90
-10	0.10874	-1.132	-10.41	-334.533	3113.60
-8	0.09031	2.120	23.47	-334.533	1525.40
-6	0.07120	5.449	76.53	-334.533	537.10
-4	0.04991	8.503	170.37	-334.533	296.40
-2	0.02760	11.358	411.52	-334.533	181.30
2	-0.01565	21.240	-1357.20	-334.533	75.35
4	-0.03611	29.269	-810.55	-334.533	58.70
6	-0.05657	37.263	-658.71	-334.533	49.20
8	-0.07738	45.202	-584.16	-334.533	42.70
10	-0.09700	52.739	-543.70	-334.533	38.50
12	-0.11815	60.351	-510.80	-334.533	34.50

As the load demand gets smaller, the error between β and an assumed linear frequency bias increases.

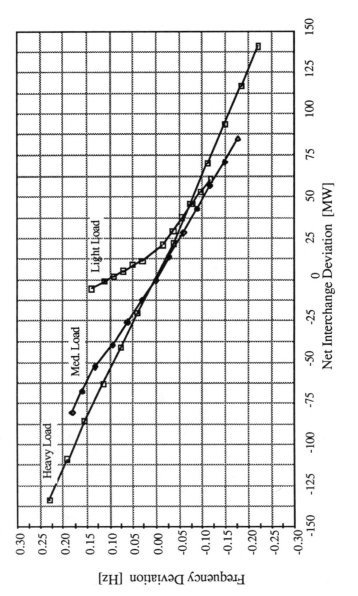

Fig. 5   Natural Response Characteristics for Heavy, Medium, and Light Loads

VI   TWO-AREA POWER SYSTEM AGC SIMULATION

In order to analyze the effects produced by three different frequency bias settings in the ACE calculation during dynamic conditions, a two-area power system has been studied in the AGCS. One of the systems, area (1), is represented in detail by eight types of generating units, an area frequency dependent load model and an energy control center. The other system, area (2), is approximated by a single-lag unit, an hourly load model and its energy control center.

By assuming the system in steady state operation, provided by constant loads in both areas, for fifty seconds, a sudden load decrease of 9% is forced out in area (2) to determine the performance of AGC with area (1) frequency bias settings larger, equal and smaller than the natural response value. The second area frequency bias is set at its natural frequency response. The following graphs show the effect of these settings on frequency deviations, ACE's, measured net interchanges, area loads, inadvertent interchanges, and area net generations. Only a five-hundred second simulation period is being shown in the figures.

One of the most prominent results is given by the system frequency. Smaller frequency deviations are observed with the larger bias setting than with the other two settings. Thus, area (1) helps the interconnection to avoid excessive excursions of frequency by assisting with supplementary control action, as shown in Fig. 6. Moreover, if $B < \beta$, the deviation is even larger, aggravating the system situation.

A. Condition 1: $B_1 > \beta_1$ and $B_2 = \beta_2$

The ACE of area (1), Fig. 7, presents large positive values calling for a decrease of area (1) generation, even though the problem was not caused within its jurisdiction. These movements will help the interconnection by absorbing the excess of area (2) generation produced by its load drop. In Fig. 9, a large flow of imported power entered area(1) due to the external unbalance. In Fig. 11, it can be seen that area (1) generation experiences large control movements. This good neighbor policy also can be observed in Fig. 13, where a large inadvertent interchange moves into area (1).

ACE (2) has also positive values, as expected, which calls for a decrease of generation, Fig. 8. Initially, it shows amounts very close to the actual load drop, since $B_2 = \beta_2$ for area (2). An increase of net interchange leaving area (2) can be seen in Fig. 10. Net generation (2) is decreased following the load drop as shown in Fig. 12. Also the inadvertent interchange (2) depicted in Fig. 14 displays a power flow leaving area (2) alleviating the excess of available generated power.

B. Condition 2: $B_1 = \beta_1$ and $B_2 = \beta_2$

The system conditions are analyzed as follows: ACE(1) fluctuates in small values but dampens out shortly as illustrated in Fig. 7. Strictly speaking it should not deviate from zero, but it does because B_1 is maintained at a constant value during the whole simulation period. Net interchange (1) -Fig. 9, area generation (1) -Fig.11, and inadvertent interchange (1) -Fig.13 experience departures from their initial values since the sudden unbalance in the system requires a dynamic adjustment to maintain equilibrium.

ACE (2), Fig. 8, displays the area (2) load-generation unbalance. In Fig. 10, the net interchange (2) presents the tie-line deviation due to its load drop. Net generation (2) shows a decrease in generated power, Fig. 12, and in Fig. 14 illustrates the inadvertent interchange leaving area (2), both with smaller values than the previous case.

C. Condition 3: $B_1 < \beta_1$ and $B_2 = \beta_2$

The system conditions are analyzed as follows: ACE (1) aggravates the interconnection situation by calling for an increase of generation when there has been a loss of load in its area, as is shown in Fig. 7. The rest of the variables can be compared to those corresponding to the other two settings in Figs. 9, 11 and 13.

ACE (2) requests a change in its generation, Fig. 8, as expected since $B_2 = \beta_2$. The rest of the variables can be compared to those corresponding to the other two settings in Figs. 10, 12 and 14.

The two-area system dynamic simulation corroborates completely the anticipated results produced in the interconnected system variables by the use of different frequency bias settings.

Area (2) will receive additional benefits from the interconnection at the expense of area (1) if $B_1 > \beta_1$, a condition which may not be clearly understood or specified in the operational agreements. On the other hand if $B_1 < \beta_1$, area (1) would act in detriment of the interconnection performance.

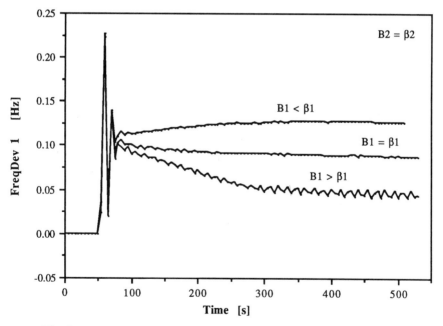

Fig.6 Frequency Deviation for Three Frequency Bias Settings

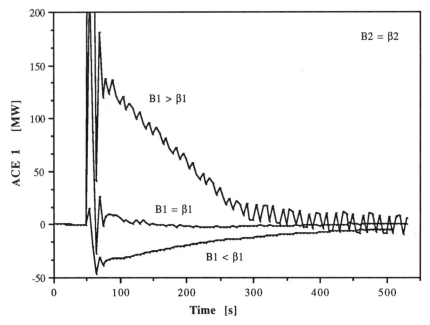

Fig. 7 ACE (1) for Three Frequency Bias Settings.

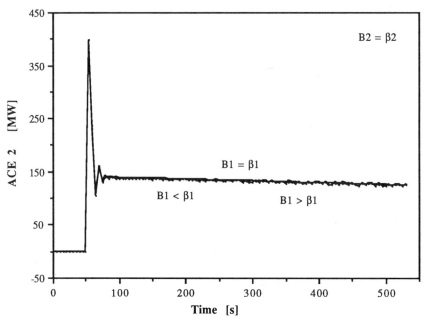

Fig. 8 ACE (2) for Three Frequency Bias Settings.

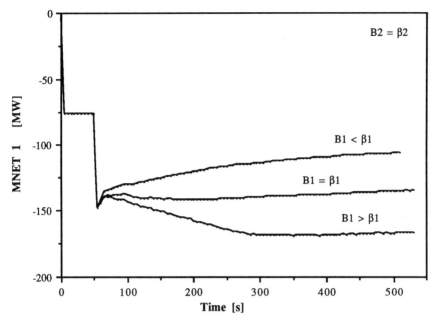

Fig. 9 MNET (1) for Three Frequency Bias Settings.

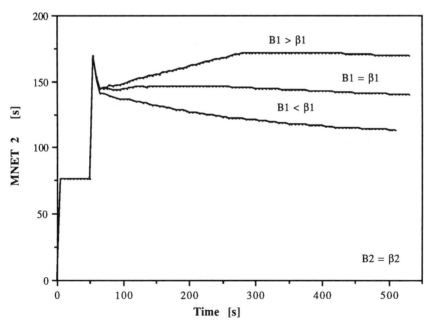

Fig. 10 MNET (2) for Three Frequency Bias Settings.

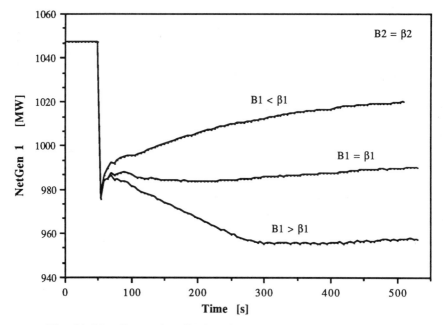

Fig. 11 Net Generation (1) for Three Frequency Bias Settings.

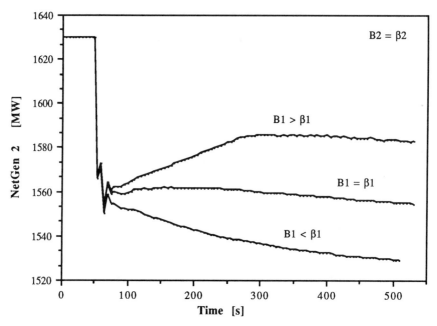

Fig. 12 Net Generation (2) for Three Frequency Bias Settings.

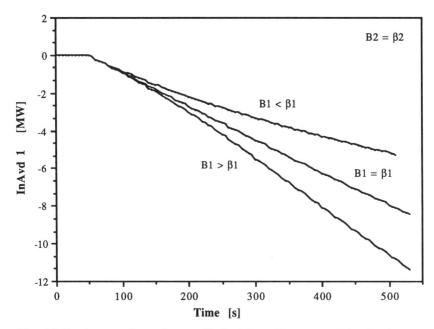

Fig. 13 Inadvertent Interchange (1) for Three Frequency Bias Settings.

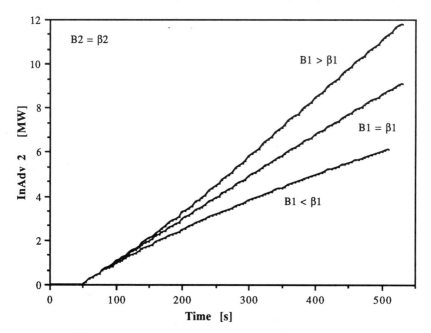

Fig. 14 Inadvertent Interchange (2) for Three Frequency Bias Settings.

VII. IMPROVEMENTS RESULTING FROM THE IMPLEMENTATION OF VARIABLE NONLINEAR FREQUENCY BIAS CHARACTERISTIC

By setting the frequency bias at the value of the area's natural frequency response, estimated from the actual on-line power area components, ACE would not experience erroneous influences in its calculation. Consequently, significant improvements in the NERC control performance criteria may be achieved.

Reduction of unnecessary generation movements, reduction of inadvertent power flow among interconnected areas, and interconnection reliability are some of the benefits. Moreover, better recognition of internal and external load-generation unbalances avoid excess of control actions in the area's generating units, less number of generating units would be needed for system regulation, and operating control performance criteria could be met more efficiently.

Since each member of a power pool represents a time-varying nonlinear system, new algorithms in the light of adaptive control theory need to be investigated. They would have to formulate a realistic decentralized approach of the control of such complex systems based upon well founded principles of actual power systems operations.

As a first step towards this research, an off-line mechanism that contains a substantial degree of realism should be considered. An automatic generation control simulator with features of adaptive control techniques, digital distributed control at power plant levels as well as short-term load forecasting would provide a suitable engineering-tool to carry out in a simulated real-time environment these innovative ideas.

REFERENCES

[1]   N. Cohn, "*Control of Generation and Power Flow on Interconnected on Interconnected Systems*," John Wiley and Sons, Inc., New York, 1961.

[2]   T. Kennedy, S. M. Hoyt, C. F. Abell, "Variable Non-Linear Tie-Line

Frequency Bias for Interconnected Systems Control," *IEEE Paper 87 SM 477-3, IEEE/PES Summer Meeting,* San Francisco, CA, 1987.

[3]   J. Jativa, *"Determination of Natural Frequency Response of a Power System for Automatic Generation Control,"* Master's Thesis, The University of Texas at Arlington, December 1988.

[4]   NERC, "Operating Guides for Interconnected Systems Operations," New Jersey, December 1987.

[5]   B. Oni, H. Graham, L. Walker, "Investigation of Nonlinear Tie Line Bias Control of Interconnected Power Systems," *IEEE Transactions on Power Apparatus and Systems,* vol. PAS-100, pp. 2350-6, May 1981.

[6]   Power Math Associates, Inc, *Automatic Generation Control Simulation,* Contractor Report SAND 86-7044, Del Mar, CA, 1986.

APPENDIX

The actual internal generation movement at point 2 is calculated from figure 2 as follows:

In the triangle (1a2), $2810.021 = \dfrac{\Delta PG}{\Delta f''}$

In the triangle (2a1'), $1000 = \dfrac{87.24484 - \Delta PG}{\Delta f''}$

By eliminating $\Delta f''$: $\Delta PG = 64.3461$ MW

And, the frequency at 2 is: $\Delta f'' = 0.0229$ Hz

f actual $= 59.77553 + 0.0229 = 59.79843$ Hz

The external generation movement at point 3 is calculated from figure 2 :

In triangle (ob3), $2810.021 = \dfrac{\Delta PG}{\Delta f'''}$

In triangle (2c3), $611.3333 = \dfrac{201.5721 - \Delta PG}{0.20157 - \Delta f''}$

By eliminating $\Delta f'''$, $\Delta PG = 64.34673$ MW

And the frequency at 3 is: $\Delta f''' = 0.0229$ Hz

f actual $= 60 + 0.0229 = 60.0229$ Hz

HIGH DYNAMIC PERFORMANCE
MICROCOMPUTER CONTROL
OF ELECTRICAL DRIVES

WERNER LEONHARD

Technical University Braunschweig,
Germany

I. INTRODUCTION

Energy is a precondition of all technical and industrial
activities; as long as it was based on human and animal
labour, only few could live well, while the rest of the
people led a short and miserable life, conditions still found
in many parts of the globe today. In industrialized countries
with ample supply of auxiliary energy the situation is quite
different. Supported by primary energy at the annual rate of
60 MWh per capita, corresponding to a continuous flow of 7 kW
per person, our lives can be relieved of physical effort to
the point, where people voluntarily take up strenuous (and
sometimes silly) exercises such as jogging long distances or
climbing steep mountains on bicycles in order to improve their
physical and mental well-being.

Most of the auxiliary energy is converted several times on
its way from the primary form, such as chemical, nuclear or
hydro, to the eventual destination, with electricity playing
an important part in transmission and distribution. Making
the energy compatible to the needs of the consumer by
converting it to its final form, be it thermal, chemical or

mechanical, is one of the outstanding features of electricity; another is efficiency of conversion and ease of control.

Much of the electrical energy ends up in mechanical work which is needed in enormous quantities wherever transportation of goods and people or industrial processes of all sorts take place. In fact, more than half the electrical energy produced in an industrial country is supplied to electrical machines for converting it back to mechanical work.

Electrical drives are being built for an extremely wide range of power, speed and torque,from micromotors in electric watches to household appliances and huge motors for driving rolling mills or pumps in hydro-storage plants. Unlike internal combustion engines they are always ready for immediate service and can withstand high temporary overload. Since they cause no environmental pollution at the point of use, apart from moderate heat losses, and produce little vibration or noise, they are ideally suited for integration into production machines. Modern machine tools no longer use complicated mechanical transmissions; instead they have multimotor drives that deliver power precisely where and when it is needed, in the best way for the particular task at hand. This is evident with industrial robots that could not be built without distributed drives. Another example of how this has changed the mechanical design is the motor vehicle, which may contain 50 or more auxiliary electrical drives, from starter motor and fuel pump to windshield wiper and seat adjustment. Unfortunately the main drive motor is still non-electric, because present day storage batteries are too bulky, heavy and expensive. A lead-acid battery containing the same amount of energy as a gasoline filled fuel tank would be fifty times as heavy, even when taking the lower efficiency of the combustion engine into account. However, as soon as the vehicle is track-bound and can be supplied with electricity from a catenary or power rail, electrical propulsion is unsurpassed.

Most electrical drives in use today employ ac-motors that are fed directly from the power line and run at about constant speed. This is the best solution for driving a grinder or an escalator, but in other cases an adjustable speed may be preferable or even necessary. Examples are electric traction, reversing rolling mills, mine hoists or boiler feed pumps. With servo drives on machine tools and robots it is necessary to closely control not only speed but acceleration and angular position as well, in order that the tool or robot hand accurately moves on the specified spatial trajectory. The excellent controllability of electrical drives is a main advantage compared with other actuators. It is estimated, that about 20% of all electrical drives are of the controlled variety but this figure is likely to rise due to the increasing automation of transport and production processes and the energy savings possible with controlled drives.

A controlled electrical drive comprises three main constituents

- the electromechanical energy converter, i.e. the electrical machine proper,
- the electronic power converter and
- the control unit.

All three components have undergone considerable evolution in the last 20 years, thanks to new materials, components and design methods. This is most pronounced with the semiconductor technology which has resulted in advanced macro- electronic switches for high power converters and in microelectronic components for their control. The design of todays controlled electrical drives calls for a systems approach based on High Tech components.

II. SPECIFICATIONS FOR CONTROLLED ELECTRICAL DRIVES

Electrical machines can be designed for translational motion but this is a rare exception, the usual drive motor is of the

rotary type. Since the mechanical power transmitted by a
shaft is the product of angular velocity (speed)ω and torque
m, an electrical machine is operable in four quadrants of the
torque-speed plane, i.e. driving and braking in both directions of rotation; in the braking mode power can be delivered
back to the electrical supply. This flexibility is characteristic of electrical drives; reversing the rotation of a shaft
driven by a steam turbine would call for a reversing gear,and
regeneration is clearly impossible.

There are many applications, where steady state operation with
adjustable speed in one or several quadrants of the torque-
speed plane suffices, for example in order to obtain the
benefits of energy saving with a centrifugal pump or compres-
sor. Similar requirements exist in process industries, such as
paper- producing machines or extruders. These specifications
can be achieved with a simple adjustable speed control law, as
seen in Fig. 1, having torque limits for protecting the
equipment; the limits may be adjustable. While it normally
is desirable to impose stringent dynamic specifications with
regard to the compensation of varying load torque, which
acts as a disturbance, changes of speed by variation of the
speed reference are often carried out gradually, i.e. through
a ramp function of the speed reference ω_{Ref}.

Different and usually much more severe dynamic specifications
are imposed on position-controlled drives used as servo motors
and actuators for positioning diverse mechanical plants such
as elevators, radar antennae, reading heads of a storage
disk, machine tools or robots. A frequent condition is not
only to quickly suppress the effects of load torques or
friction, but also to follow a given reference angle $\varepsilon_{Ref}(t)$
with utmost precision over a wide frequency band. For this
purpose it is necessary to analyse the mechanical plant in
much more detail than for a simple speed control. The
rotation of the motor can, of course, be converted to
translational motion by a mechanical linkage.

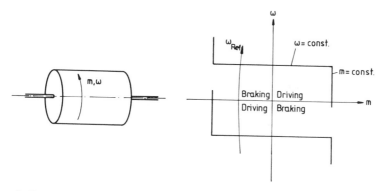

Fig. 1 Torque-speed plane with adjustable speed control curves
and torque limits

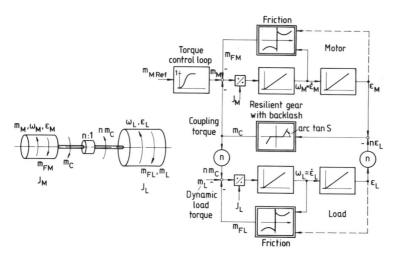

Fig. 2 Block diagram of a drive with two partial inertias

In view of the high acceleration and the peak torques that may
have to be transmitted through the drive shaft,   it is   often
important to include torsional effects in the drive model   and
to   separate   the   motor and the mechanical plant by   defining
two or more partial inertias. This is seen in Fig. 2 for a
two-mass model of a drive train consisting of   two   lumped
inertias J_M and J_L, separated by a coupling or   gear with
stiffness S; it could even exhibit backlash due to wear. Some

of the parameters are variable, for example when the load
inertia is changing with geometry, as with a winder or robot.

The motion is initiated by the inner driving torque m_M of the
motor, which is assumed to be governed by a fast torque
control loop(the subject of detailed discussion later). The
integrators describing the motion of the two inertias are
resulting from Newton's law. The effective load torque m_L
depends on the type of load; with a multi-axes robot, it
may contain various centrifugal, Coriolis and gravitational
components apart from dissipative terms.

That electrical servo drives often require mechanical gears is
due to the fact that the tangential force per unit surface of
the rotor is limited by magnetic saturation and heat losses in
the conductors; hence it is difficult to design electrical
servo motors for high specific torque. If a compact drive with
high output power per weight is needed, it must be achieved
by high motor speed, i.e. use of a gear. This may change if
superconductivity should some day become applicable to ordinary electrical machines. The situation is different with
hydraulic actuators, where the use of normal forces permits a
light-weight design for low speed without resorting to
mechanical gears; this is one of the reasons why hydraulic
actuators find wide application for moving control surfaces of
aircraft or ships or on autonomous vehicles such as fork lifts
or construction equipment.

When speaking of high dynamic performance drives, one usually
refers to the position-controlled or servo type variety, where
the power ratings are small (usually less than 10 kW) but
high dynamic accuracy and fast response to reference changes
are important. Servo drives are normally operating at intermittent duty, where intervals with high impulse torque,
necessary for overcoming friction and producing high acceleration, may alternate with periods of standstill or very low
speed. Simpler speed controlled drives can often be approximated as part of the system in Fig. 2, even though the rated
power may be in the multi-MW region.

A typical control structure that is applicable to both types of drives is shown in Fig. 3, where the coupling (or gear) is assumed to be infinitely stiff so that the motor and load inertias may be lumped together. This type of multi-loop structure has been universally accepted as suitable for position-controlled drives, be they electrical, hydraulic or pneumatic.

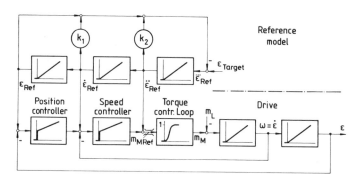

Fig. 3   Cascade control with feed forward of servo drive with lumped inertia

It consists of several nested loops, with the innermost being a torque control loop, which could be substituted by current control. This involves the fastest control action, limited only by the dynamics of the power converter and the internal transients of the machine. High dynamic performance applies to the bandwidth of the torque control; the speed control loop is often much slower, being dominated by the inertia of the mechanical plant. By electronically limiting the torque reference m_{Ref} this inner loop serves as an effective protection of the equipment against mechanical overload by essentially opening the speed loop during overload.

The superimposed control loops for speed and position are a consequence of the double integration in the mechanical plant, with the closed torque control loop serving as an actuator for the speed control and the closed speed control

loop as an actuator for position control. Hence the various
controllers can be designed step by step, i.e. in a decoupled
way, and the speed can be easily bounded. Of course, the
response to reference changes becomes slower (roughly by a
factor of two) for each loop added, but the response to
disturbances entering the plant remains fast, being a responsibility of the inner control loops.

In order to offset this cumulative effect of delayed response
to reference inputs, feed-forward is employed; by adding a
suitably shaped reference signal at the input of each controller, the superimposed controllers need not generate these
reference signals and can remain inactive. Their only task
is to issue small corrective signals to the lower level
controllers. The feed-forward reference signals are produced
by a signal generator forming a reference model for the
desired transients of the control plant. The signals should
be such that they can be followed by the physical plant
without even temporarily saturating the actuator or any of the
controllers. Hence, the transients of the complete control
system will precisely track those of the reference model,
which is a precondition for trajectory control of a multi-
drive plant such as a robot. The reference model drawn in Fig.
3 produces a set of reference signals that are continuous up
to second derivative, i.e. acceleration, because it may be
desirable to avoid a discontinuous torque reference that
could excite resonances in the mechanical plant. By modifying
the model, time optimal transients with limited jerk could
also be achieved.
Even better results, i.e. smaller dynamic position errors, are
obtained if the load torque m_L, which could be a complex
function of controlled and uncontrolled variables, is modelled
and added as a feed-forward signal $\hat{m}_L$ at the input of the
torque controller. This is called feed-forward by inverse load
model because

$$\hat{m}_L = F(\ddot{e}_{Ref}, \dot{e}_{Ref}, e_{Ref}) \qquad (1)$$

represents the inverse of the plant model

$$F(\ddot{\varepsilon}, \dot{\varepsilon}, \varepsilon) = m_M. \tag{2}$$

It is preferable to generate $\hat{m}_L$ on the basis of the signals coming from the reference generator rather than from the plant in order to avoid stability problems due to sensor errors or computational delays. Naturally, there are limits to the achievable accuracy because of the uncertainties of the mechanical plant; modelling friction effects is always questionable. If the load inertia is changing as a known function of the operating state this can also be compensated , as shown in Fig. 4 for the case of a lumped inertia drive, leading to a speed controller with feed-forward adaptation (gain scheduling). Examples are robots with changing geometry or winders with variable build-up of the coil.

As an example of a typical positioning drive the two-dimensional control system shown in Fig. 5a is considered, where a mass M_2 is to be positioned in polar coordinates $\underline{r}(t) = r(t) \, e^{j e(t)}$. There are two independently controlled servo drives, one for rotating the arm around a horizontal axis and

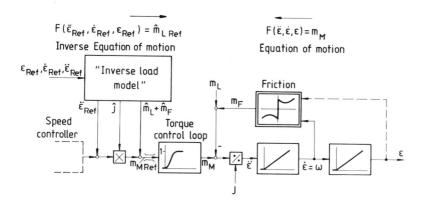

Fig. 4 Parameter adaptation and load torque compensation
        through inverse load model

the other for moving M_2 in radial direction. With some simplifications, the equations of the mechanical plant are as follows

$$(\underbrace{J_1+J_2+M_2r^2}_{J}) \frac{d\omega}{dt} = m_M - 2M_2 \, r \, \omega \, v - M_2 \, g \, r \, \cos\varepsilon - m_F - m_L, \tag{3}$$

$$\frac{d\varepsilon}{dt} = \omega \, , \tag{4}$$

$$M_2 \, \frac{dv}{dt} = f_M + M_2 \, r \, \omega^2 - M_2 \, g \, \sin\varepsilon - f_F - f_L \, , \tag{5}$$

$$\frac{dr}{dt} = v \, . \tag{6}$$

Clearly, there are nonlinear interactions between the two drive axes, caused by centrifugal, Coriolis and gravitational forces.

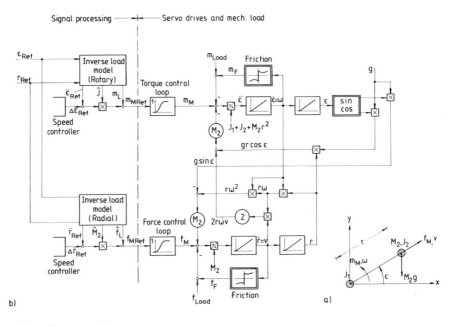

Fig. 5 Two dimensional positioning in polar coordinates; mechanical plant (a) and block diagram (b)

The pertinent block diagram and part of the control scheme are seen in Fig. 5b. Even for this simple mechanical plant, the interactions are quite complex; the relative importance of the coupling terms depends of course on the tangential and radial velocity at which M_2 is moved.

On the basis of eqs.(3-6) the interactive torques for designing the inverse load model may be computed from the reference signals e_{Ref}, r_{Ref} in order to be injected at the input of the torque controllers. This relieves the superimposed speed controllers of generating these reference signals, which would otherwise call for a temporary speed deviation. The assumed cancellation of interactive and feed-forward forces is of course only an approximation in view of the residual lag of the torque control loop, but the smaller this lag is the more effective the compensation becomes. A typical value for the lag is 0,5-2 ms for high-performance electrical servo drives in the power range to 10 kW.

As shown in Fig. 5b, varying inertia J or mass M_2 are also compensated by the inverse model in order to maintain constant loop gain. When finally adding the acceleration reference signals, there remain only friction torques and unmodelled errors caused by unknown load effects which have to be corrected by the speed controllers.

In order to obtain a quantitative estimate of how the various measures affect the positional error, a simulation has been made of the complete control system for a positioning transient as shown in Fig. 6a,b. The arm is moved by 90° and back in a time-optimal transition with limited jerk, i.e. continuous acceleration; at the same time, the radial position is changed. The two torque control loops are assumed to be identical, having second order dynamics with a residual lag of 2 ms. In Fig. 6b the reference trajectory and the computed trajectory of mass M_2 are shown when the control is activated by the positional references, without any feed-forward signals added. Clearly, there is a noticeable dynamic error, even when neglecting the exact time scale on the trajectories.

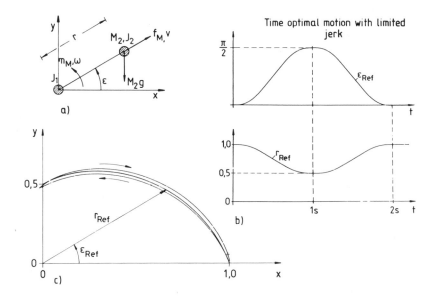

Fig. 6 Time-optimal transition of polar positioning system, no
feed-forward added

The position error $\underline{r}_{Ref}(t)-\underline{r}(t)$ is plotted in Fig. 7 on
enlarged scales for different feed-forward schemes, beginning
with the case already shown in Fig. 6c and proceeding in
several steps to the full inverse load model. The presence of
Colomb dry friction was assumed in all cases but no attempt
has been made to include it in the load model; its compensation was left to the speed controllers. The example, which of
course cannot be generalized for other plants, indicates that
an improvement of the dynamic position error by at least one
order of magnitude seems possible with well-tuned feed forward
schemes.

What is demonstrated here on a simple two dimensional arm
assumes staggering proportions on robots, where 6 or more
independent drive axes may be involved. It has been shown that
dynamic load compensation by a complete 6-axes inverse model
involves about 1400 algebraic operations that would have to be
carried out during each sampling period of the controller,
possibly 1-2 ms. Even when neglecting the interactions of the

hand   axes and reducing the model to the three   main   axes,   a
bulk   of about   hundred algebraic operations remains.   On   the
other hand,   as shown by actual measurements on robots,   there
are   substantial benefits in accuracy when compensating   these
systematic disturbance effects as closely as possible to where
they occur [23,29].

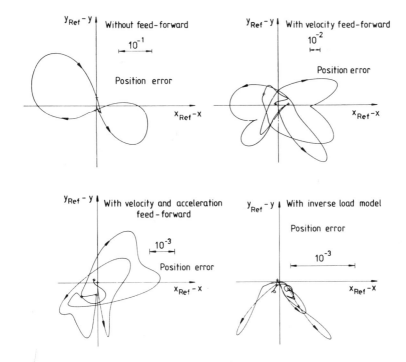

Fig. 7 Dynamic position errors of the polar positioning system
       for various feed-forward schemes

III. ENERGY CONVERSION IN ELECTRICAL MACHINES

After   discussing a suitable control structure for the   speed
and position of a shaft driven,   in principle,   by any source
of mechanical power,   let us now turn to the process of energy
conversion itself,   i.e.   the creation of torque   and   subse-

quently speed of the rotor from the voltages and currents
applied to the electrical windings of the machine. This
corresponds to the block "torque control loop" in Fig. 3 and
5b, containing additional lower level control systems.

In simplified terms, energy conversion in most electrical
machines is based on two laws of physics: Creation of
mechanical forces on current carrying conductors in a magnetic
field (Lorentz) and the law of induction, linking the change
of ·magnetic flux in a coil to a voltage across the terminals
of the coil (Faraday). The torque then results from a
summation of all the tangential forces along the circumference of the rotor, and the terminal voltage from a summation
of the voltages induced in all series connected coils,
distributed along the circumference. Electrical power, on the
other hand, is the product of current and terminal voltage.

A. MATHEMATICAL MODEL AND CONTROL OF DIRECT CURRENT (DC)-DRIVES

The oldest and until today dominant controlled electrical
drive employs dc-machines; a simplified cross section is seen
in Fig. 8.

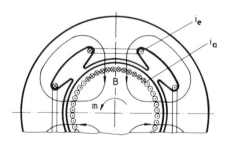

Fig. 8 Simplified cross section of a 4 pole dc-machine

A radial magnetic field, changing periodically along the circumference, is created by field windings or permanent magnets on the stator side of the airgap. Opposite the poles are axial conductors placed in slots on the rotor (armature); they carry the armature current in a direction corresponding to that of the local magnetic flux density so that all conductors produce tangential forces in the same direction. The important feature is that the spatial distribution of the armature currents remains fixed as the rotor moves; this requires that the current in any individual conductor must be alternating, having a rectangular wave shape and a frequency equal to the number of stator periods passed per second, i.e. the product of the number of pole pairs and speed. The periodic commutation of the armature current in the conductors passing through the neutral zone of the magnetic field is performed by a mechanical commutator which is part of the rotor; it serves as an inverter clocked by the angular position of the rotor, thus converting the direct voltage and current applied at the brushes to alternating voltages and currents in the armature windings. Thus the electrical power fed to the armature is converted to mechanical power at the rotor shaft (subtracting the unavoidable power losses in windings, brushes, friction etc).

Because of the fixed angular relationship of fields and currents, the machine-internal commutator causes the dc-machine to exhibit a very simple control structure; on the other hand, it becomes a limiting factor for the design, because the time available for commutation is reduced as the speed rises, and the magnetic energy contained in the inductance of the armature coil to be commutated increases with the current and the size of the machine. Therefore there are practical limits within which dc-machines can be built; at a speed of 60 min^{-1} the limit may be 10 MW but at 6000 min^{-1} only about 100 kW. Besides, there are the problems of sparking, the need for periodic maintenance and the additional axial length which is a disadvantage when space is at a premium such as on traction drives; also, there are narrow thermal limits for the time the motor can be allowed to produce torque in standstill. On the other hand, the dynamic structure of a dc-machine is straight forward because field and armature windings exhibit little magnetic coupling, which may be further reduced by a compensating winding on larger and more heavily loaded machines. Therefore it is possible to control the two important quantities, flux and armature current separately; they are the main physical quantities determining torque where the flux, controlled through the field current, has a similar effect as a continuously variable gear ratio, trading torque for speed.

The equations of a separately excited dc-machine are

$$L_a \, di_a/dt + R_a \, i_a + e = u_a \qquad \text{armature circuit,} \qquad (7)$$

$$e = c \, \Phi_e \, \omega \qquad \text{induced voltage (emf),} \qquad (8)$$

$$J_M \, d\omega/dt = m_M - m_L \qquad \begin{array}{l}\text{Newton's law assuming}\\ \text{lumped inertia,}\end{array} \qquad (9)$$

$$m_M = c \, \Phi_e \, i_a \qquad \text{motor torque,} \qquad (10)$$

$$N_e \, d\Phi_e/dt + R_e \, i_e = u_e \qquad \text{field circuit,} \qquad (11)$$

$$\Phi_e(i_e) \qquad \text{static magnetizing curve,} \qquad (12)$$

$$d\varepsilon/dt = \omega \qquad \text{rotor angle,} \qquad (13)$$

where the various parameters are defined by the equivalent circuit in Fig. 9a. The pertinent normalized block diagram is shown in Fig. 9b [16].

As mentioned before, there are two inputs for controlling the motor, field and armature voltage. The main power is supplied to the armature while the field voltage serves as an auxiliary input when higher than rated speed in the low torque region is needed; otherwise the flux is maintained at rated value. With

smaller   servo motors having permanent   magnets   for   excitation, the field windings are omitted, $\dot{\Phi}_e \sim \dot{\Phi}_{eo}$.

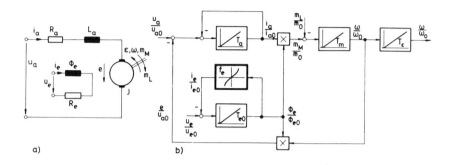

a)   b)

Fig. 9   Equivalent circuit and block diagram of separately
excited dc-machine

The   most common power actuators for dc-drives are   transistor
choppers   for   low   power servo motors   and   phase   controlled
thyristor   converters   for high   power   drives   (Fig.10).   For
reasons of efficiency,   both types of converters employ   electronic switches;   the converters are adaptable to operation of
the   drive   in all four quadrants of the   torque-speed   plane,
i.e.   driving   and   braking with   regeneration. The   use   of
transistor   choppers   for   servo motors is   motivated   by   the
higher   switching frequency,   which results in faster   current
and torque control.

The   usual   practice   is to control armature   current   as   the
innermost control function, effectively supplying the armature
with   impressed current.   This serves for limiting the   torque
and helps to protect the equipment against   overload;   current
control is included in Fig.   10.   The dc-link in Fig.   10a   is
supplied   by a diode rectifier,   thus excluding   regeneration.
The power fed back during braking is dissipated in a   resistor
R_B. This is usual practice for low power drives with intermittent duty;   at higher power ratings,   the power supply for the
dc-link can be made reversible, allowing regeneration.

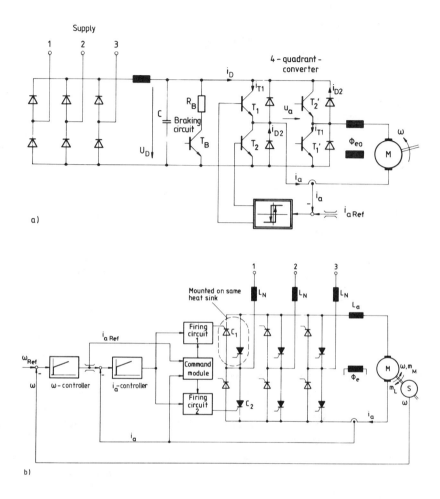

Fig. 10 4-quadrant power supplies for dc-motors
   a) Transistor chopper for servo motors
   b) 6-pulse line-commutated thyristor converter
      without circulating current

With a transistor chopper the control is performed by alterna-
tely making diagonally placed transistors conducting, at
constant frequency but varying duty cycle; this is called
pulse width modulation (PWM). Blocking a formerly conducting
transistor forces the armature current into the opposite free-

wheeling diode while the parallel transistor is made conducting. This "forced" commutation is not feasible with   normal
thyristors contained in Fig. 10b. Here the   commutation   is
achieved by sequential firing of the next thyristor, thus
creating a temporary short circuit; the outgoing thyristor is
then blocked by the line voltages during a brief overlap
period. This is called natural or line commutation. Control of
thyristor converters is achieved by shifting the firing pulses
with respect to the line voltages, which constitutes a pulse-
phase-modulation (PPM). Details of converter control cannot be
discussed here. There are more recent types of thyristors
which may also be blocked by a low power control signal (gate
turn-off-thyristors, GTO), but they are not presently applied
to dc-drives.

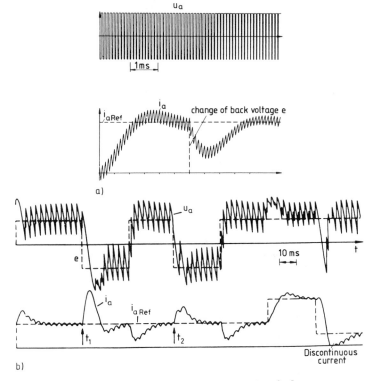

Fig. 11   Step   responses   of   current control loops
          a) Transistor chopper
          b) Line commutated 6-pulse converter

The 4-quadrant line-commutated converter shown in Fig. 10b consists of two 2-quadrant converter circuits C_1 and C_2 which alternatively carry the load current depending on its direction. Firing a thyristor at a time when the other converter is still conducting must be avoided at all costs as this would lead to a short circuit condition resulting in an interruption of service; the coordination of the two converter-halves is performed by the block "command module".

Because of the electronic power converter the current control loop is exhibiting rapid response. With a transistor chopper switching at a clock frequency of 5 kHz, the residual lag for signals of small amplitude can be less than 1 ms and with the 6-pulse converter operating on the 50 Hz-line, the response time is less than 10 ms, assuming well tuned current controllers. Typical transients are shown in Fig. 11.

The higher level control usually has a cascaded structure according to Fig. 3; this is shown in Fig. 10b for the speed control loop. The torque reference is limited in order to restrict the drive to the permitted operating range. The feed-forward signals may be produced by a reference model as in Fig. 3.

The controllers are characterized by blocks containing their step responses; the proportional plus integral (PI) controllers shown are only symbolic, they could also be other types of controllers as long as stability conditions and accuracy requirements are maintained.

On most dc-servo drives, the current- and speed-controllers are of the analogue type employing operational amplifiers, but in view of the required positional accuracy, typically 10^{-5} to 10^{-6}, digital sensors and sampled data controllers are very common with the position control loop: this also simplifies storing and processing of reference data. There now exists a general tendency to convert to fully digital control, even at the lowest control level, because of the improved flexibility of digital methods and the cost advantages made possible by the advances of microelectronics. A complete control scheme of a dc-servo drive with reference model is seen in Fig. 12.

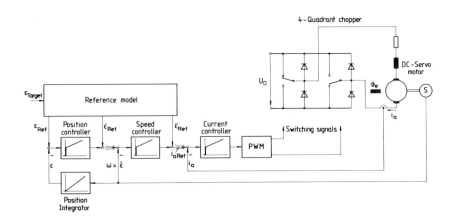

Fig. 12 Position control of a dc-servo motor with
PM-excitation

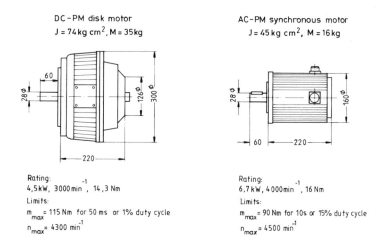

DC-PM disk motor
$J = 74\,kg\,cm^2$, $M = 35\,kg$

AC-PM synchronous motor
$J = 45\,kg\,cm^2$, $M = 16\,kg$

Rating:
4,5 kW, 3000 min^{-1}, 14,3 Nm
Limits:
m_{max} = 115 Nm for 50 ms or 1% duty cycle
n_{max} = 4300 min^{-1}

Rating:
6,7 kW, 4000min^{-1}, 16 Nm
Limits:
m_{max} = 90 Nm for 10s or 15% duty cycle
n_{max} = 4500 min^{-1}

Fig. 13 Comparison of dc- and ac-servo motors with
PM excitation

It was pointed out before that the mechanical commutator is
the weakest link of a dc-drive. Hence a considerable effort
has been spent to make controlled ac-drives competitive with
dc-drives. This has been achieved in recent years thanks to

the development of semiconductor technology. The progress
made possible also in the design of electromechanical
converters is exemplified by the two industrial servo motors
seen in Fig. 13, with the ac-motor exhibiting about half the
mass and inertia. Similar relations exist with large motors
in the MW-region needed for traction, which is greatly
accelerating the general transition from dc- to ac-drives.

B. MATHEMATICAL MODEL AND CONTROL OF ALTERNATING CURRENT (AC)-DRIVES

Since commutation of currents is necessary with all rotating
machines having periodically distributed magnetic fields,
there have been numerous attempts to perform the commutation
process in static equipment outside the machine proper. The
result are arrangements such as seen in Fig. 14, where a
variable frequency converter, preferably of solid state, is
supplying power to the stator windings of an ac-machine; the
drive is thus freed of moving contacts and the restrictions
mentioned before are lifted.

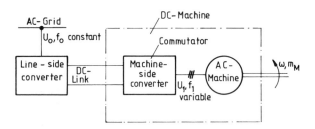

Fig. 14 Controlled ac-drive with dc-link

This considerably simplifies the design and construction of
the machine but, unfortunately, complicates its control dynamics which are much more involved than those of a dc-machine;
this is so because there no longer is a fixed angular
relationship for the distribution of fields and currents in
stator and rotor and there is variable magnetic coupling

between stator and rotor windings;   important quantities,
such as rotor currents in induction machines with cage rotor
cannot be measured at all.

In order to arrive at a mathematical model of an ac-machine,
considerable simplifications are necessary; this is acceptable
as long as the purpose of the model is not the design of the
machine itself but of the control system, where inaccurate
knowledge of the plant parameters is a way of life. On the
other hand, a machine model suitable for control design should
include all major dynamic effects and remain valid for any
voltage and current wave-form that may be impressed by the
switched converter;   this is much more demanding than the
usual assumption of sinusoidal wave-forms in steady state.

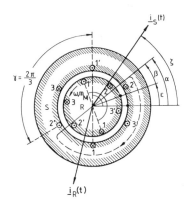

Fig. 15   Cross section of a symmetrical two-pole three phase
          ac-machine

Figure 15 depicts a simplified cross section of a general ac-
machine with circular symmetry - no salient poles, no slots -
where three phase windings with axially orientated and finely
distributed conductors are assumed on the smooth stator and
rotor surfaces.   The spatial distribution of the windings is
such that the ampere-turns waves in stator and rotor are - at
any instant - sinusoidal functions of the circumferential
angles α or β, respectively. Only the conductors at the
center of each conducting band are shown in Fig. 15. If the

instantaneous values of the stator currents are $i_{S1}(t)$, $i_{S2}(t)$, $i_{S3}(t)$ where, because of the isolated neutral point,

$$i_{S1}(t) + i_{S2}(t) + i_{S3}(t) = 0 \qquad (14)$$

holds at any time, the ampere-turns wave in the stator is, with $\gamma = 2\pi/3$ and N_S the number of turns of each stator phase,

$$\Theta_S(\alpha,t) = N_S[i_{S1} \cos\alpha + i_{S2} \cos(\alpha-\gamma) + i_{S3} \cos(\alpha-2\gamma)]$$

$$= N_S/2 \ [\underline{i}_S e^{-j\alpha} + \underline{i}_S^* e^{j\alpha}] \ . \qquad (15)$$

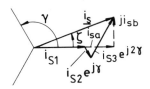

Fig. 16   Time dependent complex current vector

The current vector, also called space vector [15],

$$\underline{i}_S(t) = i_{S1}(t) + i_{S2}(t)e^{j\gamma} + i_{S3} \ e^{j2\gamma}$$

$$= i_{Sa}(t) + j \ i_{Sb}(t) = i_S(t) \ e^{j\zeta(t)} \qquad (16)$$

indicates the instantaneous magnitude and orientation of the spatially sinusoidal ampere-turns wave, Fig. 16. $\underline{i}_S^*$ is the conjugate complex current vector. A corresponding expression holds for the rotor ampere-turns wave,

$$\Theta_R(\beta,t) = N_R/2 \ [\underline{i}_R e^{-j\beta} + \underline{i}_R^* e^{j\beta}] \ . \qquad (15b)$$

The use of complex vectors is a consequence of the two-dimensional representation in Fig. 15, where the axial dimension is redundant when end effects are neglected. The current vectors are convenient mathematical abstractions but the ampere-turns waves are measurable real quantities. The time dependent vectors should not be confused with the phasors commonly employed for steady state analysis of sinusoidal currents and voltages.

Assuming infinite permeability of the iron core and neglecting eddy currents, the radial flux density at the stator side of the airgap is obtained by superposition at the stator angle α = $\varepsilon + \beta$, where ε is the mechanical angle of rotation.

$$B_S(\alpha,t) = \mu_O/2h \; [\Theta_S(\alpha,t) + \varkappa \, \Theta_R(\varepsilon+\beta,t)] \; ; \tag{17}$$

h is the lenght of the airgap and $\varkappa < 1$ a coupling factor which takes magnetic leakage into account. Integrating the flux density over the rotor surface of lenght 1 and observing the sinusoidal distribution of the windings leads to the flux linkage of the stator windings, also expressed in vectorial form

$$\underline{\Psi}_S = L_S \, \underline{i}_S + M \, \underline{i}_R \, e^{j\varepsilon} \, , \tag{18a}$$

where $\underline{i}_R \, e^{j\varepsilon}$ represents the rotor ampere-turns in stator coordinates; a similar equation holds for the rotor flux, defined in rotor coordinates

$$\underline{\Psi}_R = L_R \, \underline{i}_R + M \, \underline{i}_S \, e^{-j\varepsilon} \, . \tag{18b}$$

Assuming equal number of turns in the stator and (fictitious) rotor windings and introducing the usual leakage factors

$$M = L_O, \quad L_S = (1+\sigma_S) \, L_O, \quad L_R = (1+\sigma_R) \, L_O \, , \tag{19}$$
$$\sigma = 1-1/(1+\sigma_S)(1+\sigma_R)$$

results in the voltage equations for stator and rotor windings

$$\underline{u}_S = R_S \underline{i}_S + d \, \underline{\Psi}_S/dt = R_S \underline{i}_S + L_S \, d\underline{i}_S/dt +$$
$$+ \, L_O \, d/dt(\underline{i}_R \, e^{j\varepsilon}) \, , \tag{20}$$

$$\underline{u}_R = R_R \underline{i}_R + d \underline{\psi}_R/dt = R_R \underline{i}_R + L_R d\underline{i}_R/dt +$$

$$+ L_0 d/dt(\underline{i}_S e^{-j\epsilon}) , \tag{21}$$

where
$$\underline{u}_S(t) = u_{S1}+u_{S2} e^{j\gamma} + u_{S3} e^{j2\gamma}=u_{Sa}(t) + ju_{Sb}(t) \tag{22}$$

is a complex voltage vector combining the line-to-neutral stator voltages. In case of a cage motor, the rotor windings are short circuited, $\underline{u}_R=0$.

The electrical torque of the motor can be derived by integrating the tangential Lorentz-forces caused by the radial flux density and the axial current distribution at the rotor surface, Fig. 17. This results in a vector product for instantaneous torque

$$m_M(t) = 2L_0/3 \ \text{Im} \ [\underline{i}_S(\underline{i}_R e^{j\epsilon})^*] . \tag{23}$$

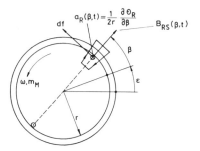

Fig. 17 Lorentz-force acting on circumference of rotor

Thus the mechanical motion of the rotor is described by

$$J_M \ d\omega/dt=m_M-m_L = 2L_0/3 \quad \text{Im}[\underline{i}_S(\underline{i}_R e^{j\epsilon})^*]-m_L, \tag{24}$$

$$d\epsilon/dt = \omega , \tag{25}$$

where J_M is the inertia of the rotor, m_L the effective load torque at the coupling of the motor, ω its angular velocity and ϵ the angle of rotation.

The four complex differential equations or the equivalent six real equations represent a mathematical model of a general symmetrical ac-machine, which is adequate for designing high performance controls. By imposing different constraints on the rotor circuit, the generalized model can be adapted to synchronous, asynchronous or doubly-fed machines. This is shown in Fig. 18. The model is valid for any waveforms of voltages and currents, as long as eq. (14) is satisfied.

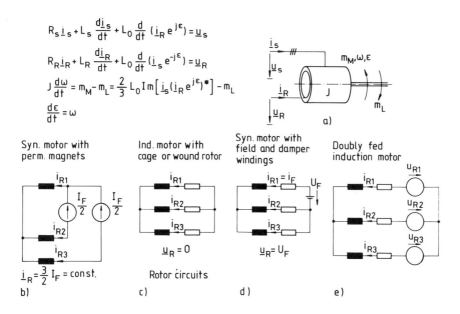

$$R_s \underline{i}_s + L_s \frac{d\underline{i}_s}{dt} + L_0 \frac{d}{dt} (\underline{i}_R e^{j\varepsilon}) = \underline{u}_s$$

$$R_R \underline{i}_R + L_R \frac{d\underline{i}_R}{dt} + L_0 \frac{d}{dt} (\underline{i}_s e^{-j\varepsilon}) = \underline{u}_R$$

$$J \frac{d\omega}{dt} = m_M - m_L = \frac{2}{3} L_0 \mathrm{Im} \left[\underline{i}_s (\underline{i}_R e^{j\varepsilon})^* \right] - m_L$$

$$\frac{d\varepsilon}{dt} = \omega$$

Syn. motor with perm. magnets

$$\underline{i}_R = \frac{3}{2} I_F = const.$$
b)

Ind. motor with cage or wound rotor

$$\underline{u}_R = 0$$

Rotor circuits
c)

Syn. motor with field and damper windings

$$\underline{u}_R = U_F$$
d)

Doubly fed induction motor

e)

Fig. 18 Model of a symmetrical ac-machine with different constraints for the rotor circuit

This indicates a variety of possible ac-drives; only the cases

a) synchronous motor with permanent magnets in the rotor
   and
b) induction motor with cage rotor

will be discussed here. They are the types of motors which, in combination with PWM-inverters, are preferred for high dynamic performance drives at lower to medium power ratings.

When splitting the complex model equations (20,21,24,25) into real and imaginary parts, six nonlinear first order equations emerge which can be represented by block diagrams involving six state variables. A typical version of such a dynamic model of a voltage supplied induction motor with short circuited rotor winding is shown in Fig. 19. It is of considerable complexity which had control engineers puzzling for a long time , even though highly sophisticated design techniques had already been available during the sixties. The problem was finally solved by introducing moving coordinates and formulating the principle of field orientation [3, 9].

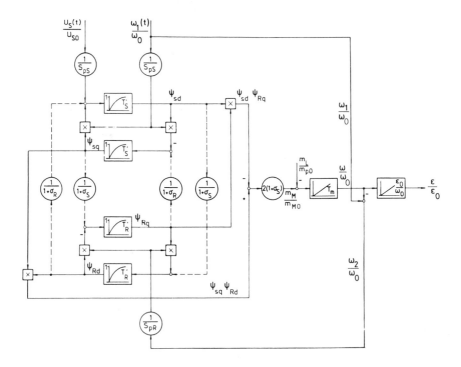

Fig. 19   Dynamic block diagram of voltage-fed   induction
           machine

There have been various attempts of controlling ac-motors on the basis of simplified steady-state models, which may at least be briefly mentioned.

In order to operate an ac-motor at an acceptable efficiency,
it is necessary to supply the stator windings with voltages
and currents of a frequency depending on the rotational   speed
of   the   machine.   Power electronic converters   are   producing
voltages which possess, in addition to the desired fundamental
component,   a multitude of higher harmonics;   the currents are
also distorted but less so than the voltages because they   are
smoothed by the inductances of the machine.

When   neglecting the harmonics and assuming in steady state   a
symmetrical set of line-to-neutral stator voltages having   the
RMS-value U_S and frequency ω_1

$$u_{S\nu}(t) = \sqrt{2}\ U_S\ \cos\ [\omega_1 t + \tau - (\nu-1)2\pi/3], \qquad \nu=1,2,3, \tag{26}$$

the complex voltage vector (eq.   22) consisting of the   fundamental components

$$\underline{u}_S(t) = \frac{3\sqrt{2}}{2}\ U_S\ e^{i\tau}\ e^{j\omega_1 t} = \frac{3\sqrt{2}}{2}\ \underline{U}_S\ e^{j\omega_1 t} \tag{27}$$

moves   on   a circular path. $\underline{U}_S$ is the phasor   of   the   stator
voltages.   A   corresponding   definition holds for   the   stator
currents

$$\underline{i}_S(t) = \frac{3\sqrt{2}}{2}\ \underline{I}_S\ e^{\ j\omega_1 t} . \tag{28}$$

Introducing   eqs.(27,28)   in eqs.(20-25) and   assuming   steady
state conditions,   i.e.   constant load torque m_L and speed ω,
results in the well-known steady-state phasor equations,

$$\underline{U}_S = (R_S + j\omega_1 L_S)\ \underline{I}_S + j\omega_1 L_o \underline{I}_R,$$

$$\underline{U}_R = (R_R + j\omega_2 L_R)\ \underline{I}_R + j\omega_2 L_o \underline{I}_S, \tag{29a-d}$$

$$m_M = 3\ L_o\ \text{Im}\ [\underline{I}_S\ \underline{I}_R^{\ *}] = \text{const.}\ ,$$

$$\omega_1 = \omega + \omega_2,$$

where ω_2 is the slip frequency and $\underline{U}_R$=o holds in case   of   an

induction motor with cage or short circuited rotor winding.
They correspond to the single phase equivalent circuit diagrams shown in Fig. 20.

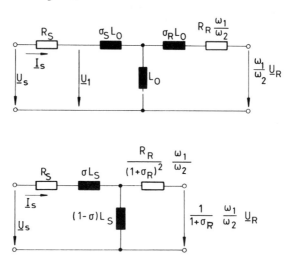

Fig. 20   Single phase equivalent circuits of ac-motor

These models are often used for designing simple adjustable
speed ac-drives such as those with constant flux, i.e. constant stator voltage to frequency ratio; however, when
aiming for high dynamic performance, the differential equations (20-25) are clearly preferable for modelling the plant
dynamics.

As a first step, considerable simplification is possible by
supplying the motor with impressed stator currents instead of
voltages; this can be achieved by again opting for current
control as the innermost control loop. If the control bandwidth is sufficiently wide, limited only by the power converter, the stator voltage equation (20), being now subject
to the current controllers, is no longer affecting the dynamics of the drive; hence the drive control is greatly
simplified. Besides, controlling the stator currents is
advantageous for controlling the torque and protecting the
inverter.

As an example of a frequently used power converter with current control, a voltage source transistor inverter with pulse-width modulation is shown in Fig. 21, where the motor terminals are sequentially switched to the upper or lower bus of a constant voltage dc-link. When employing bipolar transistors, the clock frequency of the inverter ranges up to several kHz, the power rating may exceed 200 kW. The circuit includes regeneration if the line-side converter, which is not shown, allows power to be fed back to the line.

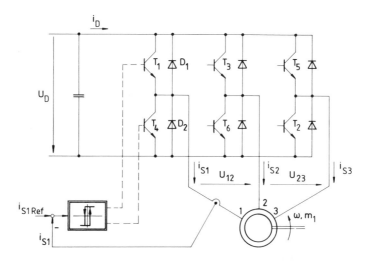

Fig. 21   Voltage source PWM transistor inverter with current
           control

The stator currents are controlled as indicated in principle for one phase. Pulse-width modulation and control may be performed in many different ways, analogue or digital, using a microcontroller or special microelectronic hardware (ASIC's). The on-off controller shown is only the simplest example. In view of the condition (14), one of the three current controllers is redundant, there should be provisions for avoiding conflict of the controllers.

When the stator currents are assumed to be impressed by fast current control, the dynamic model of the ac-motor is reduced to eqs.(21,24,25), with the stator currents serving as input quantities. In view of the rotor circuits (Fig. 18) the approach for controlling the PM-synchronous motor and the induction motor with cage rotor must be different.

IV. CONTROL OF HIGH PERFORMANCE AC-DRIVES

A. ROTOR-ORIENTATED CONTROL OF A PM-SYNCHRONOUS MOTOR

With high energy magnets, preferably of the rare earth type attached to the circumference of the rotor, an impressed and sinusoidally distributed mmf can be created; this may be modelled by current sources, Fig. 18, resulting in an impressed rotor current vector, $\underline{i}_R(t)=3/2\ I_F$ = const. so that the rotor voltage equation (21) becomes redundant and can also be removed from the drive dynamics. This leaves only the mechanical equations (24,25)

$$J_M\ d\omega/dt = L_o\ I_F\ \text{Im}\ [\underline{i}_S\ e^{-j\varepsilon}] - m_L\ , \tag{30}$$

$$d\varepsilon/dt = \omega\ , \tag{31}$$

where the term in brackets,

$$\underline{i}_S(t)\ e^{-j\varepsilon} = i_S(t)\ e^{j(\zeta-\varepsilon)} = i_S(t)\ e^{j\delta} =$$

$$= i_{Sd}(t) + j\ i_{Sq}(t)\ . \tag{32}$$

can be interpreted as the stator current vector viewed from the moving rotor; it may be called the stator current in rotor coordinates or the rotor-orientated stator current vector, $\delta = \zeta-\varepsilon$ is the load angle. The expression for the motor torque then reads

$$m_M = L_o\ I_F\ i_S(t)\ \sin\ (\zeta-\varepsilon) =$$

$$= \phi_R\ i_S(t)\ \sin\ \delta = \phi_R\ i_{Sq}(t)\ . \tag{33}$$

Hence the torque is proportional to the flux produced by the
permanent magnets and the quadrature component of the stator
current vector in rotor coordinates. The pertinent vector
diagram is shown in Fig. 22a; it should be noted that this is
an instantaneous view of time dependent vectors, not a phasor
diagram. Clearly, eq.(33) is analogous to eq.(10), with the
quadrature current component representing the torque-producing input of the synchronous motor, analogous to the armature
current of a PM-excited dc-motor. It may be recalled that
rotor coordinates are also the basis of the classical Park-
transformation for synchronous machines.

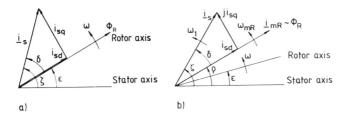

a)   b)

Fig. 22   Vector diagram of PM-synchronous motor   (a)
          and induction motor with cage rotor (b).

By assuming impressed stator currents, realized by current
control around a pulse width modulated voltage source inverter, the model of the PM-synchronous motor is reduced to
the mechanical equations (30,31). This results in a very
simple digital torque-control scheme, for example as shown in
Fig. 23, where the sampled stator currents are first converted
to polar coordinates $|\underline{i}_S(t)|$, $\zeta(t)$. Subtracting the measured
angle of rotation, $\delta = \zeta - \varepsilon$, transforms the current vector into
rotor coordinates so that the electrical torque m_M may be
computed for use as feedback quantitiy in a torque control
loop. The torque controller in turn issues a reference for
the magnitude of the current vector $|\underline{i}_{SRef}|$ and for the load
angle $\delta_{Ref} = \pm \pi/2$ which results in minimum current for a
given torque. By then adding the measured rotor angle, $\zeta_{Ref} = \delta_{Ref} + \varepsilon$, the current reference is converted back to stator
coordinates before it is split up into the three phase ac-

current references. These quantities are then supplied
through D/A-converters to the current controllers.

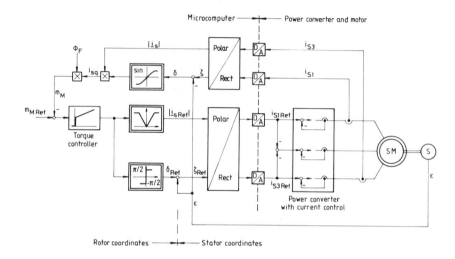

Fig. 23   Torque   control of a PM-synchronous   servo motor in
          rotor coordinates

A main feature of torque control in rotor coordinates is, of
course, that the controller processes dc-signals, while ac-
quantities occur only at the power converter and motor. The
speed and position control loops correspond to those of the
dc-servo motor, Fig. 12.

The scheme in Fig. 23 is only one of many possibilities for
realizing torque control. Another would be to supply the
motor with impressed voltages from the PWM-inverter and close
the current loops in rotor coordinates; likewise, the control
could be implemented in cartesian coordinates, commanding
$i_{SdRef}=0$. It has also been shown that a certain amount of
field weakening is feasible in order to reach higher speeds at
low torque; this would call for $|\delta_{Ref}| > \pi/2$. However, as the
required stator current rises sharply, this is only applicable
to motors operated at intermittent duty. All these schemes
are characterized by substantial complexity for real-time

signal processing so that they only became practical since microcomputers and signal processors have been available. For time - critical tasks close to the power converter, for instance PWM, special hardware components (ASIC's) or microcontrollers are increasingly being applied. As an example of the high performance that may be achieved with this type of drive, the step responses of a position-controlled PM-synchronous motor are depicted in Fig. 24a. The control scheme employed stator current control in rotor coordinates and a limited amount of field weakening [17].

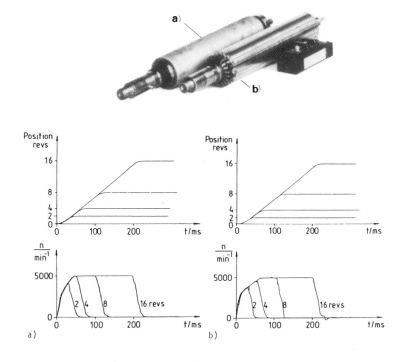

Fig. 24   Step responses of a digitally controlled 1.2kW ac-servo motor at no-load
a) PM synchronous motor
b) induction motor with cage rotor

B. FIELD-ORIENTATED CONTROL OF AN INDUCTION MOTOR

1. Principle and realization

The situation is more involved with an induction motor, because eq.(21), describing the currents in the rotor windings is still valid there. By rewriting it

$$R_R \underline{i}_R + L_o \, d/dt \, [(1+\sigma_R)\underline{i}_R + \underline{i}_S \, e^{-j\varepsilon}] = 0 \qquad (34)$$

and substituting the non-observable rotor current vector $\underline{i}_R$ by a synthetic magnetizing current vector representing rotor flux

$$\underline{i}_{mR} = i_{mR} \, e^{j\rho} = \underline{i}_S(t) + (1+\sigma_R) \, \underline{i}_R \, e^{j\varepsilon} \, , \qquad (35)$$

we obtain with $T_R = L_R/R_R$

$$T_R \, d\underline{i}_{mR}/dt + (1-j\omega T_R) \, \underline{i}_{mR} = \underline{i}_S \, ; \qquad (36)$$

this is split into real and imaginary parts

$$T_R \, di_{mR}/dt + i_{mR} = Re \, (\underline{i}_S \, e^{-j\rho}) = i_{Sd} \, , \qquad (37a)$$

$$d\rho/dt = \omega_{mR} = \omega + Im(\underline{i}_S \, e^{-j\rho}) \, /T_R \, i_{mR} =$$

$$= \omega + i_{Sq}/T_R \, i_{mR} \, , \qquad (37b)$$

where $\underline{i}_S \, e^{-j\rho} = i_{Sd} + j \, i_{Sq}$ \qquad (38)

is the stator current vector in a frame of reference defined by the rotor flux wave or, briefly, the field-orientated stator current vector. T_R is the main field time constant corresponding to the field lag of a dc-machine; it may be in the order of 0.1 sec for a small servo motor but ranging up to 1 sec for a larger machine.

When replacing the rotor current by the magnetizing current in the expression for the torque (eq.23),

$$m_M(t) = 2\ L_o/3(1+\sigma_R)\ i_{mR}\ i_{Sq} = 2/3(1+\sigma_R)\ \mathring{\Phi}_R\ i_{Sq} =$$

$$= K\ i_{mR}\ i_{Sq} \qquad (39)$$

is found, where $L_o\ i_{mR} = \mathring{\Phi}_R$ corresponds to rotor flux. There
is again a simple analogy to the torque of the separately
excited dc-machine; however, the current vectors are now
defined in a frame of reference that moves across the rotor
surface at slip frequency. The pertinent vector diagram is
seen in Fig. 22b; for sinusoidal currents and constant torque
the vectors $\underline{i}_S$ and $\underline{i}_{mR}$ rotate at the same velocity so that the
load angle δ remains constant.

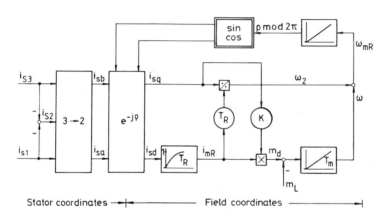

Fig. 25   Block diagram of the current-fed induction motor in
           field coordinates

A block diagram of the current-fed induction motor in field
coordinates, rendering a graphical representation of the model
equations (24,25,37,39) is shown in Fig. 25. It contains the
generation of the rotor flux and torque as well as the
coordinate transformation involving four multiplications with
$\cos\rho$, $\sin\rho$. T_R again characterizes the main field lag. Most
remarkable is the fact that, in contrast to Fig. 19, the
torque is now produced through just one channel, combining the
rotor flux $L_o i_{mR}$ and the quadrature component of the field
orientated stator current vector. This is the main idea of
controlling an ac-motor in field coordinates: By transforming

the stator variables into a coordinate system moving with the
rotor flux, the complex dynamic structure of the ac-machine is
greatly simplified becoming equivalent to that of a separately
excited dc-machine [3].

The problem is thus reduced to controlling the field orientated stator current components i_{Sd} and i_{Sq} as a first step
towards controlling flux and torque. The stator current
control loops exhibit only a short lag, caused by the
converter dynamics, leakage inductances etc.; hence, i_{Sq} is
the quantity best suited for fast torque control. On the other
hand, the flux $L_0 i_{mR}$ can only be changed through a main lag
and should therefore be maintained constant at the maximum
possible value, determined in the base speed range by iron
saturation and in the field weakening region by the voltage
ceiling of the inverter. The second problem of controlling the
motor in field coordinates is that of decoupling the nonlinear interactions caused by the coordinate transformation. If
the current controlled converter produces nearly impressed
stator currents with negligible delay, decoupling can be
achieved by an inverse transformation performed on the control
side, thus in effect cancelling the machine internal transformation, as shown in Fig. 26. There are now two control
channels for flux and torque (speed, position) with the controllers processing dc-quantities. Only in the center part,
around the converter and stator, are there ac-quantities.
Assuming fast current control, this part may be omitted or
crudely approximated when designing the drive control.

A similar solution exists when the motor is supplied with
impressed stator voltages, for instance coming from a PWM
voltage source inverter but without the current control. Of
course, this complicates the motor model because the stator
voltage equations (20) must be taken into account; otherwise
the solution of field orientated control is the same. The
resulting interactions are seen in Fig. 27. The motor currents
i_{Sd}, i_{Sq} are now controlled in field coordinates, i.e. the two

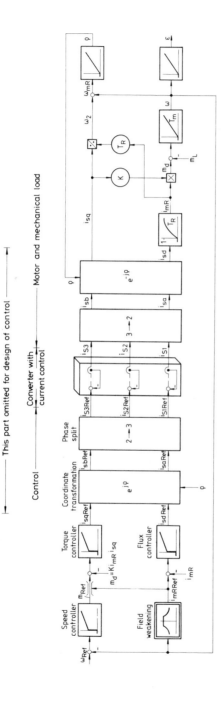

Fig. 26 AC-machine and control in field coordinates; ac-current control in stator coordinates

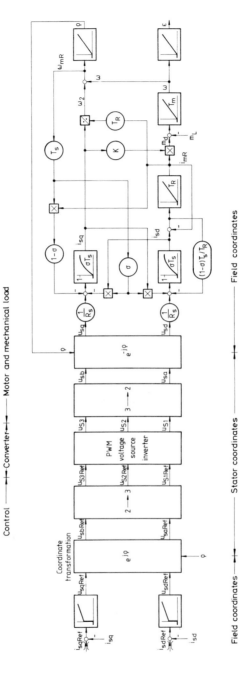

Fig. 27 AC-machine with impressed voltages; current control in field coordinates

inverse coordinate transformations become part of the current control loops. Again assuming an inverter with sufficiently high clock frequency, the voltage interactions are easily handled by the current control; they could also be compensated by adding decoupling signals at the outputs of the two current controllers.

The complete drive control scheme again consists of a flux controller prescribing i_{SdRef} and a second channel for controlling torque (speed, position), acting through i_{SqRef}. The two control schemes shown in Fig. 26, 27 are essentially equivalent but they differ with regard to practical implementation.

The main difficulties that are now left for the practical realization of field orientated control are

- the complexitiy of the signal processing required, particularly the multiplications with trigonometric functions of unbounded angles and
- the acquisition of the machine internal variables i_{mR}, ρ, i_{Sd}, i_{Sq}, m_M needed for the control.

The first aspect presented a serious obstacle as long as only analogue control hardware was available; however, with digital microelectronics becoming ever more powerful and cost effective in the last 10 years, these problems have now been overcome. Most of the complexity is embedded in software; also, the required signal processing power, i.e. accuracy and speed of computation, is no longer a limiting factor. Even a standard 16 bit microprocessor can execute the complete control program for an ac-motor in less than 1 ms; with signal processors the computations are carried out in less than 100 µs.

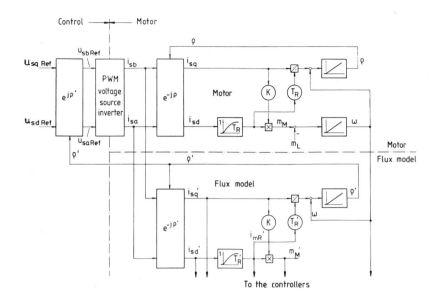

Fig. 28   Flux model of induction motor, based on rotor voltage
equation, using measured stator currents nd speed

The other problem of acquiring valid signals in a dynamic
state of the drive is still difficult to solve because of the
distributed magnetic fields and the fact that some variables
such as currents in a cage rotor winding are not directly
measureable unless telemetering equipment is built into the
machine which would defeat the purpose of having a simple and
rugged drive [4]. After various abandoned attempts at direct
field measurements it has evolved during the past 10 years
that the best solution is to compute the internal machine
quantities from measured variables such as currents, voltages
and speed, using a dynamic model of the machine. In view of
the nonlinearities of the model equations this is a task that
can be effectively handled only by digital microcomputers,
integrating the nonlinear machine equations in real time.
There are several ways in which the flux vector and the
associated signals may be computed from the model equations
(20,21,24,25); the simplest, based on the rotor voltage
equation (37) is shown in Fig. 28; it corresponds to Fig. 25,

with measurements of stator currents and speed serving as inputs. Stator resistance and primary leakage do not affect this numerical "flux model", but the rotor time constant T_R which may be changing due to heating of the rotor and desaturation in the field weakening range has to be considered. Clearly, if the model parameter T_R' is not in line with the corresponding motor parameter T_R, the flux angle ρ' and the other field orientated variables will be faulty.

A complete block diagram of an ac-servo drive with field orientated control is seen in Fig. 29, where all the control functions including pulse-width modulation have been implemented on a signal processor TMS 32010 [17]. The sampling frequency of all the controllers and the clock frequency of the PWM-inverter are identical, 4 kHz, which is adequate for servo drives operating at a stator frequency up to 100-200 Hz. The programming was in Assembler using integer arithmetic, but the same speed of computation could now be achieved with higher level languages and floating point arithmetic, using more recent microcomputers.

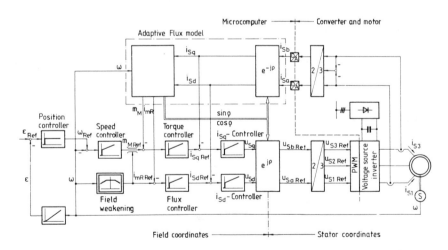

Fig. 29 Block diagram of an ac-servo drive with current control in field coordinates

With properly tuned field orientated control, an induction
motor with cage rotor can exhibit a similar dynamic
performance as a PM-synchronous motor. This was shown by
alternatively fitting the stator of a three phase servo motor
with a solid steel rotor having Sm-Co magnets on its
circumference or with a laminated rotor carrying a cage
winding. The inverter and the digital controller were the
same; however, for the PM synchronous motor the microcomputer
program was simpler because the flux model could be omitted.
Step-responses of both motors are shown in Fig. 24 for the
same test conditions except a slightly higher dc link-voltage
in case of the induction motor; there are only minor
differences in dynamic performance [17].

2. Comparison of ac-servo motors

When comparing the PM-synchronous motor and the induction
motor as a position-controlled servo drive, several additional
view points have to be considered:

- Higher efficiency of the PM-synchronous motor because no
  rotor currents are needed for generating torque,
- Simpler control algorithm for the synchronous motor because
  the torque control takes place in rotor coordinates, i.e. no
  flux model is needed,
- The induction motor requires reactive power which is to be
  supplied by the inverter.

On the other hand,

- the induction motor offers the option of field weakening
  resulting in a wider speed range,
- the cost of an induction motor should be less than that of
  the synchronous motor because of the high cost of the
  magnets.

Obviously, there is no simple answer, valid for every
situation. Experience so far indicates, that PM-synchronous
motors are preferred for positioning drives, while the

induction motor is the ideal solution for spindle drives on
machine tools.

3. Self-tuning and adaptive motor control

It was mentioned before that the need to fully utilize the
capacity of an induction motor may call for a flux model that
is responsive to changing saturation and rotor temperature
which both influence the rotor time constant T_R. The effect of
detuning the flux model is seen in the vector diagram in Fig.
30, indicating that it results in a shift between flux- and
torque-producing current components while the torque,
commanded by the speed controller, is upheld as long as no
current limits are reached.   However, if under overload
condition maximum quadrature current is called for by the
torque controller and this does not correspond to the actual
magnetic condition of the motor, there is loss of maximum
torque.

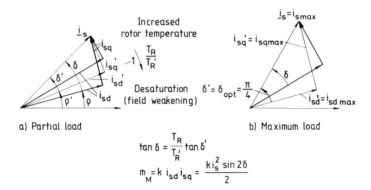

a) Partial load   b) Maximum load

$$\tan \delta = \frac{T_R}{T_R'} \tan \delta'$$

$$m_M = k\, i_{sd}\, i_{sq} = \frac{k i_s^2 \sin 2\delta}{2}$$

Fig. 30 Effect of detuning flux model of induction motor

This is demonstrated with measured reversing transients of a
22 kW wound rotor induction motor whose rotor time constant
could be reduced at will by placing external resistors in the
rotor circuit. Fig. 31 depicts current limited transients in
the base speed range with tuned and detuned flux model.
Clearly, with detuned flux model the maximum torque is
reduced, no matter whether $T_R' \gtrless T_R$.

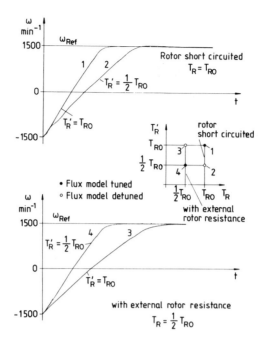

Fig. 31   Speed reversal of 22 kW wound   rotor   induction motor
          at base speed with tuned and detuned   flux model

In   view of the fact,   that temperature-induced change of   the
rotor   resistance is a slow process,   there is a   good   chance
that   the   flux model can be tuned   on-line   using   additional
measurements   such   as   stator   voltages   and   with   suitable
identification   algorithms.   Several   procedures   have   been
proposed,   but   it is not possible to discuss them   in   detail
here [8,10,11,28,32].

While a change of temperature affecting rotor resistance   can
take   minutes,   a   variation   of rotor   inductance   caused   by
saturation   occurs   instantaneously   as the   motor   enters   or
leaves the base speed range.   Identification procedures   would
be   too   slow   to   track such   changes,   hence   an   open   loop
compensation based on prior measurements seems preferable.   By
approximating   the   magnetization   curve $i_{mR}(\hat{\Phi}_R)$ with   an
empirical stored function,   the flux model   may be extended to

include both effects. The block diagram in Fig. 32 depicts a possible solution, where the input R_R/R_{Ro} for the thermal variations would be activated by one of the identification algorithms mentioned before [10].

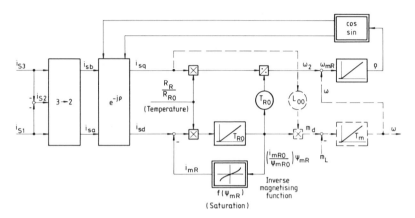

Fig. 32   Adaptive   flux model of   induction   motor   including
          thermal and saturation effects

The   advantage   of   taking   saturation   into   account   is demonstrated in Fig. 33, where a speed reversal of the 22 kW induction motor above base speed is seen. When employing the saturation dependent flux model, the time required for speed reversal with the same maximum current is markedly reduced; this indicates that the adaptive flux model is much better suited to follow the true magnetic state of the motor.

These examples, which could be extended to other types of machines and converters, indicate the enormous progress that has been made in recent years in the field of high-performance drives. The control structures, most of them realized in software, are straight forward in principle but involved in their practical realization. As the microcomputers continue to become more powerful, there are promising attempts to use this additional potential for automatic adjustment of the control; this may take the form of self-tuning during commissioning of the drive or on-line adaptive control autonomously tracking the variable operating conditions of the plant [10,26].

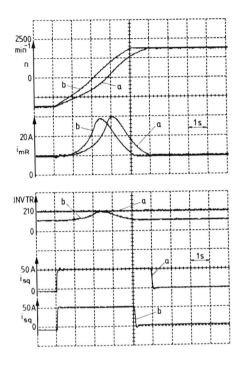

Fig. 33   Speed reversal of 22 kW induction   motor   above   base
         speed   without (a)   and with (b)   saturation dependent
         flux model

Self-tuning can be achieved by automatically performing simple
tests, using the inverter as a controllable voltage source,
with the motor at standstill. By analyzing the currents the
electromagnetic parameters of the machine are determined. As
an example, the step response of a completely self-tuned
current controller for the 22 kW motor is seen in Fig. 34.
The control scheme corresponded to that shown in Fig. 29, with
a 30 kVA voltage source transistor inverter operating at a
clock frequency of 1 kHz and a TMS 320C25 control computer
with 2 kHz sampling frequency. After tuning the inner current
control loops, the speed controller is automatically adjusted
also, but this calls for tests with the motor moving in order
to estimate the parameter of the mechanical load [10].

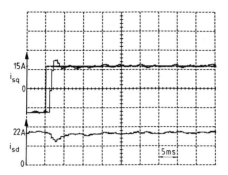

Fig. 34   Step response of self-tuned current control loops in
          field coordinates

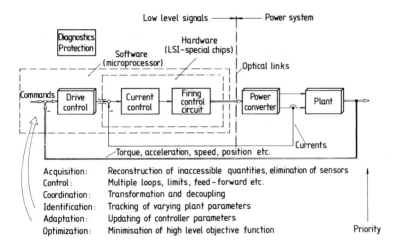

Fig. 35   Structure of future power converter and digital drive
          control systems

Much research is still needed in the field of on-line adaptive
drive control, because changes of drive parameters can occur
rapidly, calling for quick convergence of the identification
procedures. These algorithms are quite complex so that integer
arithmetic and Assembler programming are of little help; the
problems of adaptive drive control can only be solved with
higher   level   programming   languages   and   floating-point

arithmetic throughout. On the other hand, there are time-
critical control tasks close to the power converter, such as
pulse-width-modulation, where fine resolution in time is of
highest priority while the functions themselves are
repetitive and unchanging; this is the area where special LSI-
chips (ASIC's) are offering an advantage [14,31]. The likely
future structure of advanced high-performance drives is
depicted in Fig. 35.

A typical ASIC with 26 000 gates for an all digital uninter-
ruptible power supply control system with 20 kHz sampling
frequency is seen in Fig. 36 [14].

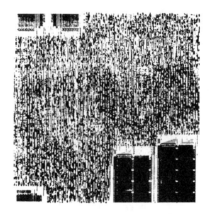

Fig. 36 ASIC for digital control system with 20 kHz sampling
        frequency

V. CONCLUSION AND OUTLOOK

Based on the evolution of macro- and micro-electronics there
is a world-wide emergence of controlled ac-drives that are no
longer subject to the limitations of their dc-predecessors.
Once commutation is separated from the electrical machine
proper and is performed in solid-state converters, there
exists a multitude of combinations for ac-drives, comprising

different types of converters, machines and control. Only two types of ac-drives have been discussed in this survey which excel for their dynamic performance and are widely employed as servo motors, for example on machine tools and robots. It is shown what the requirements for the drives are and how the multivariable machine models can be simplified by converting to moving coordinates. Such methods call for complex real-time signal processing which is best handled by microelectronics, using hardware as well as software. Research and development in the field are proceeding rapidly and on a global scale. The problems to be solved next are self-tuning and adaptive drives, where the control automatically tracks the changing needs of the mechanical load. New microprocessors, offering high-level language capability and floating point arithmetic at signal processor speed are suitable tools for solving the problems of the next generation in motion control.

SOME REFERENCES

[1] Abbondanti, A.: Method of flux control in induction motors driven by variable frequency, variable voltage supplies, Proc. IEEE/IAS Int. Semicond. Power Conv. Conf., 1977.

[2] Abraham, L.: Verfahren zur Steuerung des von einer Asynchronmaschine abgegebenen Drehmomentes, German Patent 1.563.228, 1966

[3] Blaschke, F.: The principle of field orientation, as applied to the new TRANSVECTOR-closed loop control systems for rotating field machines, Siemens Review 1972, pg. 217

[4] Blaschke, F.; Böhm, K.: Verfahren der Flußerfassung bei der Regelung stromrichtergespeister Asynchronmaschinen, 1. IFAC Symp., Control in Power Electronics and Electrical Drives, Düsseldorf 1974, Vol. 1, pg. 635

[5] Bose, B.K.: Power electronics and AC-drives, Prentice Hall, Englewood Cliffs, N.J. 1986

[6] Depenbrock, M.: Direkte Selbstregelung (DSR) für hochdynamische Drehfeldantriebe mit Stromrichterspeisung, ETZ-Archiv 1985, pg. 211

[7] Gabriel,R.;Leonhard,W.;Nordby,C.: Field orientated control of a standard AC-motor using microprocessors, IEEE-Trans. Ind. Appl. 1980, pg. 186

[8] Garces, L.J.: Parameter adaption for the speed controlled static AC drive with squirrel cage induction motor, IEEE-Trans. Ind. Appl.1980, pg. 173

[9] Hasse, K.: Zur Dynamik drehzahlgeregelter Antriebe mit stromrichtergespeisten Asynchron-Kurzschlußläufermaschinen, Diss. TH Darmstadt, 1969

[10] Heinemann, G.; Leonhard, W.: Self-tuning Field-orientated Control of an Induction Motor Drive, Intl. Power Electronics Conf., Tokyo 1990, pg. 465

[11] Irisa, T.; Takata, S.; Ueda, R.; Sonoda, T.; Mochizuki, T.: A novel approach on parameter self tuning method in AC servo systems, 3. IFAC Symp. Control in Power Electronics and Electrical Drives, Lausanne 1983, pg. 41

[12] Iwakane, I.; Inokuchi, A., Kai, T., Hirai, J.: AC servo motor drive for precise positioning control, Proc. Int. Power Electr. Conf. Tokyo 1983, pg. 1453

[13] Jönsson, R.: Measurements on a new induction motor control system, EPE 89, Aachen, pg. 17

[14] Kiel, E.; Schumacher, W.; Gabriel, R.: PWM-Gate Array for AC-Drives, EPE 87, Grenoble, pg. 653

[15] Kovacz, R.P.; Racz, J.: Transiente Vorgänge in Wechselstrommaschinen, Ung. Akad. d. Wissenschaften, Budapest 1959

[16] Leonhard, W.: Control of Electrical Drives, Springer Verlag, Berlin, Heidelberg, New York, Tokyo, 1985

[17] Lessmeier, R.; Schumacher, W.; Leonhard, W.: Microprocessor-controlled AC-servo drives with synchronous or induction motors: Which is preferable? IEEE-Trans. Ind. Appl., 1986, pg. 812

[18] Letas, H.H.; Kiel, E.: Digitales ASIC ersetzt analoge Umrichterregelung, IAM-Forum, Braunschweig, Vol.5 pg. 18

[19] Lipo, T.A.: State variable steady state analysis of a controlled current induction motor drive, IEEE-Trans. Ind. Appl. 1975, pg. 704

[20] Mokrytzki, B,: Pulse width modulated inverters for ac motor drives, IEEE Trans. Ind. Appl., pg. 312, 1968

[21] Murphy, J.M.D., Howard, L.S., Hoft, R.G.: Microprocessor control of a PWM-inverter induction motor drive, Conf. Rec. IEEE-Power Electr. Spec. Conf., 1979, pg. 344

[22] Nabae, N.; Otsuka, K.; Uchino, H.; Kurosowa, R.: An approach to flux control of induction motors operated with variable-frequency power supply, IEEE Trans. Ind. Appl., 1980, pg. 342,

[23] Olomski, J.: Bahnplanung und Bahnführung von Industrierobotern, Diss. TU Braunschweig 1989, also Vol. 4 in "Fortschritt der Technik"-Series, Vieweg

[24] Pfaff,G.; Weschta,A.; Wick,A.: Design and experimental results of a brushless AC servo drive, Conf. Rec. IEEE Ind. Appl. Ann. Meeting, 1982, pg. 692

[25] Plunkett,A.B.: Direct flux and torque regulation in a PMW-inverter induction motor drive, IEEE Trans. Ind. Appl., IA-13, no. 2, 1977

[26] Schierling, H.: Self-commissioning-A novel feature of inverter-fed motor drives, IEE-Power Electronics and Variable Speed Drives, London 1988, pg. 287

[27] Schönung, A.; Stemmler, H.: Geregelter Drehstrom-Umkehrantrieb mit gesteuertem Umrichter nach dem Unterschwingungsverfahren, BBC-Nachrichten 1964, pg. 555

[28] Schumacher, W.; Leonhard, W.: Transistor-fed AC-servo drive with microprocessor control, Intl. Power Electronics Conf., Tokyo 1983, pg. 1465

[29] Seeger, G.; Leonhard, W.: Estimation of rigid body models for a six-axes manipulator with geared electric drives, Proc. IEEE-Conf. on Robotics and Automation, 1989, pg. 1690

[30] Takahashi, I.; Noguchi, T.: A new quick-response and high efficiency control strategy of an induction motor, IEEE-Trans. Ind. Appl. 1986, pg. 820

[31] Tez, E.S.: MOTONIC-chip set: High performance intelligent controller for industrial variable speed AC drives, IEE-Power Electronics and Variable Speed Drives, London 1988, pg. 287

[32] Yoshida, Y., Ueda, R., Sonoda, T.: A new inverter-fed induction motor drive with a function of correcting rotor circuit time constant, Proc. Int. Conf. on Power Electr., Tokyo, 1983, pg. 672

INDEX